Methods in Enzymology

Volume 123
VITAMINS AND COENZYMES
Part H

METHODS IN ENZYMOLOGY

EDITORS-IN-CHIEF

Sidney P. Colowick   Nathan O. Kaplan

Methods in Enzymology

Volume 123

Vitamins and Coenzymes

Part H

EDITED BY

Frank Chytil
DEPARTMENT OF BIOCHEMISTRY
VANDERBILT UNIVERSITY SCHOOL OF MEDICINE
NASHVILLE, TENNESSEE

Donald B. McCormick
DEPARTMENT OF BIOCHEMISTRY
EMORY UNIVERSITY SCHOOL OF MEDICINE
ATLANTA, GEORGIA

1986

ACADEMIC PRESS, INC.

Harcourt Brace Jovanovich, Publishers

Orlando   San Diego   New York   Austin
London   Montreal   Sydney   Tokyo   Toronto

ACADEMIC PRESS, INC.
Orlando, Florida 32887

United Kingdom Edition published by
ACADEMIC PRESS INC. (LONDON) LTD.
24–28 Oval Road, London NW1 7DX

LIBRARY OF CONGRESS CATALOG CARD NUMBER: 54-9110

ISBN 0–12–182023–8

PRINTED IN THE UNITED STATES OF AMERICA

86 87 88 89   9 8 7 6 5 4 3 2 1

Table of Contents

Section I. Cobalamins and Cobamides (B_{12})

Section II. Vitamin A Group

v

Section III. Vitamin D Group

Section IV. Vitamin E Group, Tocopherols

Section V. Vitamin K Group

Section VI. Carnitine

Section VII. Heme Porphyrins and Derivatives

Contributors to Volume 123

Article numbers are in parentheses following the names of contributors.
Affiliations listed are current.

M. AKHTAR (43), *Department of Biochemistry, The University of Southampton, Southampton SO9 3TU, England*

JANICE R. ALMOND (17), *Department of Medicine, Duke University Medical Center, Durham, North Carolina 27710*

DAVID H. ALPERS (3), *Department of Medicine, Division of Gastroenterology, Washington University School of Medicine, St. Louis, Missouri 63110*

T. C. AW (20), *Department of Biochemistry, National University of Singapore, Singapore 0511*

WILLIAM S. BECK (1), *Department of Medicine, Harvard Medical School and Massachusetts General Hospital, Boston, Massachusetts 02114*

PANGALA V. BHAT (9), *Clinical Research Institute of Montreal and Department of Medicine, University of Montreal, Montreal, Quebec H2W 1R7, Canada*

L. L. BIEBER (28, 29), *Department of Biochemistry, Michigan State University, East Lansing, Michigan 48824*

DAVID F. BISHOP (38), *Division of Medical Genetics, Mount Sinai School of Medicine, New York, New York 10029*

IAIN A. BORTHWICK (46), *Department of Biochemistry, The University of Adelaide, Adelaide, South Australia 5000, Australia*

EJAL BOSIN (10), *National Institutes of Health Clinical Center, National Institute of Alcohol Abuse and Alcoholism, Bethesda, Maryland 20201*

C. D. B. BRIDGES (12), *Department of Ophthalmology, Cullen Eye Institute, Baylor College of Medicine, Houston, Texas 77030*

HARRY P. BROQUIST (33), *Department of Biochemistry, Vanderbilt University, Nashville, Tennessee 37232*

KENNETH L. BROWN (2), *Department of Chemistry, The University of Texas at Arlington, Arlington, Texas 76019-0065*

MICHELE BRUGH-COLLINS (7), *Differentiation Control Section, Laboratory of Cellular Carcinogenesis and Tumor Promotion, National Cancer Institute, National Institutes of Health, Bethesda, Maryland 20205*

G. L. CATIGNANI (23), *Department of Food Science, North Carolina State University, School of Agriculture and Life Sciences, Raleigh, North Carolina 27695-7624*

KIM E. CREEK (7), *Differentiation Control Section, Laboratory of Cellular Carcinogenesis and Tumor Promotion, National Cancer Institute, National Institutes of Health, Bethesda, Maryland 20205*

HARRY A. DAILEY (47, 48), *Department of Microbiology, University of Georgia, Athens, Georgia 30602*

LUIGI M. DE LUCA (7), *Differentiation Control Section, Laboratory of Cellular Carcinogenesis and Tumor Promotion, National Cancer Institute, National Institutes of Health, Bethesda, Maryland 20205*

ROBERT J. DESNICK (38), *Division of Medical Genetics, Mount Sinai School of Medicine, New York, New York 10029*

MARC K. DREZNER (17), *Departments of Medicine and Physiology, Duke University Medical Center, Durham, North Carolina 27710*

W. E. DUNCAN (20), *Endocrine-Metabolic Service, Walter Reed Army Medical Center, Washington, D.C. 20307-5001*

WILLIAM H. ELLIOTT (46), *Department of Biochemistry, The University of Adelaide, Adelaide, South Australia 5000, Australia*

C. FIOL (29), *Department of Biochemistry, Michigan State University, East Lansing, Michigan 48824*

JENNIE E. FLEMING (47), *Department of Microbiology, University of Georgia, Athens, Georgia 30602*

S.-L. FONG (12), *Department of Ophthalmology, Cullen Eye Institute, Baylor College of Medicine, Houston, Texas 77030*

NANCY E. FRANK (18), *Department of Nutrition, School of Medicine, Case Western Reserve University, Cleveland, Ohio 44106*

MARIJKE FRÀTER-SCHRÖDER (5), *Biochemistry Institute, University of Zurich, CH-8057 Zurich, Switzerland*

PETER GIMSING (1), *Department of Medicine, Gentofte Hospital, DK-2900 Hellerup, Denmark*

RALPH GREEN (2), *Departments of Laboratory Hematology and Cardiovascular Research, The Cleveland Clinic Foundation, Cleveland, Ohio 44106*

J. G. HADDAD (20), *Department of Medicine, Endocrine Section, University of Pennsylvania, Philadelphia, Pennsylvania 19104-6067*

BRUCE A. HALLEY (11), *Merck Sharp & Dohme Research Laboratories, Rahway, New Jersey 07065*

BERTILLE M. HARBIN (47), *Department of Zoology, University of Georgia, Athens, Georgia 30602*

ILMO E. HASSINEN (34), *Department of Medical Biochemistry, University of Oulu, SF-90220 Oulu, Finland*

MARK R. HAUSSLER (22), *Department of Biochemistry, The University of Arizona, College of Medicine, Health Sciences Center, Tucson, Arizona 85724*

BRUCE W. HOLLIS (18, 19), *Department of Nutrition, School of Medicine, Case Western Reserve University, Cleveland, Ohio 44106*

RONALD L. HORST (14), *Department of Physiology, National Animal Disease Center, Agricultural Research Service, United States Department of Agriculture, Ames, Iowa 50010*

F. M. HUENNEKENS (4), *Division of Biochemistry, Department of Basic and Clinical Research, Scripps Clinic and Research Foundation, La Jolla, California 92037*

DONALD W. JACOBSEN (2, 4), *Departments of Cardiovascular Research and Laboratory Hematology, The Cleveland Clinic Foundation, Cleveland, Ohio 44106*

C. JONES (43), *Division of Antibiotics and Chemistry, National Institute for Biological Standards and Control, London NW3 6RB, England*

GLENVILLE JONES (15), *Departments of Medicine and Biochemistry, Queen's University, Kingston, Ontario K7L 3N6, Canada*

PETER M. JORDAN (51, 52), *Department of Biochemistry, The University of Southampton, Southampton SO9 3TU, England*

SHOSUKE KAWANISHI (49), *Department of Public Health, Faculty of Medicine, Kyoto University, Kyoto 606, Japan*

J. KERNER (28), *Institute of Biochemistry, University Medical School, Szigeti UT 12, H-7624 Pecs, Hungary*

NICK J. KOSZEWSKI (14), *Department of Physiology, National Animal Disease Center, Agricultural Research Service, United States Department of Agriculture, Ames, Iowa 50010*

PIERRE LABBE (41), *Institut Jacques Monod, Laboratoire de Biochimie des Porphyrines, Université Paris VII, 75251 Paris Cedex 05, France*

ANDRE LACROIX (9), *Clinical Research Institute of Montreal and Department of Medicine, University of Montreal, Montreal, Quebec H2W 1R7, Canada*

AGHDAS LAGHAI-NEWTON (52), *Department of Biochemistry, The University of*

Southampton, Southampton SO9 3TU, England

C. K. LIM (44, 45), *Division of Clinical Cell Biology, Medical Research Council Clinical Research Center, Harrow, Middlesex HA1 3UJ, England*

G. I. LIOU (12), *Department of Ophthalmology, Cullen Eye Institute, Baylor College of Medicine, Houston, Texas 77030*

BRUCE LOBAUGH (17), *Departments of Medicine, Physiology, and Surgery, Duke University Medical Center, Durham, North Carolina 27710*

MOMOYO MAKINO-TASAKA (6), *Department of Pharmacology, Hyogo College of Medicine, Nishinomiya, Hyogo 663, Japan*

STAVROS C. MANOLAGAS (21), *Department of Medicine, Division of Endocrinology and Metabolism, University of California, San Diego, San Diego, California 92161*

BRIAN K. MAY (46), *Department of Biochemistry, The University of Adelaide, Adelaide, South Australia 5000, Australia*

BARBARA E. MILLER (16), *Norwich Eaton Pharmaceuticals, Inc., Norwich, New York 13815-0231*

NOBUO MONJI (10), *Genetic Systems Corporation, Seattle, Washington 98121*

JOSEPH L. NAPOLI (13, 14), *Department of Biochemistry, University at Buffalo, State University of New York, Buffalo, New York 14214*

ELDON C. NELSON (11), *Department of Biochemistry, Oklahoma State University, Stillwater, Oklahoma 74078*

ANTHONY W. NORMAN (16), *Department of Biochemistry, University of California, Riverside, Riverside, California 92521*

RICHARD ODESSEY (30), *Department of Physiology, Louisiana State University Medical Center, New Orleans, Louisiana 70112*

KENNETH W. OLSEN (36), *Department of Chemistry, Loyola University of Chicago, Chicago, Illinois 60626*

SHRI V. PANDE (27, 32), *Laboratory of Intermediate Metabolism, Clinical Research Institute of Montreal, Affiliated to the University of Montreal, Montreal, Quebec, Canada H2W 1R7*

T. J. PETERS (44, 45), *Division of Clinical Cell Biology, Medical Research Council Clinical Research Center, Harrow, Middlesex HA1 3UJ, England*

J. WESLEY PIKE (22), *Department of Biochemistry, The University of Arizona, College of Medicine, Health Sciences Center, Tucson, Arizona 85724*

BYRON A. PIROLA (46), *Department of Biochemistry, The University of Adelaide, Adelaide, South Australia 5000, Australia*

CHARLES J. REBOUCHE (31), *Department of Pediatrics, The University of Iowa, Iowa City, Iowa 52242*

TIMOTHY A. REINHARDT (19), *Department of Physiopathology, National Animal Disease Center, Agricultural Research Service, United States Department of Agriculture, Ames, Iowa 50010*

DONATA RIMOLDI (7), *Differentiation Control Section, Laboratory of Cellular Carcinogenesis and Tumor Promotion, National Cancer Institute, National Institutes of Health, Bethesda, Maryland 20205*

MARK G. ROCKLEY (11), *Department of Chemistry, Oklahoma State University, Stillwater, Oklahoma 74078*

NATALIE L. ROCKLEY (11), *Department of Chemistry, Oklahoma State University, Stillwater, Oklahoma 74078*

A. CATHARINE ROSS (8), *Department of Physiology and Biochemistry, The Medical College of Pennsylvania, Philadelphia, Pennsylvania 19129*

SEIYO SANO (49), *Department of Public Health, Faculty of Medicine, Kyoto University, Kyoto 606, Japan*

TSUTOMU SASA (42, 50), *Division of Biology, Miyazaki Medical College, Kiyotake, Miyazaki 889-16, Japan*

JASBIR S. SEEHRA (51), *Genetics Institute, Cambridge, Massachusetts 02140*

BELLUR SEETHARAM (3), *Department of*

Medicine, Division of Gastroenterology, Washington University School of Medicine, St. Louis, Missouri 63110

YOSHIHIKO SEKI (49), Department of Public Health, Faculty of Medicine, Kyoto University, Kyoto 606, Japan

ANDREAS SEUBERT (39), Department of Dermatology, University of Göttingen, D-3400 Göttingen, Federal Republic of Germany

SIGRID SEUBERT (39), Department of Dermatology, University of Göttingen, D-3400 Göttingen, Federal Republic of Germany

MARTIN J. SHEARER (24, 25), Department of Haematology, Clinical Science Laboratories, Guy's Hospital, London SE1 9RT, England

YUZO SHIOI (42, 50), Division of Biology, Miyazaki Medical College, Kiyotake, Miyazaki 889-16, Japan

GOPESH SRIVASTAVA (46), Department of Biochemistry, The University of Adelaide, Adelaide, South Australia 5000, Australia

JAMES G. STRAKA (40), Department of Medicine, University of Minnesota College of Medicine, Minneapolis, Minnesota 55455

TATSUO SUZUKI (6), Department of Pharmacology, Hyogo College of Medicine, Nishinomiya, Hyogo 663, Japan

JIN TAMAOKA (26), Institute of Applied Microbiology, University of Tokyo, Bunkyo-ku, Tokyo 113, Japan

KEN TSUTSUI (37), Department of Biochemistry, Cancer Institute, Okayama University Medical School, Okayama 700, Japan

SERGE N. VINOGRADOV (35), Department of Biochemistry, Wayne State University, School of Medicine, Detroit, Michigan 48201

Preface

Since Volumes 62, Part D, 66, Part E, and 67, Part F of "Vitamins and Coenzymes" were published as part of the *Methods in Enzymology* series considerable new information has become available. Advances continue to be made, rapid in some cases, on techniques and methodology attendant to assays, isolations, and characterization of vitamins and those systems responsible for their biosynthesis, transport, and metabolism. In some instances, additional coenzymic forms of principal groups, e.g., pterins, are now known to participate in roles not heretofore recognized, e.g., methane formation. In others, e.g., porphyrins and hemes, the coenzyme-like function of many warrants inclusion in current volumes.

Because of the new information on vitamins and coenzymes and the certainty that methods attendant to this area continue to provide impetus for further research, we have sought to provide investigators with more current modifications of earlier procedures as well as with those that have newly evolved. The collated information is divided into two new volumes that are the result of our efforts in soliciting contributions from numerous, active experimentalists who have published most of their findings in the usual, refereed research journals. Volume 122, Part G, covers most of the classically considered "water-soluble" forms including ascorbate, thiamin, pantothenate, and biotin. Volume 123, Part H, covers the "fat-soluble" groups of A, D, E, and K, and also covers B_{12}, carnitine, heme, and porphyrins.

We should like to express our gratitude to the contributors for their willingness to supply the information requested and for their tolerance of our editorial revisions. There has been an attempt to allow such overlap as would offer flexibility in the choice of method, rather than presume any one is best for all laboratories. Omissions may be deliberate where modification of a "tried and true" technique is quite minor, but in some cases may be attributed to inadvertent oversight of the editors. We believe these volumes will provide useful and fairly replete addendums to the six earlier ones on this subject: 18 A, 18 B, 18 C, 62 D, 66 E, and 67 F.

We wish to acknowledge the encouragement of the founding editors of the *Methods in Enzymology* series, Dr. Nathan O. Kaplan and the late Dr. Sidney P. Colowick. It is with sadness that we join our colleagues in no longer having the kind and sapient advice of Dr. Colowick. Continued thanks are due to the staff of Academic Press for their help.

FRANK CHYTIL
DONALD B. McCORMICK

METHODS IN ENZYMOLOGY

EDITED BY

Sidney P. Colowick and Nathan O. Kaplan

VANDERBILT UNIVERSITY
SCHOOL OF MEDICINE
NASHVILLE, TENNESSEE

DEPARTMENT OF CHEMISTRY
UNIVERSITY OF CALIFORNIA
AT SAN DIEGO
LA JOLLA, CALIFORNIA

I. Preparation and Assay of Enzymes
II. Preparation and Assay of Enzymes
III. Preparation and Assay of Substrates
IV. Special Techniques for the Enzymologist
V. Preparation and Assay of Enzymes
VI. Preparation and Assay of Enzymes (*Continued*)
Preparation and Assay of Substrates
Special Techniques
VII. Cumulative Subject Index

METHODS IN ENZYMOLOGY

EDITORS-IN-CHIEF

Sidney P. Colowick and Nathan O. Kaplan

VOLUME XVIII. Vitamins and Coenzymes (Parts A, B, and C)
Edited by DONALD B. MCCORMICK AND LEMUEL D. WRIGHT

VOLUME XIX. Proteolytic Enzymes
Edited by GERTRUDE E. PERLMANN AND LASZLO LORAND

VOLUME XX. Nucleic Acids and Protein Synthesis (Part C)
Edited by KIVIE MOLDAVE AND LAWRENCE GROSSMAN

VOLUME XXI. Nucleic Acids (Part D)
Edited by LAWRENCE GROSSMAN AND KIVIE MOLDAVE

VOLUME XXII. Enzyme Purification and Related Techniques
Edited by WILLIAM B. JAKOBY

VOLUME XXIII. Photosynthesis (Part A)
Edited by ANTHONY SAN PIETRO

VOLUME XXIV. Photosynthesis and Nitrogen Fixation (Part B)
Edited by ANTHONY SAN PIETRO

VOLUME XXV. Enzyme Structure (Part B)
Edited by C. H. W. HIRS AND SERGE N. TIMASHEFF

VOLUME XXVI. Enzyme Structure (Part C)
Edited by C. H. W. HIRS AND SERGE N. TIMASHEFF

VOLUME XXVII. Enzyme Structure (Part D)
Edited by C. H. W. HIRS AND SERGE N. TIMASHEFF

VOLUME XXVIII. Complex Carbohydrates (Part B)
Edited by VICTOR GINSBURG

VOLUME XXIX. Nucleic Acids and Protein Synthesis (Part E)
Edited by LAWRENCE GROSSMAN AND KIVIE MOLDAVE

VOLUME XXX. Nucleic Acids and Protein Synthesis (Part F)
Edited by KIVIE MOLDAVE AND LAWRENCE GROSSMAN

VOLUME XXXI. Biomembranes (Part A)
Edited by SIDNEY FLEISCHER AND LESTER PACKER

VOLUME LX. Nucleic Acids and Protein Synthesis (Part H)
Edited by KIVIE MOLDAVE AND LAWRENCE GROSSMAN

VOLUME 61. Enzyme Structure (Part H)
Edited by C. H. W. HIRS AND SERGE N. TIMASHEFF

VOLUME 62. Vitamins and Coenzymes (Part D)
Edited by DONALD B. MCCORMICK AND LEMUEL D. WRIGHT

VOLUME 63. Enzyme Kinetics and Mechanism (Part A: Initial Rate and Inhibitor Methods)
Edited by DANIEL L. PURICH

VOLUME 64. Enzyme Kinetics and Mechanism (Part B: Isotopic Probes and Complex Enzyme Systems)
Edited by DANIEL L. PURICH

VOLUME 65. Nucleic Acids (Part I)
Edited by LAWRENCE GROSSMAN AND KIVIE MOLDAVE

VOLUME 66. Vitamins and Coenzymes (Part E)
Edited by DONALD B. MCCORMICK AND LEMUEL D. WRIGHT

VOLUME 67. Vitamins and Coenzymes (Part F)
Edited by DONALD B. MCCORMICK AND LEMUEL D. WRIGHT

VOLUME 68. Recombinant DNA
Edited by RAY WU

VOLUME 69. Photosynthesis and Nitrogen Fixation (Part C)
Edited by ANTHONY SAN PIETRO

VOLUME 70. Immunochemical Techniques (Part A)
Edited by HELEN VAN VUNAKIS AND JOHN J. LANGONE

VOLUME 71. Lipids (Part C)
Edited by JOHN M. LOWENSTEIN

VOLUME 72. Lipids (Part D)
Edited by JOHN M. LOWENSTEIN

Section I

Cobalamins and Cobamides (B_{12})

[1] Determination of Cobalamins in Biological Material

By PETER GIMSING and WILLIAM S. BECK

Cobalamins are present in all animal tissues and many microorganisms (but are lacking in plants).[1] In animal tissues, five different cobalamins have been identified in cells and extracellular fluids: MeCbl,[2] AdoCbl, OH-Cbl, CN-Cbl, and HSO$_3$-Cbl. In addition, several cobalamin-like substances (i.e., other corrinoids and cobalamin analogs) are found in animal and bacterial materials under various conditions. We are concerned in this chapter mainly with "true cobalamins."

Within organisms these compounds are bound to specific proteins: transport proteins [e.g., intrinsic factor (IF) and transcobalamin]; cobalamin-dependent enzymes (e.g., N^5-methyltetrahydrofolate-homocysteine methyltransferase, EC 2.1.1.13; and methylmalonyl-CoA mutase, EC 5.4.99.2) or proteins of unknown function (e.g., haptocorrin or R-protein). Thus when complete data are desired, it may be necessary to resolve multiple molecular species from multiple compartments before assaying individual cobalamins. In less demanding circumstances, total cobalamin content in biological preparations may be assayed without prior separation of the different molecular forms. We discuss here the determination of cobalamins from certain compartments and speculatively extrapolate these principles to other compartments. Our experiences have related mainly to human plasma, blood cells, bile, and liver.

The major aims of a cobalamin determination are (1) the accurate estimation of total cobalamin content, and (2) estimation of the relative amount of each cobalamin form. Since only two cobalamins (MeCbl and AdoCbl) can be determined individually by specific assays, existing methodologic limitations make this approach necessary. The determination of total cobalamin has been described in an earlier volume of this

[1] Original experiments were supported by Research Grant AM-26755 and International Research Fellowship TWO3341 from the National Institutes of Health, Bethesda, Maryland, by a grant from the P. Carl Petersen Foundation, Copenhagen, Denmark, and by the John Phyffe Richardson Fund.

[2] Abbreviations for corrinoids are as recommended by IUPAC-IUB Commission on Biochemical Nomenclature of Corrinoids (1973 Recommendations) as summarized in "Cobalamin: Biochemistry and Pathophysiology" [(B. M. Babior, ed.), p. 462. Wiley, New York and London]: CN-Cbl, cyanocobalamin; MeCbl, methylcobalamin; AdoCbl, 5'-deoxyadenosylcobalamin; OH-Cbl, hydroxo/aquacobalamin; HSO$_3$-Cbl, sulfitocobalamin; (CN)$_2$-Cbl, dicyanocobalamin; (Ade)CN-Cba, pseudovitamin B$_{12}$; (CN)$_2$-Cbi, dicyanocobinamide.

series.[3,4] Hence, this methodology will not be given here, although it will be the object of discussion. Determination of the relative amounts of cobalamins will be given in detail.

Principle

The assay depends on a preliminary extraction of cobalamins from tissue or fluid, further isolation and/or purification of the cobalamins before resolution of the different forms, and quantification of each form. Two methods will be presented. One, using thin-layer chromatography (TLC), is a refinement[5] of earlier methods.[6–8] The other employs high-performance liquid chromatography (HPLC).

Characteristics of Ideal Assay

Extraction and purification procedures should achieve a full recovery of all cobalamins, regardless of their biological compartment of origin. All procedures should be mild enough to avoid chemical modifications of cobalamins e.g., conversion of one form (such as AdoCbl) to another (such as OH-Cbl) or degradation to noncobalamin corrinoids (e.g., conversion of CN-Cbl to cyanocobinamide). The purification step should remove all substances that could interfere with the separation and quantification of cobalamins, so that the behavior of forms present in biological materials resembles that of pure standards. Resolutions should be complete, so that each cobalamin form is well separated from every other form. Finally, the assay should detect and quantify only cobalamins and all cobalamins should be equally responsive to the assay system.

Assay Method

Preparation of Samples

All specimens are handled in the dark or in dim-red illumination. Serum or plasma is separated from whole blood within a few hours after venepuncture to minimize liberation of cobalamins from granulocytes, which contain substantial amounts of cobalamin. Tissue specimens are

---

[3] U.-H. Stenman and L. Puutula-Räsänen, this series, Vol. 67, Part F, p. 24.
[4] R. Green, this series, Vol. 67, Part F, p. 99.
[5] P. Gimsing, E. Nexø, and E. Hippe, *Anal. Biochem.* **129**, 296 (1983).
[6] J. E. Ford and E. Holdsworth, *Biochim. J.* **53**, 22 (1953).
[7] K. Lindstrand and K.-G. Stahlberg, *Acta Med. Scand.* **174**, 665 (1963).
[8] J. C. Linnell, M. MacKenzie, J. Wilson, and D. M. Matthews, *J. Clin. Pathol.* **22**, 545 (1969).

washed with cold saline and homogenized or sonicated before analysis. Liberation of cobalamins from proteins by papain proteolysis has been recommended[4]; however, in our experience when this procedure is used with liver tissue or bile, recovery of total cobalamin is increased but mobility of cobalamin forms on HPLC is altered, perhaps by peptide residues remaining attached to cobalamins.

Assay of Total Cobalamin

Accurate assay of total cobalamin is achieved only with isotope dilution methods employing pure IF as binder[9] or with microbiological assays using *Lactobacillus leichmannii*, *Euglena gracilis*, or *Ochromonas malhamensis*.[10]

Determination of Relative Amount of Various Cobalamins

Although two different separation methods can be used, the same techniques are used for extraction and partial purification prior to either separation procedure.

Extraction of Cobalamins

Principle. Cobalamins are extracted with hot ethanol, which denatures and precipitates cobalamin-binding proteins. Since OH-Cbl binds nonspecifically to histidine residues in these proteins, its recovery on ethanol extraction is somewhat lower than that of the other cobalamins.[11,12] However, preincubation with an excess of cadmium ions competitively inhibits this nonspecific binding.[12,13] Thiol-blocking agents like *N*-ethylmaleimide have similar effects.[14]

Materials

Glass tubes and acetylation flasks. Use disposable washed glassware, or glassware soaked in chromic-sulfuric acid before rinsing in distilled water
Absolute ethanol
80% ethanol (v/v)

[9] J. F. Kolhouse, H. Kondo, N. C. Allen, E. Podell, and R. H. Allen, *N. Engl. J. Med.* **299**, 785 (1978).
[10] W. S. Beck, *Methods Hematol.* **10**, 31 (1983).
[11] W. R. Bauriedel, J. C. Picken, and L. A. Underkofler, *Proc. Soc. Exp. Biol. Med.* **91**, 377 (1956).
[12] P. Gimsing, *Anal. Biochem.* **129**, 288 (1983).
[13] E. L. Lien and J. M. Wood, *Biochim. Biophys. Acta* **264**, 530 (1970).
[14] J. van Kapel, L. J. M. Spijkers, J. Lindemans, and J. Abels, *Clin. Chim. Acta* **131**, 211 (1983).

Deionized and distilled water
Cadium acetate
80° water bath
Rotatory evaporator
Centrifuge

Procedure. Excess cadmium acetate is added to the sample and then incubated for 2 hr at room temperature, or 16 hr at 2 to 8°. The amount of cadmium to be added depends on the anticipated protein content (i.e., number of histidine residues) of the sample. For serum samples the final cadmium concentration should be about 0.2 mol/liter. Following this addition, the sample is added with vigorous mixing to 4 vol absolute ethanol preheated to 80°, and the mixture is incubated for 20 min at this temperature. After cooling in an ice bath the sample is centrifuged (10 min, 2000 *g*). The supernatant fraction is saved, and the precipitate is mixed with 2 vol cold 80% ethanol (v/v). After another centrifugation the second supernatant fraction is combined with the first and evaporated to dryness in a rotatory evaporator at 30 to 40° (vacuum 1.8 to 8.0 kPa).

Desalting and Purification

Principle. Most biological samples contain salts and other materials that may interfere with the resolution of cobalamins. Therefore an effort is made to eliminate these substances. Cobalamins possess both polar and nonpolar groups. In aqueous solution cobalamins are adsorbed to nonpolar materials such as silanized silica gel, Amberlite XAD-2, or alkyl-bonded phase columns like C$_8$ and C$_{18}$ reverse phase. This type of adsorption is a satisfactory substitute for the phenol extractions used in the past; moreover it saves time and improves recovery. Cobalamins are adsorbed to the columns and less hydrophobic compounds are washed away by water or weak organic solvent solutions. The cobalamins are then eluted with higher concentrations of organic solvent solutions that are sufficient to elute cobalamins but not more hydrophobic compounds.

Eluted samples are concentrated on rotary evaporator (as described above) or by lyophilization. Our experience has been chiefly with two kinds of columns: Amberlite XAD-2, which is used before TLC, and Sep-Pak C$_{18}$, which is used before HPLC. It is likely that both columns can be used interchangeably.

Procedure with Amberlite XAD-2 Columns

Materials

Amberlite XAD-2 (50–100 μm) (Servachrom, Germany)
Double-distilled water

Methanol, HPLC grade
Acetone, analytical grade
Glacial acetic acid, HPLC grade
Potassium hydroxide, analytical grade
Glass columns with polyethylene discs (8 × 330 mm)
−80° freezer (Forma Bio-Freezer, Forma Scientific)
Lyophilizer
Solvent A: 1% glacial acetic acid in double-distilled water (v/v)
Solvent B: 1% glacial acetic acid, 10% methanol in double-distilled
    water (v/v/v)
Solvent C: 1% glacial acetic acid, 50% methanol in double-distilled
    water (v/v/v)
Solvent D: KOH in methanol (0.1 mol/liter)
Note: All four solvents are degassed by suction filtration through a
    0.2-μm filter (Rainin Instrument Co.)

Preparation of XAD-2 Columns. A 30 g portion of Amberlite XAD-2 is suspended in 50 ml acetone, and washed with another 50 ml acetone on a porcelain filter. After drying at 80° it is suspended in 200 ml methanol and allowed to sediment in a 250-ml cylinder. The supernatant portion containing fines is discarded. This procedure is repeated 2 to 4 times until the supernatant fraction is clear. From a final suspension of Amberlite in 100 ml methanol, glass columns are packed to a height of 40 mm. After washing with 12 ml solvent C and 12 ml solvent A, columns are ready for use. After use columns may be regenerated by cleaning with 8 ml solvent D before the washing procedure described above. Columns may be reused 4 to 5 times.

Desalting the Samples. Evaporated extracts are dissolved in 2 ml solvent A, applied to the columns, and allowed to flow in spontaneously. After preliminary washing with 12 ml solvent A and 12 ml solvent B, cobalamins are eluted with 8 ml solvent C. The eluate is frozen at −80° and lyophilized in conical tubes.

Procedure with Sep-Pak C$_{18}$ Cartridges

Materials

Sep-Pak C$_{18}$ cartridges (Waters Associates)
Double-distilled water
Acetonitrile, HPLC grade
t-Butanol, analytical grade

Procedure. Sep-Pak C$_{18}$ cartridges are prepared by prewetting with 2 ml acetonitrile, and rinsing with 6 ml of double-distilled water. Evaporated extracts are dissolved in 2 ml of double-distilled water and applied to columns at a flow rate of 2 ml/min. At the same flow rate cartridges are washed with 12 ml of double-distilled water and then eluted with 6 ml 20% *t*-butanol (v/v). Eluates are lyophilized in conical tubes.

TLC and Quantification by Bioautography

Principle. The five cobalamins can be separated by one-dimensional ascending TLC. Tailing is minimized if TLC plates are equilibrated in an atmosphere saturated with the solvent. After development, known amounts of CN-Cbl and MeCbl standards are applied to calibrate the bioautographic quantification of the different cobalamin forms. The growth response of the cobalamin-dependent *E. coli* strain used for bioautography varies with the cobalamin form. However, the conversion of AdoCbl, MeCbl, and OH-Cbl to HSO_3-Cbl makes it possible to derive results for these compounds from a MeCbl calibration curve. A tetrazolium indicator is added to agar, which has been seeded with microorganisms, to facilitate identification of zones of growth and to increase sensitivity.[6] Following growth, indicator is extracted and the relative amount that was reduced by bacterial growth is measured spectrophotometrically.

Separation Procedure

Materials

TLC aluminum sheets silica gel 60 F_{254} (Merck)
TLC solvent: *sec*-butanol : 2-propanol : water : ammonia, 30 : 45 : 25 : 2 (v/v/v/v)
Cobalamin standards
Stock solutions (about 1 mmol/liter)
CN-Cbl, MeCbl, AdoCbl, and OH-Cbl (Sigma)
HSO_3-Cbl prepared from OH-Cbl according to Hill *et al.*[15]
Actual concentrations of these solutions are determined by converting aliquots to $(CN)_2$-Cbl in 1 mmol/liter KCN, and estimating absorbance at 367.5 nm. The molar extinction coefficient is 30,800
Identification standard: CN-Cbl, 20 nmol/liter; 40 nmol/liter of other four cobalamins are prepared in double-distilled from stock solutions
Calibration standards: CN-Cbl, 20 nmol/liter; MeCbl, 40 nmol/liter

[15] J. A. Hill, J. M. Pratt, and R. J. P. Williams. *J. Theor. Biol.* **3,** 423 (1962).

60-W lamp

Incubation chamber

Procedure. Lyophilized, partially purified extracts are dissolved in 100 μl of double-distilled water. To TLC plates are applied under air streams 8 μl identification standards and about 50 and 250 fmol (calculated from total cobalamin and assuming an estimated recovery of 50%) of each unknown specimen. Plates are incubated for 2 hr in a chamber saturated with TLC solvent. Plates are then developed (6 to 20 hr) by ascending TLC at room temperature (18 to 25°). Calibration standards of CN-Cbl (10 to 80 fmol) and MeCbl (20 to 240 fmol) are applied on top of the dried TLC plates, which are then exposed to light (60 W for 1 hr at a distance of 0.5 m).

Bioautography and Quantification

Materials

Escherichia coli strain NCIB 9270 (ATCC 14169) (Glaxo Lab. Ltd., England) is maintained in 1% agar (w/v) in tryptic soy broth. Other cobalamin-dependent *E. coli* strains like NCIB 8134 (ATCC 10799) (Glaxo) or *E. coli* 113-3 (ATCC 11105), may also be used. However, each strain has a somewhat different specificity for cobalamins and other corrinoids. A working inoculum is obtained by isolating a pure colony of the bacteria after seeding on blood agar plates. The colony is grown in 4% peptone water (w/v), pH 7.2, with D-glucose (11 mmol/liter) and sodium chloride (85 mmol/liter) at 30° for 2 hr. This solution is stored at 2 to 8° and can be used for up to 1 month.

Agar solution: 100 ml 3% solution (w/v) is autoclaved in 250-ml glass flasks and stored at room temperature.

Buffer solution: K_2HPO_4, 80 mmol/liter; KH_2PO_4, 44 mmol/liter; sodium citrate, 4.7 mmol/liter; magnesium sulfate heptahydrate, 1.7 mmol/liter; ammonium sulfate, 15 mmol/liter; and sodium chloride, 1.7 mmol/liter, pH 7.0

D-Glucose (Merck)

2,3,5-Triphenyltetrazolium chloride (Merck)

Sodium bisulfite (Merck)

Glass plates (200 × 200 mm)

Plexiglas frames (180 × 180 mm)

Whatman filter paper No. 1 (180 × 180 mm)

Bunsen burner

30° incubator

Procedure. In a boiling water bath 100 ml agar solution is mixed with 100 ml buffer solution in which has been dissolved 0.18 mmol 2,3,5-triphenyltetrazolium chloride and 3.3 mmol D-glucose. After the mixture cools to 47 to 43°, 0.8 ml of *E. coli* working inoculum is added. After cautious mixing that avoids air bubbles two agar plates are poured onto flame-sterilized glass plates in Plexiglas frames on a horizontal table. A filter paper wetted with 5 ml sterile filtered sodium bisulfite (2 mol/liter) is placed on each of the cooled agar plates. The agar plate is placed next to the TLC plate with the filter paper between them to form a sandwich. The

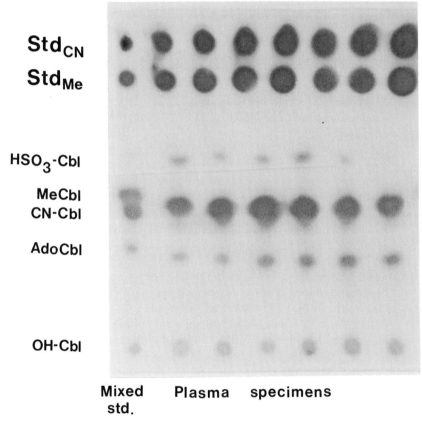

FIG. 1. Thin-layer chromatography followed by bioautography of a mixed standard (left channel) containing OH-Cbl, AdoCbl, CN-Cbl, MeCbl, and HSO$_3$-Cbl. Results of analyses of six normal plasma specimens appear on the right. Calibration standards of CN-Cbl (Std$_{CN}$) (10–140 fmol) and of MeCbl (Std$_{Me}$) (20–200 fmol) appear at the top.

TABLE I
R_f VALUES OF COBALAMINS AND COBINAMIDE
SEPARATED BY TLC[a]

Compound	R_f value
OH-Cbl	0.00 ± 0.00
(CN)$_2$-Cbi	0.003 ± 0.003
AdoCbl	0.23 ± 0.01
CN-Cbl	0.35 ± 0.02
MeCbl	0.40 ± 0.02
HSO$_3$-Cbl	0.58 ± 0.02

[a] Data are mean values and standard deviations
from 12 estimates on 4 different days.

glass plate and frame are removed, and the sandwich is placed in a humid incubator at 30° for 16 to 20 hr. Cobalamin spots are located as crimson areas of reduced tetrazolium (Fig. 1). R_f values are compared with those of standards (Table I). Colored spots are cut out and dye is extracted in 2 ml absolute ethanol (for 2 hr at 37° or 16 hr at room temperature). Amounts of cobalamin are determined by reading CN-Cbl spots against the CN-Cbl standard curve; other cobalamin spots are read against a MeCbl standard curve. Relative amounts in each spot are calculated. The actual amount of each cobalamin may be calculated as the product of relative amounts and total cobalamin in the sample.

HPLC and Quantification

Principle. Cobalamins and other corrinoids may be readily separated by reverse phase HPLC.[14,16–18] This method has been used mainly for separating radioactive cobalamins or nonlabeled cobalamins in large quantities. Although the cobalamin content of biological material is usually too small to allow direct spectrophotometric localization of HPLC peaks derived from reasonable amounts of starting material, the determination of total cobalamin concentration in each fraction makes the HPLC method practical. Since most solvents used in HPLC analysis of cobalamins are acidic, fractions must be neutralized during collection to prevent

[16] J. W. Jacobson, R. Green, E. V. Quadros, and Y. D. Montejano, *Anal. Biochem.* **120,** 394 (1982).
[17] E. P. Frenkel, R. L. Kitchens, and R. Prough, *J. Chromatogr.* **174,** 393 (1979).
[18] R. A. Beck and J. J. Brink, *Environ. Sci. Technol.* **10,** 173 (1976).

degradation of eluted cobalamins. When radioactive cobalamins are analyzed, known standards must be added.

Materials

HPLC gradient system (Rabbit HP) (Rainin Instrument Co.)
RP-8 Spheri-5 OS-GU precolumn (Brownlee Lab. Inc.)
Microsorb C_8 (4.6 × 250 mm) (Rainin Instrument Co.)
250 μl injection loop (Rainin Instrument Co.)
Acrodisc filter (0.2 μm) (Gelman)
Double-distilled water
Phosphoric acid, HPLC grade
Triethanolamine, analytical grade
Acetonitrile, HPLC grade
Fraction collector (ISCO)
Polypropylene tubes (10 × 75 mm)
Helium gas
Human serum albumin (HSA), free of cobalamin
Solvent A: 0.085 mol/liter phosphoric acid; pH set to 3 with triethanolamine
Solvent B: acetonitrile
Note: both solvents are degassed by suction filtration through a 0.2-μm filter
Borate/HSA: borate, 0.25 mol/liter, pH 9.5 with 8% HSA (w/v)

Procedure. Partly purified extracts are dissolved in 2 ml of double-distilled water and filtered through 0.2-μm Acrodisc filters. After a final lyophilization, samples are dissolved in 300 μl solvent A just before injec-

TABLE II
RETENTION TIME OF COBALAMINS AND
COBINAMIDE ON HPLC ANALYSIS[a]

Compound	Retention time (min)
OH-Cbl	15.2 ± 0.5
CN-Cbi	16.6 ± 0.4
CN-Cbl	19.6 ± 0.2
(CN)₂-Cbi	22.2 ± 0.2
AdoCbl	23.2 ± 0.2
MeCbl	27.5 ± 0.4

[a] Data are means and standard deviations from 48 determinations on 18 different days. Times are adjusted for retention time of mobile phase.

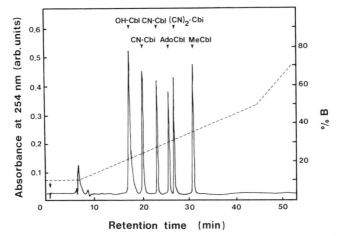

Fig. 2. HPLC separation of a mixture of cobalamins (OH-Cbl, CN-Cbl, AdoCbl, and MeCbl) and $(CN)_2$-Cbi on a Microsorb C_8 column. Cyanocobinamide appears to present as two peaks, which as indicated are probably CN-Cbi and $(CN)_2$-Cbi. The dashed line indicates a gradient of increasing concentration of acetonitrile from pump B.

tion into the HPLC. Samples are introduced through a 250 μl injection loop; 25 μl is saved for recovery calculations. The HPLC gradient is formed by mixing solvent A and B from pumps A and B, which are controlled by a microcomputer. Solvents are sparged with helium for at least 20 min before the start of a run and continuously throughout the run. The gradient starts at 10% solvent B (v/v) (i.e., 90% A) and increases linearly to 50% B at 40 min. Flow rate is 0.5 ml/min; pressure is 900 psi. The gradient continues linearly to 70% B in the next 8 min and is kept at this concentration for 5 min while the flowrate increases to 1.0 ml/min. Starting at 3.6 min after injection of the sample, 0.4 min fractions are collected into polypropylene tubes, each containing 200 μl borate/HSA. A total of 120 fractions is collected. The total cobalamin concentration of each fraction is determined by the radioisotope dilution method IF as binder. Cobalamins are identified by their retention times (Table II). When radioactive cobalamins are being analyzed, 4 μl of the following mixed standard is added to the sample before injection (concentrations in mmol/liter): CN-Cbl, 0.1; MeCbl, 0.2; AdoCbl, 0.3; OH-Cbl, 0.15; and $(CN)_2$Cbl, 0.2. Absorbance at 254 nm is recorded continuously (Fig. 2).

General Comments

The described methods are acceptable for the analysis of biological materials in which expected cobalamin types are known. In this setting it is reasonable to *detect* cobalamins from their growth-promoting activity with cobalamin-dependent microorganisms or their binding behavior in the radioisotope dilution assay—and to *identify* them by comparison of TLC R_f values or HPLC retention times with those of cobalamin standards. New cobalamins with the same mobility as expected known ones will go unrecognized. Although the radioisotope dilution assay using pure IF as binder is specific for cobalamins, cobalamin-dependent *E. coli* mutants may respond to such cobalamin-like substances as pseudovitamin B$_{12}$ [(Ade)CN-Cba], a compound whose mobility matches that of AdoCbl in our TLC system. It is emphasized therefore that analyses of materials likely to contain unusual or unknown cobalamins (or corrinoids) should utilize at least two different separation systems, e.g., both TLC and reverse-phase HPLC or ion-exchange chromatography (like SP-Sephadex[19]).

As stressed above in discussion of the "ideal assay," recovery of each cobalamin form from each compartment should be complete. Since unknown biological materials may contain cobalamin-binding substances of varying stability in the extraction procedure, some compartments may be overestimated at the expense of other compartments. For example, cobalamins bound to transcobalamin are liberated easily by acidification, while their liberation from haptocorrin requires boiling in addition. Thus recovery studies are essential when examining biological materials that might contain uncharacterized cobalamin-binding proteins.

[19] J. A. Begley and C. A. Hall, *J. Chromatogr.* **177,** 360 (1979).

[2] Analysis of Cobalamin Coenzymes and Other Corrinoids by High-Performance Liquid Chromatography

By Donald W. Jacobsen, Ralph Green, and Kenneth L. Brown

Most body fluids and tissues contain low concentrations ($<10^{-6}$ M) of corrinoids.[1,2] The most prevalent forms of corrinoids found in fluid and tissue extracts from normal individuals not receiving vitamin B$_{12}$ supple-

[1] J. C. Linnell, A. V. Hoffbrand, H. A.-A. Hussein, I. J. Wise, and D. M. Mathews, *Clin. Sci. Mol. Med.* **46,** 163 (1974).
[2] E. V. Quadros, D. M. Mathews, I. J. Wise, and J. C. Linnell, *Biochim. Biophys. Acta* **421,** 141 (1976).

ments include the cobalamin coenzymes adenosylcobalamin (AdoCbl),[3] methylcobalamin (MeCbl), and aquacobalamin (AqCbl). Other minor forms have also been identified including sulfitocobalamin (HSO_3-Cbl) and cyanocobalamin (CN-Cbl, vitamin B_{12}). In addition, a number of unidentified corrinoids have been detected by bioautography.[2] This chapter describes methods for rapid analysis of naturally occurring corrinoids and corrinoid analogs based on high-performance liquid chromatography.[4]

Principle

Since corrinoids have dissimilar amphipathic solubility properties, complex mixtures can be resolved on reverse-phase HPLC columns using either isocratic or gradient elution techniques. Body fluid and tissue corrinoids, which are usually associated with specific binding proteins, must be extracted and desalted prior to HPLC. Resolved corrinoids are detected directly by UV-VIS flow-cell absorbance spectrophotometry if chemical levels exceed 1 pmol per compound. Lower levels of radiolabeled corrinoids (e.g., ^{57}Co or ^{60}Co) can be resolved, collected, and counted.

Reagents and Columns

[^{57}Co]CN-Cbl (Amersham) or *Streptomyces griseus* fermentation[5]
Acetonitrile, HPLC grade (Burdick and Jackson)
CN-Cbl, crystalline (Sigma)
Reverse-phase columns
    LiChrosorb RP-8, 10 μm (EM Laboratories)
    Ultrasphere C_8, 5 μm (Beckman/Altex)

---

[3] AdoCbl, 5'-deoxy-5'-adenosylcob(III)alamin (B_{12} coenzyme); AqCbl, aquacob(III)alamin (B_{12a}); MeCbl, methylcob(III)alamin; HSO_3-Cbl, sulfitocob(III)alamin; CN-Cbl, cyanocob-(III)alamin (vitamin B_{12}); ForCbl, 5'-deoxy-5'-formycinylcob(III)alamin; ε-AdoCbl, 5'-deoxy-5'-(1,N^6-etheno)adenosylcob(III)alamin; DNCbl, 5'-deoxy-5'-(2,6-diamino)nebularinylcob(III)alamin; AraACbl, 5'-deoxy-(9β-D-arabinofuranosyl)-5'-adenylcob(III)-alamin; AC_2Cbl, 2-aminoethylcob(III)alamin; AC_3Cbl, 3-aminopropylcob(III)alamin; AC_5Cbl, 5-aminopentylcobalamin; AC_8Cbl, 8-aminooctylcob(III)alamin; AC_{11}Cbl, 11-aminoundecylcob(III)alamin; PrCbl, propylcob(III)alamin; DapCbl, dansylamidopropylcob-(III)alamin; (Aq)$_2$Cbi, diaquacob(III)inamide; (CN)$_2$-Cbi, dicyanocob(III)inamide; CN-Cbi, monocyanomonoaquacob(III)inamide (Factor B); AdoCbi, mono-5'-deoxy-5'-adenosylmonoaquacob(III)inamide; MeCbi, monomethylmonoaquacob(III)inamide; AC_3Cbi, 3-aminopropylcob(III)inamide; DapCbi, dansylamidopropylcob(III)inamide; HPLC, high-performance liquid chromatography.

[4] D. W. Jacobsen, R. Green, E. V. Quadros, and Y. D. Montejano, *Anal. Biochem.* **120**, 394 (1982).

[5] H. H. Thornberry and H. W. Anderson, *Arch. Biochem.* **16**, 389 (1948).

Ultrasphere C$_{18}$, 10 μm (Beckman/Altex)
Bondapak C$_{18}$, 10 μm (Millipore/Waters)
Sep-Pak C$_{18}$ cartridges (Millipore/Waters)

Synthesis of Cobalamin Coenzymes and Other Corrinoids

Methylcobalamin, PrCbl, and DapCbl were synthesized from cob-(I)alamin (obtained by reduction of CN-Cbl with powdered Zn) and methyl iodide, propyl iodide, and dansylaminopropyl chloride, respectively, as described previously.[6–8] The aminoalkylcobalamins AC$_2$Cbl, AC$_3$Cbl, AC$_5$Cbl, AC$_8$Cbl, and AC$_{11}$Cbl were synthesized from cob-(I)alamin and 2-chloroethylamine hydrochloride, 3-chloropropylamine hydrochloride, 5-bromopentylamine hydrobromide, 8-bromoctylamine hydrobromide, and 11-bromoundecylamine hydrobromide, respectively, according to procedures described previously.[6,9] AdoCbl and the upper-axial nucleoside analogs of the coenzyme (ForCbl, ε-AdoCbl, DNCbl, TubCbl, and AraACbl) were prepared from cob(I)alamin and the appropriate 5'-chloro-5'-deoxynucleoside.[7,8] AqCbl and CN-Cbl were prepared from CN-Cbl by the procedures of Hogenkamp and Rush[10] and Friedrich and Bernhauer,[11] respectively. AdoCbi, MeCbi, AC$_3$Cbi, and DapCbi were synthesized from cob(I)inamide (obtained by reduction of CN-Cbi with powdered Zn) and the corresponding alkylchloride according to the general procedure for the synthesis of cobalamin coenzyme analogs.[8]

Extraction of Corrinoids from Prokaryotic and Eukaryotic Cells

Cell Production. Lactobacillus leichmannii (ATCC 7830) was maintained on *Lactobacillus* broth (Difco 0901-15) and grown in medium[12] containing 1 nM [^{57}Co]CN-Cbl to early stationary phase (absorbance = 0.90 at 660 nm) at 37° for 16–18 hr. Murine leukemia L1210 cells were cultured in RPMI 1640 medium containing 10% human serum saturated with [^{57}Co]CN-Cbl and antibiotics (penicillin, 100 units/ml; streptomycin,

[6] D. W. Jacobsen, P. M. DiGirolamo, and F. M. Huennekens, *Mol. Pharmacol.* **11,** 174 (1975).

[7] D. W. Jacobsen, R. J. Holland, Y. D. Montejano, and F. M. Huennekens, *J. Inorg. Biochem.* **10,** 53 (1979).

[8] D. W. Jacobsen, this series, Vol. 67, Part F, p. 12.

[9] D. W. Jacobsen, Y. D. Montejano, and F. M. Huennekens, *Anal. Biochem,* **113,** 164 (1981).

[10] H. P. C. Hogenkamp and J. E. Rush, *Biochem. Prep.* **12,** 121 (1968).

[11] W. Friedrich and K. Bernhauer, *Chem. Ber.* **89,** 2507 (1956).

[12] R. L. Blakley, *J. Biol. Chem.* **240,** 2173 (1965).

100 μg/ml). Cells were grown to late-log phase (~2 × 10⁶ cells/ml) for 48 hr at 37° in a humidified atmosphere of 95% air–5% CO_2.

Cell Extraction. The following procedures were performed under dim-red illumination to avoid photolysis of light-sensitive corrinoids. Washed *L. leichmannii* cells were disrupted by sonication in water; ethanol was then added to a final volume of 80%. After 30 min at 80°, cell debris was removed by centrifugation and subjected to a second heat extraction with 80% ethanol. The combined extracts were evaporated to dryness *in vacuo* at 37°. The residue was dissolved in 5 ml of water and passed through a Sep-Pak C_{18} cartridge to remove salts and other hydrophilic contaminants. After washing the cartridge with water, adsorbed corrinoids were eluted with 50% acetonitrile and concentrated to dryness *in vacuo* at 40°. The residue was dissolved in water (≤500 μl) and stored in the dark at −80°.

L1210 cells were collected by centrifugation, washed with phosphate-buffered isotonic saline, and extracted with 80% ethanol at 70° for 20 min. Cell debris was removed by centrifugation and subjected to a second heat extraction with 80% ethanol. The combined extracts were evaporated to dryness *in vacuo* at 45°. The residue was dissolved in water (≤5 ml) and passed through a Sep-Pak C_{18} cartridge and processed as described above for *L. leichmannii*.

Resolution of Corrinoids by HPLC

Isocratic Method. Naturally occurring cobalamins and other corrinoids were resolved on a 4.6 × 250 mm organosilane C_8 reverse-phase column (LiChrosorb RP-8, 10 μm particle size) using 30% acetonitrile in 1 mM ammonium acetate, pH 4.40. The solvent system was prepared by adding 1.0 ml of 1.0 M ammonium hydroxide to 1.0 liter of 30% acetonitrile in water, titrating to pH 4.0 with glacial acetic acid, and filtering through a solvent-inert filter (Millipore, Type UR). The column was developed at 2.0 ml/min (~900 psi), and corrinoids, as they eluted, were detected spectrophotometrically by measuring absorbance at 254 nm in a flow cell. Retention times for corrinoids resolved by isocratic analysis are reported in Table I.

Gradient Method. Complex mixtures of corrinoids were better resolved on reverse-phase columns using a gradient development method. The solvent system for Pump A consisted of 0.05 M phosphoric acid titrated to pH 3.0 with concentrated ammonium hydroxide and filtered through a type HA Millipore filter. Degassed acetonitrile was used in Pump B. Columns were developed by increasing acetonitrile concentration linearly from 5 to 30% (routine analysis, see below) or 5 to 50%

TABLE I
RETENTION TIMES OF COBALAMIN COENZYMES AND RELATED
CORRINOID ANALOGS ON A LICHROSORB RP-8 COLUMN

Corrinoid	Gradient elution (min)	Isocratic elution (min)
A. Naturally occurring		
AqCbl	9.0	4.6–4.8[a]
HSO$_3$-Cbl	9.7	1.1
CN-Cbl	10.2	1.6
AdoCbl	11.4	2.8–3.0[a]
MeCbl	12.8	2.2
B. Coenzyme analogs		
DNCbl	10.9	5.2
TubCbl	11.1	9.6
AraACbl	11.2	3.3
ForCbl	11.1	4.3
ε-AdoCbl	11.7	3.5
PrCbl	14.5	9.1$_s$[b]
DapCbl	16.8	27.1$_s$
C. Aminoalkylcobalamins		
AC$_2$Cbl	9.8	5.9
AC$_3$Cbl	10.5	10.4$_s$
AC$_5$Cbl	11.2	nd
AC$_8$Cbl	12.8	nd
AC$_{11}$Cbl	14.9	nd
D. Cobinamides		
(Aq)$_2$Cbi	7.8	33.8
(CN)$_2$-Cbi	—[c]	1.6
CN-Cbi	9.0, 10.5[d]	4.9, 5.7[d]
AC$_3$Cbi	9.2	nd
AdoCbi	10.2	6.1
MeCbi	12.7	9.6
DapCbi	17.0	23.9

[a] Sharp peaks but positions somewhat variable.
[b] Spreading.
[c] Compound unstable at pH 3.0; monocyano isomers observed.
[d] Isomeric forms.

(extended run) at the rate of 1.25%/ml flow volume. For routine analysis of chemically synthesized AdoCbl, MeCbl, and other synthetic naturally occurring cobalamins (e.g. AqCbl and HSO$_3$-Cbl) on 4.6 × 250 mm columns, the acetonitrile concentration was increased from 5 to 30% in 10 min at a flow rate of 2.0 ml/min and then recycled to initial conditions as shown in Fig. 1A. Program time was 18 min. For analysis of more hydro-

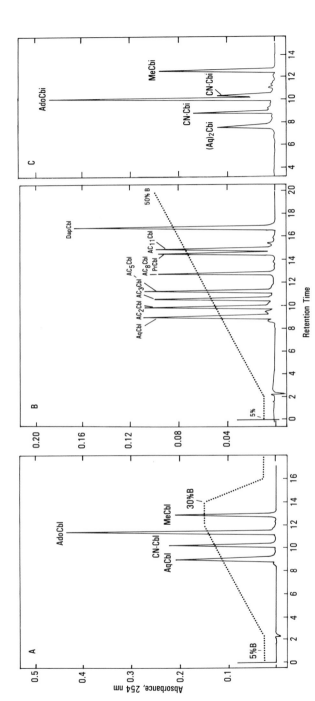

Fig. 1. Gradient HPLC separation of cobalamin and cobinamide standards on a 4.6 × 250-mm LiChrosorb C$_8$ column. The gradient profile is indicated in A and B by the dotted line. (A) Separation of four naturally occurring cobalamins (0.02 ml, each at 0.25 mM); 254 nm full-scale absorbance = 0.50. (B) Resolution of a mixture of cobalamins containing different upper-axial ligands. In this chromatogram, the gradient increased to 50% B and then decreased (not shown) to 5% B (initial conditions) over 8 min; 254 nm full-scale absorbance = 0.20. (C) Separation of cobinamide analogs (0.02 ml, each at 0.25 mM); 254 nm full-scale absorbance = 0.20. A gradient identical to that in A was used to develop this chromatogram. Each chromatogram was developed at 2.0 ml/min, and the chart speed was 1.0 cm/min.

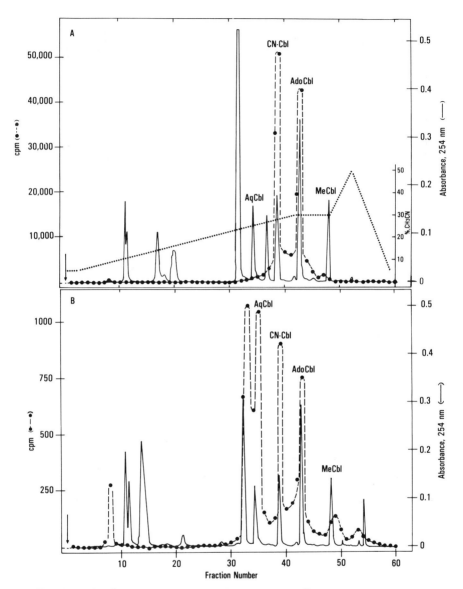

Fig. 2. (A) Gradient HPLC separation of *L. leichmannii* [57]Co-labeled cobalamins on a 4.6 × 250-mm LiChrosorb RP-8 column. Cell extract (100 μl) was mixed with 100 μl of an aqueous cobalamin standard mixture (AqCbl, CN-Cbl, AdoCbl, and MeCbl, each at 0.25 m*M*) and clarified by centrifugation at 12,000 *g* for 5 min. The supernatant (0.02 ml) was injected and developed by gradient elution at 0.50 ml/min; dotted line, percentage Pump B. The peaks generated by the four cobalamin standards are identified. Fractions (0.50 ml) of the column effluent were collected and their [57]Co content determined. (B) Gradient HPLC separation of L1210 [57]Co-labeled cobalamins L1210 cell extracts were prepared for HPLC injection as described in A. The chromatogram was developed, and fractions were collected as described in A.

phobic corrinoids (e.g., $AC_{11}Cbl$ and DapCbl), the gradient was extended to 50% acetonitrile and then recycled (Fig. 1B). Program time was 30 min.

Cell extracts containing ^{57}Co-labeled corrinoids were analyzed by gradient elution at a flow rate of 0.5 ml/min. The column was developed by a gradient from 5 to 30% over a 40-min period and then recycled as shown in Fig. 2A. Program time was 60 min.

Comments

Isocratic Method. The primary naturally occurring cobalamins were resolved on a reverse-phase column (LiChrosorb RP-8) in less than 5 min using 30% acetonitrile containing 1.0 mM ammonium acetate (pH 4.4). As shown in Table I, the order of elution was HSO_3-Cbl (1.1 min), CN-Cbl (1.6 min), MeCbl (2.2 min), AdoCbl (2.9 min), and AqCbl (4.7 min). AdoCbl and AqCbl eluted as broad peaks and their retention times were highly dependent upon the pH of the solvent system (Table I). The sensitivity of the isocratic elution method using UV absorbance at 254 nm was ~1 pmol for CN-Cbl. This system was particularly useful for resolving closely related upper-axial ligand analogs of AdoCbl (e.g., AraACbl, 3.3 min; ε-AdoCbl, 3.5 min; ForCbl, 4.3 min; and DNCbl, 5.2 min). Cobinamide analogs which lack the lower-axial nucleotide were also resolved by isocratic elution and retention times are reported in Table I. CN-Cbi (Factor B) eluted in two peaks (retention times = 4.9 and 5.7 min) which probably correspond to the axial position isomers (i.e., upper or lower cyanide). The cobinamide analogs of the cobalamin coenzymes, AdoCbi and MeCbi, separated as sharp peaks with retention times at 6.1 and 9.6 min, respectively.

Gradient Method. Improvements in corrinoid peak resolution, shape, and reproducibility were achieved by developing reverse-phase columns with a gradient of acetonitrile. The naturally occurring cobalamins (Fig. 1A) eluted in the order of AqCbl, HSO_3-Cbl, CN-Cbl, AdoCbl, and MeCbl with retention times (Table I) of 9.0, 9.7, 10.2, 11.4, and 12.8 min, respectively. Order of elution and retention times were very reproducible (±0.05 min) for reverse-phase columns of similar packing and dimensions (three 4.6 × 250 mm LiChrosorb RP-8 columns have been tested).

A mixture of eight cobalamins, including a series of five aminoalkylcobalamin homologs (AC_2Cbl, AC_3Cbl, AC_5Cbl, AC_8Cbl, and $AC_{11}Cbl$), was resolved into its components using a gradient of 5 to 50% acetonitrile (Fig. 1B).

The cobinamides were separated as well-resolved sharp peaks by gradient development as shown in Fig. 1C. $(Aq)_2Cbi$ eluted early (7.8 min) in the gradient system but late (33.8 min) in the isocratic system. When a

TABLE II
COBALAMIN RETENTION TIMES ON REVERSE-PHASE COLUMNS
DEVELOPED BY GRADIENT ELUTION

	Column		
Compound	LiChrosorb RP-8 (min)	Ultrasphere C_8 (min)	Ultrasphere C_{18} (min)
AqCbl	9.0	11.8	14.4
CN-Cbl	10.2	14.1	16.9
AdoCbl	11.4	16.6	19.5
MeCbl	12.8	19.6	22.4
Column size	4.6 × 250 mm	4.6 × 150 mm	4.6 × 250 mm
Flow rate	2.0 ml/min	1.0 ml/min	1.0 ml/min

neutral aqueous solution of $(CN)_2$-Cbi was injected, only the two CN-Cbi isomers were observed. This is probably due to the acidic pH of the gradient system (3.0) in which $(CN)_2$-Cbi is unstable. The gradient elution method has been used successfully with a number of different reverse-phase columns (Table II).

Gradient Elution Analysis of Intracellular Corrinoids. Extracts obtained from *L. leichmannii* and L1210 cells grown in the presence of [57Co]CN-Cbl were analyzed by the gradient method (Fig. 2A and B). The column was developed at 0.5 ml/min (5 to 30% in 40 min) which allowed for more efficient collection of the column effluent. Retention times of standards added to cell extracts were 32.9, 35.0, 37.4, 41.4, and 46.8 min for AqCbl, HSO_3-Cbl, CN-Cbl, AdoCbl, and MeCbl, respectively. Columns were regenerated by increasing the gradient to 50% and then decreasing it to starting conditions (Fig. 2A).

Extracts of early stationary-phase *L. leichmannii* cells contained the substrate [57Co]Cn-Cbl and [57Co]AdoCbl. No radioactivity was associated with AqCbl, HSO_3-Cbl, or MeCbl. In contrast extracts of late-log L1210 cells contained, in addition to 57Co-labeled CN-Cbl and AdoCbl, significant amounts of labeled AqCbl, MeCbl, and at least three other unidentified peaks which did not correspond to the cobalamin standards. The major peak centered in fraction 33 has a mobility identical to that of authentic GSCbl and thus, may be a newly identified intracellular form of cobalamin.

[3] Isolation of Native and Proteolytically Derived Ileal Receptor for Intrinsic Factor–Cobalamin

By Bellur Seetharam and David H. Alpers

The gastrointestinal absorption of cobalamin (vitamin B_{12}; Cbl) in mammals is initiated by two binding reactions. First, after being released from food by acid and pepsin the dietary cobalamin complexes with salivary Cbl binding proteins (nonIF). Cobalamin is subsequently released in the intestinal lumen after nonIF hydrolysis by serine proteases secreted from the pancreas.[1] The cobalamin thus released then binds to intrinsic factor (IF) secreted by the gastric mucosa. In the second binding reaction, the IF–Cbl complex thus formed binds, via a calcium dependent step, to the IF–Cbl receptor located in the ileum.[2]

Among all the mammalian ileal tissues tested canine ileal mucosal homogenates had the largest number of binding sites for human or hog IF–Cbl complex per g of mucosa.[3] This fact prompted us to use this tissue to isolate the receptor in milligram quantities.

The catalytic domain of most functional proteins of the brush border is on the lumenal side, thus facilitating hydrolytic or binding reactions involving nutrients.[4] These proteins can be isolated either in their native form by an initial solubilization using a nonionic detergent (e.g. Triton X-100), or they can be obtained in a hydrophilic form after solubilization using proteases (e.g. papain). The IF–Cbl receptor obtained by either method contained the ligand-binding sites. However, the receptor obtained by papain solubilization lacks the membrane anchor peptide.[5]

Methods

Assay for the Ileal Receptor

Receptor activity during the purification was assessed by measuring the binding of the hog IF–[^{57}Co]Cbl complex made with purified hog IF. Hog IF can be replaced by human IF. However the important consider-

---

[1] R. H. Allen, B. Seetharam, N. Allen, E. Podell, and D. H. Alpers, *J. Clin. Invest.* **61**, 1628 (1978).

[2] V. Herbert, *J. Clin. Invest.* **38**, 102 (1959).

[3] D. Hooper, D. H. Alpers, R. L. Burger, C. S. Mehlman, and R. H. Allen, *J. Clin. Invest.* **13**, 3074 (1973).

[4] D. H. Alpers and B. Seetharam, *N. Engl. J. Med.* **396**, 1047 (1977).

[5] B. Seetharam, S. S. Bagur, and D. H. Alpers, *J. Biol. Chem.* **256**, 9813 (1981).

METHODS IN ENZYMOLOGY, VOL. 123

ation is that IF preparations should be free of any possible contamination of other Cbl-binding proteins (nonIF). Purification of hog IF and its separation from non IF can be achieved by the method of Allen and Mehlman.[6]

The receptor activity in homogenates was determined by the Millipore filter assay.[3] However, once the receptor is solubilized, the separation of bound and free IF–[^{57}Co]Cbl can be achieved using precipitation with neutral ammonium sulfate. Forty percent neutral ammonium sulphate precipitates over 95% of receptor bound IF–[^{57}Co]Cbl, but only 3% of free IF–[^{57}Co]Cbl. In the standard assay 1 ml of reaction mixture contains Tris–HCl buffers (10 μmol), sodium chloride (140 μmol), CaCl$_2$ or Na$_2$EDTA (10 μmol), hog IF–[^{57}Co]Cbl complex (0–1.2 pmol), Triton X-100 (0.1%), and various amounts of solubilized receptor. After incubating for 60 min at 22° 50 μl of normal human serum is added, and the tube is cooled to 4°. Then 660 μl of 3.9 M (NH$_4$)$_2$SO$_4$ is added with mixing. After incubation for 10–15 min at 4°, the tube is centrifuged at 10,000 g for 15 min. The supernatant fraction containing unbound IF–[^{57}Co]Cbl and the pellet containing receptor bound IF–[^{57}Co]Cbl are assayed for radioactivity. Receptor specific binding is calculated as the difference between the binding observed in the presence of CaCl$_2$ and in the absence of calcium with EDTA.

Purification of the Ileal Receptor

Principle. The major step in the purification of the ileal receptor is immunoabsorbent chromatography. Solubilized extracts are first complexed with excess IF–Cbl in the presence of CaCl$_2$ at pH 7.5. The reaction mixture is passed over a CNBr-activated Sepharose column to which has been bound antiserum to IF. After extensive washing of the column, receptor can be eluted using pH 5.0 buffer containing EDTA and using Triton X-100. This treatment removes the receptor, leaving behind IF still attached to the anti-IF-Sepharose. The eluted receptor is neutralized immediately to pH 7.4 and CaCl$_2$ is added to complex the EDTA present in the elution buffer.

Tissues, Proteins, and Reagents. Dog ileums can be purchased from Pel Freeze Biologicals Company (Rogers, AR). Usually they are shipped in fresh ice, and can be obtained on the day of sacrifice. The mucosa is scraped and can be stored at $-70°$ for several weeks prior to use. Purified hog intrinsic factor prepared from hog stomach was used to raise antibody in rabbits as described by Allen and Mehlman.[6] Two milliliters of antiserum was coupled to 1 ml of CNBr-activated Sepharose 4B as described by Burger *et al.*[7]

---

[6] R. H. Allen and C. S. Mehlman, *J. Biol. Chem.* **248,** 3670 (1973).
[7] R. L. Burger, C. S. Mehlman, and R. H. Allen, *J. Biol. Chem.* **250,** 770 (1975).

[^{57}Co]CN-Cbl (specific activity 20 miCi/mmol) was purchased from Amerhsam Chemicals, Arlington Heights, IL). Crystalline CN-Cbl and papain (2× crystallized) were obtained from Sigma Chemical Company, St. Louis, MO. All other chemicals were of analytical grade.

Preparation of the Native IF–Cbl Receptor. Frozen mucosa (2.5 kg) was thawed and homogenized in a Waring Blender for 2–3 min at top speed using 5 mM potassium phosphate buffer (pH 7.4) containing 140 mM NaCl, 5 mM KCl, and 0.2 mM phenylmethylsulfonyl fluoride (Solution A). The 25% homogenate thus obtained is treated with Triton X-100 (1% v/v), stirred overnight (16 hr) at 4°, and then centrifuged at 20,000 g for 30 min. The supernatant fraction (10 liters) is now treated with 4.3 liters of absolute ethanol (stored at −20°) and stirred at 5° for 1–2 hr after the addition of ethanol. The turbid solution is centrifuged at 20,000 g for 30 min, and the pellet is suspended in Solution A, homogenized, and dialyzed against Solution A. This treatment results in the precipitation of the receptor but the volume is decreased from 10 to 3 liters.

The dialyzed material (ethanol free) is again treated with Triton X-100 1% (v/v) to extract the receptor. The solubilized receptor is now separated by centrifugation for 1 hr at 150,000 g_{max}. Of the clear supernatant 2.5 liters contains most of the receptor activity. Further reduction in volume can be achieved by a second ethanol precipitation and reextraction by Triton X-100. The receptor activity is fairly stable to repeated treatment with ethanol, which has the advantage of concentrating the receptor to a smaller volume (<1 liter). It is important to minimize the volume, since it is preferable to maintain low flow rates (20–25 ml/hr) during subsequent fractionation by affinity chromatography.

Solubilization of the Ileal IF–Cbl Receptor Using Papain. Intestinal tissue contains adsorbed pancreatic proteases, whose action on membrane proteins cannot be prevented during purification steps, but can only be minimized. When commercial tissues are used this can be a more serious problem because of the shipping time. It is important to remove pancreatic proteases first if one is to obtain a controlled proteolytic release of receptor from the brush border membrane. The use of homogenization buffer (0.25 M mannitol in 10 mM K-phosphate) containing 0.5 mM phenylmethylsulfonyl fluoride still results in the loss of nearly 20% of receptor into the supernatant fraction. The particulate fraction can be further washed at least twice in the same buffer, yielding a membrane preparation which had lost more than 99% of its endogenous proteases. However, such complete washing also removes close to 35–40% of the homogenate receptor activity.

The remaining receptor can be solubilized by incubating washed membranes (suspended and homogenized in 10 mM potassium phosphate buffer (containing 140 mM NaCl, 5 mM KCl, and 2 mM dithiothreitol) for

1 hr at 37° in the presence of papain (1 mg papain to 100 mg of homogenate protein). The reaction can be stopped by adding 0.5 mM phenylmethylsulfonyl fluoride. The reaction mixture is then centrifuged for 1 hr at 100,000 g_{max}, and the pellet discarded. The supernatant fraction containing the receptor activity (800 ml) is dialyzed against 24 liters of 5 mM potassium phosphate buffer, pH 7.4 containing 140 mM NaCl and 5 mM KCl. Solid ammonium sulfate is added to the dialyzed fraction to give a final concentration of 55%. The precipitate containing the receptor activity is recovered by centrifugation at 20,000 g, suspended, and dialyzed against the same buffer used for the first dialysis.

Immunoaffinity Chromatography on Anti-Intrinsic Factor-Sepharose. The soluble fraction (obtained either with Triton X-100 or papain extraction) is assayed for receptor activity. Based on the actual binding obtained at this stage of purification a 2-fold excess of hog intrinsic factor [^{57}Co]Cbl is added and incubated for 1 hr at 22° in the presence of 2.5 mM CaCl$_2$ and 1.25 mM MgSO$_4$ in 5 mM potassium phosphate buffer, pH 7.4.

Rabbit anti-hog intrinsic factor-Sepharose 4B, with a capacity to bind at least a 2-fold excess of intrinsic factor used in the previous incubation, is poured into a short column (2.5 × 1–2 cm). Before the application of the sample, the column is washed in sequence as follows: 250–500 ml of 5 mM potassium phosphate buffer, pH 5.0 containing 140 mM NaCl, 5 mM KCl, 5 mM EDTA, and 1% Triton X-100 (wash 1); 250–500 ml of potassium phosphate buffer, pH 7.4 containing 140 mM NaCl and 1% Triton X-100 (wash 2); and 250–500ml of 5 mM potassium phosphate buffer, pH 7.4 containing 140 mM NaCl, 2.5 mM CaCl$_2$, 1.25 mM MgSO$_4$, and 1% Triton X-100 (wash 3).

When the supernatant fraction from the papain solubilization is subjected to affinity chromatography the same sequence of washing is carried out, but in the absence of Triton X-100.

The IF–Cbl receptor now complexed with hog IF–[^{57}Co]Cbl and unbound IF–[^{57}Co]Cbl is passed on the column at a flow rate not exceeding 20–25 ml/hr. At this flow rate, 95–98% receptor complexes to the antibody via the bound IF–[^{57}Co]Cbl. Separation of unbound IF–[^{51}Co]Cbl can be monitored by measuring the decrease in [^{57}Co]Cbl radioactivity of the column breakthrough. If the flow rate decreases due to clogging of the column, this can be reversed by stopping the column, disconnecting the reservoir, gently stirring the slurry, allowing it to settle, and washing with 25–30 ml of buffer (pH 7.4). This will remove the small particles which are clogging the column. The receptor sample can now be added again.

After the entire sample passes through the column, the column is sequentially washed with 50 ml of wash 3 and 500 ml of wash 2 at a flow rate of 30–40 ml/hr. It is important during these washes to gently stir the

TABLE I

PURIFICATION OF THE TRITON-SOLUBILIZED CANINE ILEAL
RECEPTOR FOR INTRINSIC FACTOR–COBALAMIN[8]

Step	Volume (ml)	Total protein (mg)	Total IF–Cbl binding activity (nmol)	Specific activity (pmol/mg protein)
Triton-treated homogenate	12,300	600,000	23.2	0.039
First ethanol precipitation	3,800	30,000	17.7	0.59
Second ethanol precipitation	900	9,700	14.7	1.51
Immunoaffinity chromatography	65	2.4	6.05[a]	2,520

[a] Twenty-six percent recovery, 65,000-fold purification.

resin with a glass rod, in order that the washing will be complete. The elution of the bound receptor can be efficiently carried out with 50–60 ml of wash 1 containing 0.1% Triton X-100 at a flow rate of 50 ml/hr. It is important to titrate the pH to 7.4 immediately with 1 M NaOH, and to remove the effect of EDTA with addition of $CaCl_2$ to a final concentration of 5 mM. A typical purification for Triton and papain solubilized receptor is given in Tables I[8] and II[9]. The receptor can be stored at 5° for at least 4–6 weeks with no loss of activity. The receptor purified by the above procedures is very pure as judged by electrophoresis under both native and denaturing conditions. The Triton X-100 solubilized receptor has an M_r of 180,000 and consists of two subunits of M_r 59,000 and 42,000. Based on amino acid composition has 222×10^3 g of amino acids per mole of IF–[^{57}Co]Cbl binding activity. The receptor obtained with limited proteolysis is heterogeneous comprising 4 to 6 bands, but each protein band binds IF–[^{57}Co]Cbl. The M_r of these active bands varied from 100,000 to 170,000.

Comments on the Purification

The most important step in the purification is the immunoaffinity chromatography using rabbit anti-hog intrinsic factor bound to Sepharose 4B. For the purification of the dog receptor attempts to use either intrinsic factor alone or the intrinsic factor–cobalamin complex as an affinity absorbent resulted in very poor binding of the receptor compared to rabbit anti-hog intrinsic factor antiserum. The advantage of this method is the

[8] B. Seetharam, D. H. Alpers, and R. H. Allen, *J. Biol. Chem.* **256,** 3785 (1981).
[9] B. Seetharam, S. S. Bagur, and D. H. Alpers, *J. Biol. Chem.* **257,** 183 (1982).

TABLE II
PURIFICATION OF PAPAIN-SOLUBILIZED RECEPTOR FOR INTRINSIC FACTOR–COBALAMIN[9]

Step	Volume (ml)	Total protein (mg)	Total IF–Cbl binding activity (nmol)	Specific activity (pmol/mg protein)
Crude homogenate	2,000	75,000	3.0	0.04
0.25 M mannitol particulate fraction	1,330	25,562	1.6	0.061
Papain-solubilized 105,000 g supernatant	810	4,000	1.5	0.385
0–55% ammonium sulfate fraction	170	440	1.08	2.45
Immunoaffinity chromatography	20	0.10	0.191[a]	1910

[a] Six percent recovery; 47,000-fold purification.

recycling of both the intrinsic factor–cobalamin complex and the affinity ligand. After elution of the receptor the column can be washed with 0.2 M glycine–HCl buffer, pH 3.0, to remove the intrinsic factor–cobalamin complex which was bound to the antibody. The eluate is neutralized to pH 7 and can be reused after assessing the amount of complex present. The column is washed and stored in buffer used for wash 3.

[4] Purification of B$_{12}$-Binding Proteins Using a Photodissociative Affinity Matrix

By D. W. JACOBSEN and F. M. HUENNEKENS

B$_{12}$-binding proteins[1] play essential roles in the absorption, distribution, and retention of vitamin B$_{12}$ (cyanocobalamin; CN-Cbl) and its coenzyme forms, adenosylcobalamin (AdoCbl) and methylcobalamin (MeCbl). The Cbl-binding proteins are widely distributed among body tissues and fluids in concentrations ranging from 10^{-12} to 10^{-6} M. Although conventional methods have been used to purify intrinsic factor (IF) and transco-

[1] Supported by Grants AM35080 and CA6522 from the National Institutes of Health and Grant CH-31U from the American Cancer Society.

balamin II (TC-II) from hog pylorus[2] and human plasma,[3] respectively, affinity chromatographic techniques have proven to be more rapid and efficient for the isolation of Cbl-binding proteins[4-6] and Cbl-dependent enzymes.[7] This chapter outlines a general method for the affinity purification of Cbl-binding proteins from body fluids or crude extracts utilizing a photodissociative Cbl matrix.[8]

Principle

The unique organometallic carbon–cobalt bond of alkylcorrinoids (e.g. AdoCbl and MeCbl) homolyzes when exposed to light in the UV-visible range. In most alkylcorrinoids this bond is located in the β or upper-axial ligand position. Analogs containing aminoalkyl groups in the β position readily react with activated solid-phase matrices to give immobilized corrinoids that retain their high avidity for apoCbl-binding proteins. When exposed to visible light, the organometallic bond photodissociates and releases an aquacorrinoid–protein complex to solution.

Reagents

CN-Cbl, crystalline (Sigma)
Monocyanocobinamide (from CN-Cbl)[9] or dicyanocobinamide (Calbiochem)
3-Chloropropylamine hydrochloride (Aldrich)
SP-Sephadex C-25 (Pharmacia)
Sephacryl S-200 (Pharmacia)
Sephadex G-25-80 (Pharmacia)
Liquified phenol, reagent (J. T. Baker)
Zinc, powder, CP (#4280, J. T. Baker)

Preparation of Photodissociative Corrinoid Matrices

Synthesis of ω-Aminoalkylcorrinoids. The general procedure of Jacobsen[10] for the synthesis of Cbl coenzyme analogs is used for preparing ω-

[2] L. Ellenbogen and D. R. Highley, *J. Biol. Chem.* **242**, 1004 (1967).
[3] L. Puutula and R. Gräsbeck, *Biochim. Biophys. Acta* **263**, 734 (1972).
[4] R. H. Allen, R. L. Burger, C. S. Mehlman, and P. W. Majerus, this series, Vol. 34, p. 305.
[5] E. Nexø, *Biochim. Biophys. Acta* **379**, 189 (1975).
[6] J. Lindemans, H. van Kapel, and J. Abels, *Biochim. Biophys. Acta* **579**, 40 (1979).
[7] T. Toraya and S. Fukui, this series, Vol. 67, Part F, p. 57.
[8] D. W. Jacobsen, Y. D. Montejano, and F. M. Huennekens, *Anal. Biochem.* **113**, 164 (1981).
[9] P. Renz, this series, Vol. 18, p. 82.
[10] D. W. Jacobsen, this series, Vol. 67, Part F, p. 12.

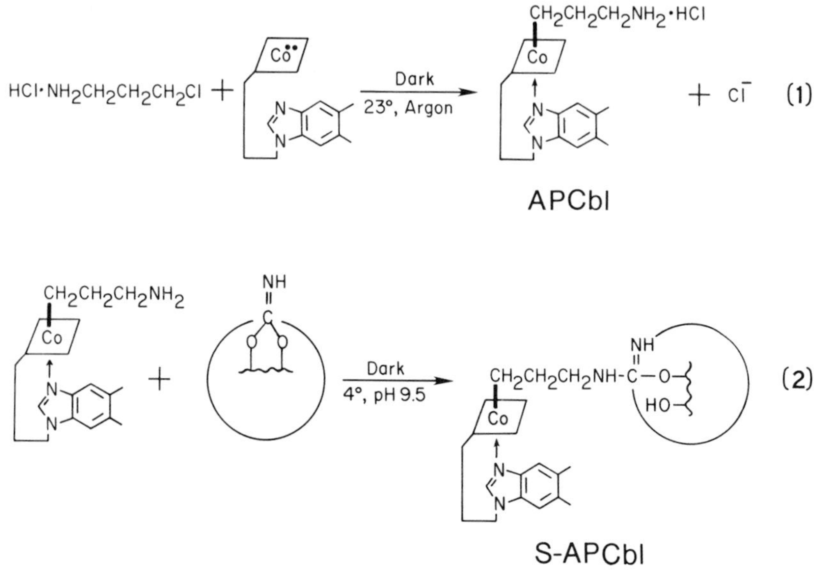

FIG. 1. Synthesis of the photodissociative matrix S-APCbl. Reaction (1): Cob(I)alamin, produced by reducing CN-Cbl with Zn, reacts with 3-chloropropylamine hydrochloride to form APCbl. Reaction (2): APCbl reacts with CNBr-activated Sephacryl S-200 to produce S-APCbl.

aminoalkylcorrinoids. A solution containing 1.00 g of CN-Cbl (0.70 mmol) and 50 g of NH$_4$Cl in 500 ml of water is purged with argon for 20 min at room temperature to remove dissolved O$_2$. With vigorous stirring under argon, 25 g of powdered Zn is added. During the next 30 min, the deep-red solution of CN-Cbl changes to a brown cob(II)alamin intermediate and then to the gray-black 2-electron reduced product cob(I)alamin. Although gray-black in appearance under the conditions described, dilute solutions of cob(I)alamin are actually green.

The following steps are carried out under dim-red illumination. A 5-fold molar excess (relative to CN-Cbl) of 3-chloropropylamine hydrochloride, dissolved in 50 ml of argon-purged distilled water, is added dropwise over a 10-min period to the stirring Zn/cob(I)alamin mixture [Reaction (1), Fig. 1]. Vigorous purging with argon is maintained during these steps. Fifty minutes after the addition of 3-chloropropylamine, the reaction is terminated by filtration through Whatman No. 1 paper. The product, aminopropyl-Cbl (APCbl), is extracted from the filtrate 3 times with 50 ml of liquified phenol : chloroform (1 : 1),[11] and the combined extracts are

[11] A. W. Johnson, L. Mervyn, N. Shaw, and E. L. Smith, J. Chem. Soc. p. 4146 (1963).

washed 5 times with equal volumes of water. After adding 3 volumes of n-butanol : chloroform (1 : 1) and 0.70 ml of 1 N HCl to the organic phase, APCbl is back-extracted into 50 ml of water 2 times. (Protonation of the primary amine on AP-Cbl facilitates its back-extraction into water.) The combined aqueous extracts are washed 3 times with equal volumes of ether to remove traces of organic solvents and then concentrated to dryness by rotary evaporation in vacuo at 45°. The residue is dissolved in 10–15 ml of water and applied to a 4 × 45 cm column of SP-Sephadex (C-25, Na$^+$ form in water).[12] After washing the column with 1 liter of water and 0.5 liter of 0.05 M sodium acetate (pH 5.0), APCbl is eluted with 0.10 M sodium acetate (pH 7.0). Fractions containing product are combined and desalted by extraction into phenol : chloroform as described above. APCbl is crystallized from aqueous acetone at 4° with a final yield of ≥80% based on CN-Cbl starting material.

The product migrates as a single red spot on Whatman No. 1 paper in the previously described solvent systems A, B, and C[13] with R_f values (relative to CN-Cbl) of 0.58, 0.70, and 0.56, respectively. It elutes from a C$_8$ reverse-phase HPLC column as a single symmetrical peak in 10.5 min using the gradient system described by Jacobsen et al.[14] Principal absorbance bands: 520 nm (ε_{mM} = 8.73), 374(10.8), 340(13.1), 315(13.7), 282(19.2), and 265(20.6) in 0.1 M sodium phosphate (pH 7.0); 452(9.52), 376(8.91), 303(24.6), 284(23.1), and 263(28.7) in 0.1 N HCl. The crystallization step can be omitted (and usually is) when APCbl is to be used primarily for the preparation of affinity matrices. With only minor modifications of the above procedures, the following ω-aminoalkylcorrinoids have been prepared and coupled to activated Sephacryl matrices: aminoethyl-Cbl, aminopentyl-Cbl, aminooctyl-Cbl, aminoundecyl-Cbl, aminopropylcobinamide, and aminoundecyl-Cbi. In the latter two preparations monocyanocobinamide replaces CN-Cbl.

Immobilization of ω-Aminoalkylcorrinoids. Washed Sephacryl S-200 beads (50 ml settled volume) are activated with CNBr (10 g in 5 ml of acetonitrile) according to the method of March et al.[15] and added to 100 ml of 2.5 mM APCbl in 0.2 M NaHCO$_3$ [Reaction (2), Fig. 1]. After agitation on a low-speed shaker for 16 hr at 4°, 100 ml of 1.0 M sodium glycinate (pH 9.0) is added to block residual coupling sites, and stirring is continued for an additional 2 hr at room temperature. The Sephacryl-APCbl beads

[12] G. Tortolani, P. Bianchini, and Y. Mantovani, *J. Chromatogr.* **53,** 577 (1970).
[13] D. W. Jacobsen, R. J. Holland, Y. D. Montejano, and F. M. Huennekens, *J. Inorg. Biochem.* **10,** 53 (1979).
[14] D. W. Jacobsen, R. Green, E. Quadros, and Y. D. Montejano, *Anal. Biochem.* **120,** 394 (1982).
[15] S. C. March, I. Parikh, and P. Cuatrecasas, *Anal. Biochem.* **60,** 149 (1974).

(S-APCbl) are collected on Whatman No. 1 paper by suction filtration, washed 3 times with 100 ml of 0.05 M potassium phosphate (pH 7.0) containing 0.5 M NaCl (hereafter referred to as "PS buffer"), and 3 times with 100 ml of water. The coupling efficiency of APCbl to Sephacryl under the conditions described is about 20% (i.e., ~1 μmol of APCbl/ml settled beads). This is determined indirectly by measuring spectrophotometrically the recovery of unreacted APCbl which is repurified for future S-APCbl preparations, or directly by photolyzing S-APCbl and measuring released AqCbl as shown in Fig. 2. The deep-red colored beads are stored in the dark at 4° in 5.0 mM sodium EDTA (pH 7.0) for up to 1 year.

Affinity Purification of Cbl-Binding Proteins

Preparation of Body Fluids and Tissue Extracts. Particulates are removed from crude starting material by filtration (if feasible) or by centrifugation. It is important that the starting material containing apoCbl binding protein be free of suspended matter.

Frozen rabbit serum (Type III, nonfiltered, nonsterile) is thawed and

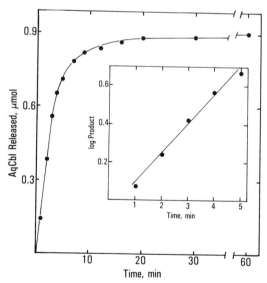

FIG. 2. Photodissociation of S-APCbl during exposure to visible light. S-APCbl (1 ml) was suspended in 30 ml of PS buffer and exposed to visible light as described under "Photodissociation of Cbl-binding proteins from S-APCbl." At the indicated times, the suspension was centrifuged and the amount of AqCbl in the supernatant was determined spectrophotometrically at 350 nm ($\varepsilon_{mM} = 24.4$). Inset: first-order plot of data (initial five points); rate constant = 0.32/min. From Jacobsen *et al.*[8]

suction filtered through a layer of cotton gauze over Whatman No. 4 paper. Pepsin-inactivated human gastric juice and human saliva are thawed and filtered in a similar fashion. Extracts from rat stomachs are prepared by homogenization in a blender followed by centrifugation at 4° until a clear supernatant is obtained. Human neutrophils are lysed by sonication in the presence of 0.5% Triton X-100 and debris is removed by centrifugation. The unsaturated Cbl-binding capacity (UCBC; equivalent to apobinder concentration) of the clarified starting material is then determined by the method of Gottlieb[16] modified as follows: samples are diluted with 0.10 M potassium phosphate (pH 7.5) containing 1 mg/ml bovine serum albumin to an approximate UCBC of 5 ng CN-Cbl/ml or less. Aliquots (0.25 ml) are then incubated with 5.0 ng of [^{57}Co]CN-Cbl (0.50 ml containing 15,000 dpm) for 15 min at 37°. Hemoglobin-coated charcoal[16] (0.50 ml) is added and the mixture is incubated for another 5 min at 37°. After centrifugation, ^{57}Co radioactivity is determined in 0.50 ml of the supernatant. The R-binder content of human gastric juice can be determined prior to IF purification by the method of Begley and Trachtenberg[17] with slight modification (the ratio of CN-Cbi to CN-Cbl is 100:1 and hemoglobin-coated charcoal replaces albumin-coated charcoal). Approximately 50% of the human gastric juice specimens screened in this laboratory have significant R-binder present which may account for up to 60% of the total UCBC. Since the present methodology is not selective, both Cbl-binding proteins will copurify; pure IF is obtained if R-binder is absent from the starting material.

Adsorption of apoCbl-Binding Proteins by S-APCbl. It is essential that the following procedures be conducted under dim-red illumination. Because S-APCbl beads slowly hydrolyze during storage (<2% of their Cbl content/month), it is necessary to remove free AqCbl by washing the beads on a Millipore filter 3 times with 25 ml of PS buffer and 3 times with 25 ml of water.

A volume of S-APCbl beads with a Cbl content 100 times the total UCBC of the starting material is added and the suspension is stirred at room temperature until the UCBC drops to near zero (Fig. 3). In representative experiments 2.0, 2.7, and 0.6 ml of S-APCbl beads (1.0 μmol Cbl/ml) are added to 1 liter of rabbit serum (UCBC = 27 ng CN-Cbl/ml; total UCBC = 20 nmol), human gastric juice (UCBC = 36.6 ng Cbl/ml; total UCBC = 27 nmol), and human saliva (UCBC = 7.3 ng Cbl/ml; total UCBC = 5.4 nmol), respectively. After adsorption is complete, the beads are allowed to settle at 0° for 60 min. The bulk of the supernatant is

[16] C. Gottlieb, K.-S. Lau, and L. R. Wasserman, and V. Herbert, *Blood* **25**, 875 (1965).
[17] J. A. Begley and A. Trachtenberg, *Blood* **53**, 788 (1979).

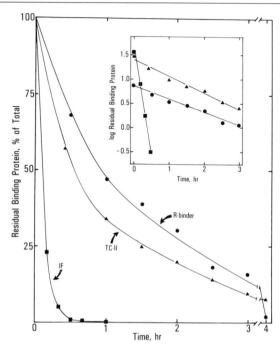

FIG. 3. Adsorption of rabbit serum TC-II (▲), human gastric juice IF (■), and human salivary R-binder (●) by S-APCbl. Experimental conditions are described under "Adsorption of apoCbl-binding proteins by S-APCbl." At the indicated times, 0.5-ml aliquots were removed, centrifuged, and the supernatants analyzed for UCBC. Inset: first-order plot of the data; rate constants are 0.013, 0.107, and 0.011/min for TC-II, IF, and R-binder, respectively. From Jacobsen et al.[8]

removed by aspiration, and the beads are transferred to a 50 ml centrifuge tube, pelleted, and washed several times with 25 ml aliquots of PS buffer at 4°. For S-APCbl–IF from human gastric juice and S-APCbl–R-binder from human saliva, a minimum of 3 washes are required to remove non-specifically adsorbed protein. For rabbit serum TC-II, it is necessary to wash more extensively: 5 times with 25 ml of PS buffer, 2 times with 25 ml of 0.05 M sodium acetate (pH 4.0) containing 0.5 M NaCl, 2 times with 25 ml of PS buffer, 2 times with 25 ml of 0.05 M sodium glycinate (pH 9.0), and 3 times with 25 ml of PS buffer. The washed bead preparations containing immobilized binder can be stored in the dark in PS buffer at 4° for up to 6 months.

Photodissociation of Cbl-Binding Proteins from S-APCbl. The washed beads containing adsorbed protein (e.g., S-APCbl-TC-II) are suspended in

10 volumes of PS buffer and irradiated with visible light (300-W bulb at 10 cm) for 30 min. The temperature of the photolysis tube is maintained at ~25° with an air blower. The beads are collected by centrifugation and the supernatant is recovered. The beads are resuspended in 5 volumes of PS buffer and photolyzed for 15 min. This last step is again repeated, and the three supernatants are combined and concentrated to 1–2 ml over a 43-mm Amicon YM10 membrane at 4°. The concentrate (in 1-ml aliquots) is applied to a 1 × 25 cm column of Sephadex G-25-80 in PS buffer and eluted in 1 ml fractions with the same buffer at 4°. The Cbl-binding proteins elute in fractions 8 and 9 while free AqCbl elutes as a broad band in fractions 16–18. Any remaining free AqCbl in the protein pool is removed by dialysis in PS buffer at 4° (3 × 1 liter, 24 hr each).

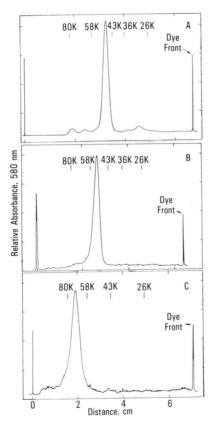

FIG. 4. SDS–PAGE of 5 μg of purified rabbit serum TC-II (A), 14 μg of human IF (B), and 3 μg of human R-binder (C). From Jacobsen et al.[8]

Comment

Recovery of rabbit serum TC-II is $\geq$ 90%, and purity, as judged from SDS–polyacrylamide gel electrophoresis[18] (Fig. 4) and spectrophotometric criteria, is usually greater than 80% if the stringent wash protocol is performed. Recovery of human IF recovery is consistently $\geq$95% with a purity of >95%. A similar result has been obtained with R-binder from of human saliva. In addition, near-homogeneous preparations of the following Cbl-binding proteins have been obtained: transcobalamin I (chicken serum); granulocyte Cbl-binder (human neutrophils); IF (rat stomachs); IF and non-IF (hog pylorus); and transcobalamins X, Y, and Z (salmon serum). In the above preparations, the Cbl-binding protein is obtained as a complex with AqCbl. The latter can be removed by dialyzing the complex against guanidinium hydrochloride as described by Allen *et al.*[4] or by reacting the complex with sodium ascorbate.[19] The apo-binding proteins can then be treated with [^{57}Co]CN-Cbl to form radiolabeled complexes for binding and transport studies.

---

[18] K. Weber, J. R. Pringle, and M. Osborn, this series, Vol. 26, Part C, p. 3.
[19] D. W. Jacobsen, Y. D. Montejano, and L. L. Zak, *Blood* **58**, Suppl. 1, 29a.

---

[5] Solid-Phase Immunoassay for Human Transcobalamin II and Detection of the Secretory Protein in Cultured Human Cells

By Marijke Fràter-Schröder

Vitamin B$_{12}$ (cobalamin, Cbl) in blood is transported by specific carrier proteins, transcobalamin II (TC2) and R-binder.[1] The two forms of transport proteins are immunologically and functionally distinct.[2,3] Plasma TC2 (MW 38,000, not glycosylated) is essential for absorption, transport, and promotion of cellular uptake of Cbl in man.[1] In contrast with the well-defined role in Cbl distribution, attributed to TC2, the function of the R-binder type of Cbl-binding proteins, also named haptocorrin, is not clear.[3] Studies on the origin of TC2 have been performed predominantly in animals. Various organs and tissues, including liver and kidney, release or

---

[1] C. A. Hall, ed., "The Cobalamins, Methods in Hematology," Vol. 10. Churchill-Livingstone, Edinburgh and London, 1983.
[2] M. Fràter-Schröder, *Mol. Cell. Biochem.* **56**, 5 (1983).
[3] E. Nexø and H. Olesen, *in* "B12" (D. Dolphin, ed.), Vol. 2, Chapter 3, p. 57. Wiley, New York, 1982.

produce TC2.[4] *De novo* synthesis has been confirmed in cell culture, using rat hepatocytes,[4] mouse fibroblasts,[4] and mouse mononuclear phagocytes.[5] The actual sites of TC2 synthesis in man are becoming apparent. *In vivo* investigations suggested that bone marrow[6] and intestine[7] are possible sources of human TC2. Another report indicated that low levels of intracellular TC2 can occur in human fibroblasts.[8] More recently, these findings were supported by the fact that both cultured skin fibroblasts and cultured bone marrow cells synthesize and secrete substantial quantities of TC2.[9]

A series of assays based on immunological principles have been reported for the quantification of human TC2 in serum.[10–16] Total, immunoreactive TC2 was determined using various modifications of the classical RIA procedure.[10,11,13–15] Attempts to quantitate holo-TC2 (approximately 10% of total), and apo-TC2 (a corresponding 90% of total TC2) in serum selectively have also been described.[12,13,15,16] The method for the quantification of total TC2 in human serum presented here uses immobilized antiserum[15] and is preferred by the author, because it is suitable for routine measurements, allowing 100 samples per day to be processed. It requires only 25 μl serum, employs extremely low levels of a stable radioactive label, and can be applied to the detection of low secretory TC2 levels in the medium of cultured human cells.[2,9] A solid-phase RIA for TC2, based on the principles and on the purification schemes presented below, has recently become commercially available.[16a]

[4] P. D. Green and C. A. Hall, this series, Vol. 67, Part F, p. 89.

[5] B. Rachmilewitz, M. Rachmilewitz, M. Chaoat, and M. Schlesinger, *Blood* **52**, 1089 (1978).

[6] M. Fràter-Schröder, C. Nissen, J. Gmür, L. Kierat, and W. H. Hitzig, *Blood* **56**, 560 (1980).

[7] I. Chanarin, M. Muir, A. Hughes, and A. V. Hoffbrand, *Br. Med. J.* **1**, 1453 (1978).

[8] N. Berliner and L. E. Rosenberg, *Metab., Clin. Exp.* **30**, 230 (1981).

[9] M. Fràter-Schröder, H. J. Porck, J. Erten, M. R. Müller, B. Steinmann, L. Kierat, and F. Arwert, *Biochim. Biophys. Acta* **845**, 421 (1985).

[10] R. J. Schneider, R. L. Burger, C. S. Mehlman, and R. H. Allen, *J. Clin. Invest.* **57**, 27 (1976).

[11] T. A. Morelli, C. R. Savage, J. A. Begley, and C. A. Hall, *J. Lab. Clin. Med.* **89**, 645 (1977).

[12] E. Nexø and J. Andersen, *Scand. J. Lab. Invest.* **37**, 723 (1977).

[13] M. Fràter-Schröder, P. Vitins, and W. H. Hitzig, in "Vitamin B12" (B. Zagalak and W. Friedrich, eds.), p. 877. de Gruyter, Berlin, 1979.

[14] P. A. Seligman, L. LaDonna, B. S. Steiner, and R. H. Allen, *N. Engl. J. Med.* **303**, 1209 (1980).

[15] M. Fràter-Schröder, L. Kierat, R. Y. Andres, and J. Römer, *Anal. Biochem.* **124**, 92 (1982).

[16] J. Lindemans, M. Schoester, and J. van Kapel, *Clin Chim. Acta* **132**, 53 (1983).

[16a] Eurodiagnostics BV, Apeldoorn, The Netherlands.

Purification of TC2

Preparation 1: TC2 Concentrate, R-Binder-Free for Tracer Preparation

The TC2 concentrate described here can be obtained from the Swiss Red Cross Central Laboratory, Berne, Switzerland. The human plasma fraction NB is the starting material. Fraction NB is similar to Cohn fraction III paste in its protein composition, but it is obtained by a slightly different fractionation technique.[17] An ion-exchange procedure, CM-Sephadex separation[18–20] was used to concentrate TC2 and also to remove the last trace of R-binder from NB. A recent modification[20] was optimized. At 4°, 6–8 kg of frozen NB (chopped) was stirred overnight and dissolved in 30 liters of 75 mM sodium phosphate, 600 mM sodium chloride, pH 5.2 buffer. Insoluble material was removed by depth filtration through cellulose pads (Filtrox AF9, Switzerland). The filtrate was diluted 3-fold and 30 g dry CM-Sephadex (C50 Pharmacia, Sweden) was added and stirred for 12 hr, in order to absorb TC2. The CM-Sephadex was washed repeatedly with 25 mM sodium phosphate, 200 mM sodium chloride, pH 5.2 buffer. Resuspension in 400 ml elution buffer (50 mM sodium phosphate, 1 M sodium chloride, pH 6.0) stirring for 1 hr, and filtration gave the first eluate with 75% of TC2. A second treatment with 200 ml buffer produced the final eluate (25% of TC2). The combined eluates (550 ml) were clarified by filtration through a 3-μm Millipore membrane and stored frozen. The resulting TC2 concentrate contained 40,000–60,000 ng/liter apo-TC2 measured by its unsaturated vitamin B$_{12}$ binding capacity (UBBC)[21] or by gel filtration[22] using G-150 Sephadex and 0.1 M phosphate buffer, pH 7.4. The specific protein activity was approximately 0.1 μg TC2 per mg protein. This preparation was stable at −20° for several years. This batch separation can be performed in 4-fold to yield 2200 ml as starting material for the affinity chromatography.

Preparation 2: Affinity Purified TC2 for Antiserum Preparation

Thermolabile ligand affinity chromatography,[23] based on the thermolabile immobilization of cyanocobalamin (CN-Cbl)[20] was used for further

[17] P. Kistler and H. Nitschmann, *Vox Sang.* **7**, 414 (1962).
[18] R. H. Allen and P. W. Majerus, *J. Biol. Chem.* **247**, 7709 (1972).
[19] C. R. Savage, A. M. Meehan, and C. A. Hall, *Prep. Biochem.* **6**, 99 (1976).
[20] J. van Kapel, B. G. Loef, J. Lindemans, and J. Abels, *Biochim. Biophys. Acta* **676**, 307 (1981).
[21] H. S. Gilbert, *J. Lab. Clin. Med.* **89**, 13 (1977).
[22] U. H. Stenman, K. Simons, and R. Gräbeck, *Scand. J. Clin. Lab. Invest.* **21**, 202 (1968).
[23] E. Nexø, H. Olesen, D. Bucher, and J. Thomsen, *Biochim. Biophys. Acta* **379**, 189 (1975).

purification of TC2. The procedure presented here is preferred over other purification schemes[18,19] because it avoids the relatively drastic exposure to guanidinium chloride to release the binding proteins. A more recent interesting modification of the affinity procedure employs photorelease of Cbl–protein complexes,[24] but this has not yet been put to use for the purification of human TC2. In the thermolabile ligand affinity chromatography procedure, unsaturated Cbl-binding proteins, retained on CN-Cbl-Sepharose at 4°, are dissociated together with bound Cbl by increasing the temperature to 37°.

Starting material was either the commercially available concentrate of preparation 1, derived from plasma fraction NB, or the equivalent preparation from Cohn III.[20] The affinity column was prepared with a spacer as described by van Kapel *et al.*[20]: 3 g activated CH-Sepharose 4B (Pharmacia, Sweden), washed and swollen according to the instructions of the manufacturer, was coupled to 3,3'-diaminodipropylamine, using 30 ml 10% (v/v) amine in 0.1 M sodium bicarbonate, 0.5 M sodium chloride buffer, pH 8.2, and overnight rotation at room temperature. Excess spacer was removed by washing with coupling buffer. Hydroxocobalamin (OH-Cbl) was added (12 mg in 10 ml coupling buffer) and incubated overnight with the CH-Sepharose conjugate in order to saturate the column with Cbl. The prepared column can be stored at 4°, protected from light, for several days in the presence of excess OH-Cbl. The suspension was transferred to the cold room (4°) and poured into a glass column (2.4 × 7.5 cm), washed with coupling buffer to remove excess OH-Cbl, resuspended in 30 ml of coupling buffer containing 0.2 M KCN, and incubated for 1 hr in order to convert bound OH-Cbl to CN-Cbl. This step is important in order to prevent unspecific absorption of other proteins to the affinity ligand.

Unbound CN⁻ was removed by washing with coupling buffer and the column was equilibrated with 0.1 M sodium phosphate, 1 M sodium chloride buffer, pH 6.0, and used directly for affinity chromatography of apo-TC2 as described.[20] The column contains approximately 2.7 mg cyanocobalamin, enough to bind 70 mg of TC2 protein. Eight 2.2 liter batches of the CM-Sephadex eluate (derived from a 4-fold batch of preparation 1 or Cohn III extract) were thawed at 4° and successively applied to the affinity column (flow rate 150 ml/hr) in a period of 8 days. Between each batch, the column was rigorously washed with 50 ml portions of (1) 50 mM sodium phosphate, 1 M NaCl, pH 6.0, (2) 0.1 M Tris–HCl, 1 M NaCl, pH 8.0, and (3) 25 mM sodium phosphate, 0.2 M NaCl, pH 5.2. The rinses

[24] D. W. Jacobsen, Y. D. Montejano, and F. M. Huennekens, *Anal. Biochem.* **113**, 164 (1981).

were repeated twice in reversing order. After the final washing, the column was resuspended in 15 ml of 0.1 M Tris–HCl, 1 M NaCl, 0.02% NaN$_3$ buffer, pH 8.0, transferred to a clean glass column, and TC2-cobalamin was eluted from the affinity adsorbent with twice 25 ml of warmed Tris–HCl buffer after 5 and 10 hr of incubation at 37°. The cooled effluents were directly concentrated to about 20 ml in an Amicon ultrafiltration cell (model 52), with a YM10 membrane (Amicon). The concentrate was centrifuged at 105,000 g for 1 hr and the supernatant was chromatographed in two portions on a Sephacryl S-200 column with the pH 8.0 Tris–HCl buffer. The pure TC2-cobalamin fractions were pooled and concentrated as above and stored at −20°. The purification scheme and yield of TC2 is outlined in Table I.

Immunization Procedure

Two rabbit anti-human TC2 antiserum preparations derived from two different sources of purified TC2 were equally suitable in the radioimmunoassay (RIA) described here.

Approximately 1 mg of pure holo-TC2 in 2.2 ml, prepared according to Savage et al.,[19] was mixed with an equal volume of Freund's adjuvant. One-half of this preparation was injected via a footpad into each of two male New Zealand White rabbits. A booster injection of the same amount was given to each rabbit 18 days later. The first blood sample was taken 18 days after the booster injection. Subsequent samplings were conducted at 7 day intervals. Whole serum was either frozen at −20° or lyophilized for storage. The second anti-human TC2 antiserum[20] was prepared from affinity purified TC2 as described. Rabbits were immunized by intramuscular injections of 0.25 mg pure human holo-TC2 with 1 ml complete Freund's adjuvant. Immunization was continued every 2 weeks by subcutaneous injections of 0.1 mg human holo-TC2. Blood was drawn in one batch by heart puncture, 3 months after the first injection. The second antiserum had similar specificities as the first antiserum, but it could be used at higher dilutions in the assay, indicating that it was four times more active than the first sample.

Preparation of Immobilized Anti-Human
   TC2 Antiserum (Immunosorbent)

A commercial substitute for the solid support used here is named "immunobead reagent" and can be obtained from Bio-Rad Laboratories, Richmond, California. The solid support consisted of copolymer beads of

TABLE I

PURIFICATION OF HUMAN TRANSCOBALAMIN II–CYANOCOBALAMIN[a]

Step of purification	Volume (ml)	Total protein (mg)	Total cobalamin binding capacity or bound cobalamin (μg)	Specific cobalamin or binding capacity		Purification factor	Yield (%)
				(μg/mg)	(mol/mol)		
Cohn fraction III (122.5 kg) solubilized	720,000	27.76×10^6	1015^b	$3.66 \times 10^{-5\,b}$	$1.03 \times 10^{-6\,b}$	1	100
CM-Sephadex G-50 batchwise elution	17,025	0.32×10^6	919^b	$2.87 \times 10^{-3\,b}$	$8.05 \times 10^{-5\,b}$	78	90.5
Affinity chromatography on CN-Cbl Sepharose	20.9	31.5	864	27.4	0.77	7.49×10^5	85.1
Sephacryl S-200 gel filtration	32.4	21.7	742	34.2	0.96	9.34×10^5	73.1

[a] Reproduced with permission from van Kapel et al.[20]
[b] Binding capacity.

acrylamide and acrylic acid.[25] The antiserum was coupled to the —COOH groups of the polyacrylamide beads as described for trypsin and ribonuclease A[25] with appropriate modifications presented before[15]: 50 μl of lyophilized reconstituted anti-TC2 antiserum, cleared by centrifugation, was mixed with 100 mg swollen polyacrylamide beads in 30 ml 0.003 M KH$_2$PO$_4$ buffer, pH 6.3, and left for 1 hr at 4°. Ten milligrams carbodiimide [1-ethyl-3-(3-dimethylaminopropyl)carbodiimide · HCl from Sigma, Germany) was added and the suspension was rotated for 3 hr at 4°. The insoluble matrix was collected by centrifugation, aspirated, and washed alternatively with phosphate buffer, pH 6.3, phosphate buffers with increasing and decreasing NaCl content, 4.5 M urea, assay buffer omitting albumin, finishing with assay buffer (0.05 M sodium, potassium phosphate, 0.1% NaN$_3$, 1% Tween 80, and 0.5% human serum albumin, pH 7.4). The final suspension was made up to 10 ml with assay buffer and lyophilized. The lyophilized material could be used for at least 2 years when stored at 4°.

Radiolabeling of TC2 (Tracer)

The partially purified TC2 concentrate from preparation 1 above (40,000 ng/liter Cbl equivalent apo-TC2) was diluted 20-fold with 0.05 M phosphate, 0.1% NaN$_3$ buffer, pH 7.4 containing radioactive cyanocobalamin (specific activity of 200 mCi/mg from Radiochemical Center, Amersham, England) in order to saturate 80–90% of apo-TC2. After 15 min incubation at 37°, traces of free radioactivity were removed by treatment with albumin-coated charcoal and centrifugation. The final tracer solution was filtered over a 0.45-μm Millipore sterile filter, and contained approximately 1500 pg/ml labeled TC2. This tracer solution was stored frozen, and was used for up to 6 months. The stability of the TC2-[^{57}Co]CN-Cbl complex was tested by gel filtration after incubation for 30 min at 37°, and 17 hr at 4°. The total radioactivity remained bound to TC2.[11]

Preparation of Serum Controls

High control (3000–4000 ng/liter apo-TC2). The partially purified TC2 (UBBC 40,000 ng/liter) from preparation 1 was diluted 20-fold with normal serum which was also used for the standard curve. The apo-TC2 content of this serum control was thus enhanced 3-fold, without appreciably changing the protein content of the serum. The total TC2 content of

---

[25] R. Mosbach, A. C. Koch-Schmidt, and K. Mosbach, this series, Vol. 44, p. 53.

the high control was enhanced 3-fold as well, because both the TC2 concentrate from preparation 1 and the whole serum contain holo-TC2 amounting to 10–12% of total TC2.[15] The serum was stored frozen.

Low control (500–600 ng/liter apo-TC2). *In vivo* parental administration of CN-Cbl causes an increased flux of TC2 into tissues and a corresponding drop of serum level.[26] Blood was withdrawn 30 min after 200 μg CN-Cbl was injected intramuscularly in a healthy person, and it was noticed that the serum level had dropped to 55% of the preinjection TC2 level. The resulting TC2 level (500–600 ng/liter) was in the order of magnitude observed for heterozygous-deficient relatives of TC2-deficient patients.[2] The serum was stored frozen.

TC2-free control (<25 ng/liter apo-TC2). TC2-free serum was used to test interference of tracer binding to serum components. It was made by treating normal serum with immobilized anti-TC2 antiserum.[15] One hundred microliters of anti-TC2 antiserum bound to 400 mg polyacrylamide beads was rotated with 5 ml serum from a healthy individual for 1 hr at 22° and overnight at 4°.

Centrifugation yielded serum with reduced TC2 which was again incubated with the same amount of regenerated immobilized anti-TC2 antiserum (bound TC2 was removed to regenerate the solid support by washing in 0.2 M glycin, 0.5 M NaCl at pH 2.8, final washing with 0.05 M sodium phosphate, 0.5% human serum albumin buffer, pH 7.4). Subsequent gel filtration (Sephadex G-150 as in Ref. 22) disclosed lack of apo-TC2 (<25 pg/ml), whereas the unsaturated R-binder fraction in the treated serum was retained. The total TC2 content of two "TC2-free" serum preparations measured by the RIA presented here was found to be 80–100 pg/ml Cbl equivalent. This indicates that slight interference from serum components does occur, and that this interference is in the order of 5–10% of normal total TC2 levels.

Standard Curve and the Determination of Total TC2 in the Reference Sample

Fresh serum from whole clotted blood, drawn from 5 healthy donors, was mixed and frozen in aliquots after the addition of 0.1% NaN_3; it could be used for at least 6 months for the standard curve described in Table II. Heparin which complexes with TC2 alters the antigen behavior of TC2 in the RIA and should be avoided. The total TC2 level of the mixed serum was calculated by comparison with total TC2 in the TC2 concentrate (preparation 1) as follows. The endogenous Cbl content of the TC2 con-

[26] J. A. Begley, T. A. Morelli, and C. A. Hall, *N. Engl. J. Med.* **297**, 614 (1977).

TABLE II
ASSAY OF IMMUNOREACTIVE TRANSCOBALAMIN II[a]

	Tube number	Serum	Buffer[b]	CN-Cbl[c]	Tracer[d]	Immuno-sorbent[e]
Total counts	1	—	—	—	50	—
Blank	2	—	800	10	50	—
Maximal binding	3	—	300	10	50	500
Standard curve[f]	4–11	1.56–200	298.5–100	10	50	500
Unknown sample	12, etc.	25	275	10	50	500

[a] All volumes are in microliters; all tubes in duplicate. Reproduced with permission from Fràter-Schröder et al.[15]

[b] Assay buffer (see section on immunosorbent).

[c] 300 ng/ml.

[d] Labeled TC2: 330 ng/liter [^{57}Co]CN-Cbl–TC2 in assay buffer.

[e] Immobilized anti-TC2 (1 : 10,000) in assay buffer.

[f] Except for tube 10 (200 μl serum and 100 μl buffer), use 100 μl of appropriately diluted reference serum (1 : 64–1 : 1 for tubes 3–9) and add 200 μl of assay buffer, in order to obtain TC2 levels ranging from 1.87 to 240 pg/tube.

centrate was high (3500 ng/liter Cbl) and could be accurately measured by means of radioisotope dilution technique.[27] This value together with the known apo-TC2 content of the sample gave an accurate estimate of the total TC2 level. Absolute TC2 levels in the standard curve serum were then determined using this known sample as a cross reference.

Procedure for Solid Phase RIA

Assay Conditions[15]

Immunoreactive TC2 was measured by competitive inhibition of labeled TC2, using the suspension of immobilized antiserum as a specific binder. Optimal assay conditions for the determination of TC2 in serum are compiled in Table II. The reagents and serum samples were diluted with assay buffer as indicated. Final dilutions were prepared 60 min before use, and kept at room temperature. The 60-min period seemed to be necessary to obtain the optimal stabilizing effect of the nonionic detergent Tween 80 in the assay.

Step 1. Eight reference samples for the standard curve, controls, and unknown samples were diluted as shown in Table II.

[27] R. P. Britt, F. G. Bolton, C. A. Cull, and G. H. Spray, *Br. J. Haematol.* **16,** 457 (1969).

Step 2. An excess amount of 3 ng CN-Cbl in 10 μl was added to each tube, mixed, and incubated for 15 min, to saturate all binding sites to eliminate any interference from traces of free radioactive CN-Cbl.

Step 3. Tracer (TC2-[^{57}Co]CN-Cbl) was added and mixed well.

Step 4. Immunosorbent, kept in suspension by gentle vortexing, was added, mixed once, and incubated overnight at 4°.

Step 5. The tubes were centrifuged (10 min at 2500 g at 4°) and the supernatant was immediately removed. The pellet was hardly visible at this point and the supernatant could be aspirated 2 mm from the bottom of the tube. The pellet was counted directly.

Step 6. A standard curve was made by plotting the percentage binding against TC2 concentrations of the standard on semilogarithmic paper or logit–log paper (Fig. 1).

$$B_0 = \text{cpm max. binding} - \text{cpm blank}$$

$$\%B/B_0 = \frac{\text{cpm standard or sample} - \text{cmp blank}}{\text{cpm max. binding} - \text{cpm blank}}$$

The corresponding TC2 value of each unknown sample could be read from the standard curve obtained. Calculation of the unknown sample could also be conducted with the aid of a computer program using logit–log linear regression and iterative optimization of the correlation. Multiplication by the dilution factor gave the concentration per ml of original material.

Sensitivity and Specificity of the Assay and Total TC2 Levels Measured in Serum

The sensitivity of this assay is 1.87 pg TC2 per tube (Fig. 1). To show that the anti-TC2 antiserum discriminates between human R-binder and TC2, normal serum was treated with excess immobilized anti-TC2 antiserum, as described above in the section on preparation of serum controls. It was then demonstrated by gel filtration that only TC2 is removed, whereas the R-binder activity was unchanged. This confirmed that the antiserum was indeed selective for TC2. The displacement curve for TC2 in a fraction of 10^7 lysed human granulocytes per ml containing only R-binder, shown in Fig. 1, confirmed that the anti-TC2 antiserum does not react with R-binder.

Species specificity was demonstrated when mouse serum (70,000 ng/liter mouse transcobalamin II) was tested against the immobilized anti-human TC2 antiserum in this assay. An apparent TC2 level of 176 ng/liter Cbl equivalent was measured. Thus, only 0.3% cross-reactivity between

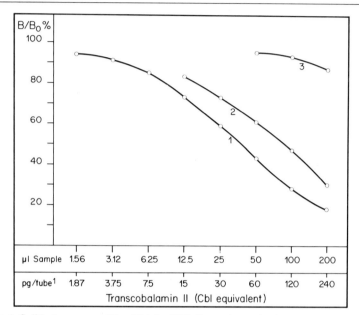

FIG. 1. Solid-phase competitive RIA for TC2. Curve 1: standard displacement calibration curve, based upon the competition between increasing amounts of unlabeled TC2 and its ^{57}Co-labeled analog (TC2–Cbl) for binding to antibody. Curve 2: displacement of labeled TC2 from antiserum by increasing volumes of low control serum containing 534 ng/liter TC2. Curve 3: lack of displacement of labeled TC2 by increasing volumes of a sample prepared from lysed granulocytes with a UBBC of 8470 ng/liter due to R-binder and no detectable TC2. Reproduced with permission from Fràter-Schröder *et al.*[15]

mouse and man was observed.[15,28] The same antiserum against human TC2 exhibited further specificities when reacted with TC2 in serum from eight other species of animals.[29]

Reactivity of holo-TC2 versus apo-TC2 was tested using the first antibody preparation in this report. The displacement curves were superimposable, indicating that the immunoreactivities of holo-TC2 and apo-TC2 were identical.[11]

Reproducibility was tested as follows. Repeat assays of a high control sample on 35 different days yielded 3570 ± 260 ng/liter ($\pm$SD) Cbl equivalent TC2. The coefficient of variation was 7% and the corresponding variation for $N = 35$ was 1.2%.[15] Both apo- and holo-TC2 are stable in serum (either human or fetal calf serum) for at least 2 weeks at room

[28] M. Fràter-Schröder, O. Haller, R. Gmür, L. Kierat, and G. Anastasi, *Biochem. Genet.* **20,** 1001 (1982).

[29] M. Haus, P. D. Green, and C. A. Hall, *Proc. Soc. Exp. Biol. Med.* **162,** 295 (1979).

temperature in a closed tube.[15] The unknown factor responsible for stabilization of apo- and holo-TC2 in serum is partially removed by the purification procedure and purified TC2 is degraded rapidly in the absence of high salt concentrations, but CN-Cbl complexation stabilizes purified TC2 adequately at assay conditions.[15]

Normal levels for total TC2 in serum from 100 healthy blood donors, using the assay presented here, have been reported.[15] The mean level and SD were 1140 ± 260 ng/liter TC2. The levels in males ($N = 61$, 1150 ± 240 ng/liter) and females ($N = 39$, 1130 ± 280 ng/liter) did not differ. Three patients with congenital TC2 deficiency had immunoreactive TC2 levels varying from 22 to 39% of the normal mean, and lacked apo-TC2 activity.[2,15] Nine healthy relatives of these three patients, shown by other methods to be heterozygous for the deficiency,[2] had immunoreactive TC2 levels ranging from 31 to 79% of the normal mean.

Quantification of Secretory TC2 in Human Cell Cultures

Synthesis and Secretion of TC2 by Cultured Skin Fibroblasts

The solid-phase RIA presented here has been applied to identification and quantification of TC2 which is secreted by fibroblasts.[9,30] Immunoreactive TC2 levels in the cell culture medium, which contains no free Cbl, were practically the same as apo-TC2 levels measured by G-150 gel filtration, indicating that TC2 is secreted predominantly in its apo form.[9]

Fibroblast cultures were derived from skin biopsies obtained from healthy individuals. Confluent cultures ($2-5 \times 10^6$ cells per flask) were maintained in 15 ml Eagle's minimal essential medium (MEM) supplemented with 10% fetal calf serum (Gibco) and the usual antibiotics plus fungizone. The medium was replaced when the cells were confluent and collected after a fixed number of days (periods from 7 to 14 days were found to be convenient) in order to measure the secretion rate of TC2.

Medium sample preparation for investigation of the TC2 content was conducted as follows[9]: an equal volume of saturated solution of ammonium sulfate (4 M) was added to medium from one culture flask, to concentrate and precipitate TC2 selectively, at 50% saturation. After 16 hr at 4° and subsequent centrifugation (8000 g, 15 min) the precipitate was reconstituted in RIA assay buffer (omitting albumin as above) equivalent to 10% of the original medium volume, and dialyzed against this buffer for 6 hr at 4° with two buffer changes (absence of NH_4^+ was tested with Nessler's reagent). Samples were stored at −20° or assayed directly.

[30] M. Sacher, F. Paky, and M. Fràter-Schröder, Helv. Paediatr. Acta 38, 549 (1983).

TC2 detection in medium samples: The RIA procedure in Table II was adapted for the investigation of cell culture medium by using larger unknown sample volumes (200 µl) and correspondingly less assay buffer (100 µl) to detect low TC2 concentrations. It was also possible to analyze more dilute samples using volumes of up to 500 µl, by increasing the immunosorbent concentration, or by increasing the total assay volume, accordingly. A background control for supplemented medium without cell contact was necessary to determine the residual fetal calf serum-derived cross-reacting material, which had to be subtracted from the TC2 levels obtained in cell culture medium.[9,30]

Synthesis and Secretion of TC2 by Cultured Bone Marrow

Liquid bone marrow cultures were derived from spongiosa residues taken from otherwise healthy individuals, undergoing orthopedic surgery and prepared as described[9]: 5×10^6 dextran sedimented bone marrow cells were maintained in 6 ml supplemented MEM, containing 20% fetal calf serum and antibiotics. The cell-free medium was investigated after a culture period of approximately 7 to 14 days, in the same way as described for skin fibroblasts, to quantify the secretory capacity of the cultured bone marrow.

Both the cultured skin fibroblasts and bone marrow cells (with continuously changing cell populations) were found to secrete increasing quantities of TC2 which accumulates in the medium. The amount of TC2 secreted per flask by 2–5 million cells for both types of cultures ranged from 200 to 2000 pg Cbl equivalent TC2, depending on the culture period and on culture conditions.[9]

Applications and General Comments

Significant elevations of apo-TC2 concentrations in serum have been observed in Gaucher's disease,[31] in multiple myeloma,[32] in certain other forms of cancer,[33] after bone marrow transplantation,[34] and in various forms of active autoimmune disease,[35] but not in rheumatoid arthritis.[36]

[31] H. S. Gilbert and N. Weinreb, *N. Engl. J. Med.* **295,** 1096 (1976).
[32] R. Carmel and D. Hollander, *Blood* **51,** 1057 (1978).
[33] H. S. Jensen, P. Gimsing, F. Pedersen, and E. Hippe, *Cancer* **52,** 1700 (1983).
[34] E. Naparstek, B. Rachmilewitz, M. Rachmilewitz, Z. Fuchs, and S. Slavin, *Br. J. Haematol.* **55,** 229 (1983).
[35] M. Fràter-Schröder, W. H. Hitzig, P. J. Grob, and A. B. Kenny, *Lancet* **2,** 238 (1978).
[36] M. Fràter-Schröder, A. Fontana, K. Fehr, and L. Kierat, *Schweiz. Med. Wochenschr.* **114,** 1396 (1984).

However, elevated levels of TC2, measured in synovial fluid, can help to discriminate rheumatoid arthritis patients from those with osteoarthrosis.[36,37] Subsequent clinical investigations have confirmed the usefulness of determining total immunoreactive TC2 levels as well, in the clinical course of systemic lupus erythematosus (SLE) and dermatomyositis.[38]

Diagnostic criteria of abnormal and absent gene products in TC2-dependent inborn errors of Cbl utilization have been defined by measurements of TC2 in serum and in cell culture. Three genetic subgroups have been proposed.[2] Early diagnosis of suspected TC2 deficiency can be performed in a cord serum sample with the RIA for TC2, because it has been shown that the newborn's own genetic TC2 type is already expressed in cord blood.[39]

Now that measurement of TC2 biosynthesis by quantification of TC2 in human cell culture medium has become accessible, it may be possible to learn more about the natural dynamics of TC2 in man in the future. It has been reported that several other human cell types, not mentioned here, secrete relatively small quantities of TC2.[9,40] Prenatal diagnosis of TC2 deficiency may be possible by investigating amniotic fluid cell cultures, where TC2 synthesis has been demonstrated.[41] The rate of TC2 secretion by fibroblasts can be influenced in various ways: CN-Cbl addition to the medium caused a decrease of net secretion (presumably as a consequence of increased uptake of the TC2–Cbl complex), whereas chloroquine and NH_4Cl, inhibitors of lysosomal proteolysis, significantly increased the quantity of TC2 recovered in the cell culture medium (presumably as a consequence of inhibition of intracellular degradation).[9]

Acknowledgments

The author is indebted to Dr. C. A. Hall, Dr. J. Lindemans, and Dr. H. J. Porck for generously supplying the antiserum preparations described above, and to Ms. L. Kierat and Dr. W. H. Hitzig for helpful discussions and support.

Supported by research Grants 3.023-81 from the Swiss National Science Foundation, the Kantonal-Zürcher Liga für Krebsbekämpfung, and the Prof. Dr. Max Cloëtta Foundation, Switzerland.

[37] P. A. Christensen, Y. Brynskov, P. Gimsing, and J. Petersen, *Scand. J. Rheumatol.* **12,** 268 (1983).
[38] U. Lässer, L. Kierat, P. J. Grob, W. H. Hitzig, and M. Fràter-Schröder, *Clin. Immunol. Immunopathol.* **36,** 345 (1985).
[39] H. J. Porck, M. Fràter-Schröder, R. R. Frants, L. Kierat, and A. W. Eriksson, *Blood* **62,** 234 (1983).
[40] R. Rabinowitz, B. Rachmilewitz, M. Rachmilewitz, and M. Schlesinger, *Isr. J. Med. Sci.* **18,** 740 (1982).
[41] M. Fràter-Schröder, P. Krieg, L. Kierat, J. Erten, and W. H. Hitzig, *Helv. Paediatr. Acta Suppl.* **50,** Abstr. 115 (1984).

Section II

Vitamin A Group

[6] Quantitative Analysis of Retinal and 3-Dehydroretinal by High-Pressure Liquid Chromatography

By MOMOYO MAKINO-TASAKA and TATSUO SUZUKI

Animals use vitamin A aldehyde as a chromphore of the visual pigment. Retinal (vitamin A_1 aldehyde) and 3-dehydroretinal (vitamin A_2 aldehyde) are two kinds of chromophore and a third chromophore found in the fly is probably 3-hydroxyretinal.[1] Generally, terrestrial and marine animals have visual pigment with retinal as its chromophore (rhodopsin). Visual pigment with 3-dehydroretinal (porphyropsin) is known in many freshwater vertebrates, e.g., teleosts, amphibians, and turtles.[2] Often, the vitamin A_2-based visual pigment coexists with the vitamin A_1-based pigment to form a paired rhodopsin/porphyropsin pigment system. The rhodopsin/porphyropsin system has been studied using partial bleaching[3] and the Carr–Price reaction (antimony chloride test).[4] High-pressure liquid chromatography (HPLC) has been used to study visual pigments and a sensitive analytical method (retinaloxime method) was developed by Groenendijk et al.[5] to determine geometric configuration of chromophore of the pigment. We have further developed this analytical method to determine retinal and 3-dehydroretinal (and their geometric isomers) in the retina.[6] The method is so sensitive that 5 pmol of retinal (or 3-dehydroretinal) can be quantitatively analyzed. The extraction procedure, HPLC, comparison with a conventional method, and applications to biological materials are described below.

Extraction of Retinal and 3-Dehydroretinal as Oximes

The extraction procedure is essentially the same as that reported by Groenendijk et al.[5] All operations are carried out under a safety light, a dim deep-red light for visual pigments and a yellow light for oximes. We

[1] K. Vogt and K. Kirschfeld, Naturwissenschaften **71,** 211 (1984).
[2] A. Knowles and H. J. A. Dartnall, in "The Eye" (H. Davson, ed.), 2nd ed., Vol. 2B, p. 581. Academic Press, New York, 1977.
[3] H. J. A. Dartnall, "The Visual Pigment." Methuen, London, 1957.
[4] C. D. B. Bridges and R. A. Alverez, this series, Vol. 81, Part H, p. 463.
[5] G. W. T. Groenendijk, P. A. A. Jansen, S. L. Bonting, and F. J. M. Daemen, this series, Vol. 67, Part F, p. 203.
[6] T. Suzuki and M. Makino-Tasaka, Anal. Biochem. **129,** 111 (1983).

describe here a standard procedure.[6a] Volumes of reagents may vary depending on the sample volume, but the volume ratio should not be changed (and the concentration of NH_2OH must be in more than 1000-fold molar excess of the amounts of retinal and 3-dehydroretinal). Tissue homogenate (100–200 μl) is mixed with 100 μl of 1.0 M NH_2OH (stock solution of 3.0 M NH_2OH is diluted with 1.0 M NaOH before use), and then 1 ml of methanol is added to form retinaloxime and 3-dehydroretinaloxime. After the mixture has stood on ice for 5 min, 1 ml dichloromethane is added and the mixture is shaken vigorously. Then 0.5 ml distilled water is added and shaken. To separate the dichloromethane layer, 2 ml of *n*-hexane is added, stirred, and the mixture is centrifuged at 3000 rpm for 5 min at 4°. The dichloromethane/hexane layer above the aqueous methanol is transferred to another tube with a pipette. The extraction with dichloromethane/hexane is repeated three times. The extracts are combined, evaporated under reduced pressure, and dissolved in an elution solvent for HPLC. When the tissue contains free retinol or carotenoids, it should be freeze-dried and washed with petroleum ether before extraction of retinal and 3-dehydroretinal.

Preparation of Standard Oximes

In order to quantitate retinal and 3-dehydroretinal, a standard mixture of retinaloxime and 3-dehydroretinaloxime at known concentration is required. In the retina, usually, 11-*cis* and all-*trans* isomers of retinal (and 3-dehydroretinal) are present as the chromophores of the visual pigment. Therefore, a standard mixture of 11-*cis*-retinal, 11-*cis*-3-dehydroretinal, all-*trans*-retinal, and all-*trans*-3-dehydroretinal is sufficient for the usual analysis of visual pigment. These four compounds are now easily obtained in high purity by HPLC using a silica gel column.[4,5] The purified isomer is dissolved in ethanol and the concentration is determined by a spectrophotometer using the value of the absorption coefficient reported.[7] Standard oximes are prepared as follows. Synthesized phosphatidylcholine dipalmitoyl is suspended in 0.1 M phosphate buffer (pH 6.8) in the final concentration of 250 μg/ml (0.34 μM). The lipid suspension is sonicated until it becomes clear. The ethanol solution of 50 μl containing 800 pmol each of 11-*cis*-retinal, all-*trans*-retinal, 11-*cis*-3-dehydroretinal, and all-*trans*-3-dehydroretinal is added to the lipid suspension of 500 μl and

---

[6a] Recently, we reexamined the extraction procedure and found that the addition of dichloromethane was not necessary. Oximes were completely recovered by hexane alone when the extraction was repeated three times.

[7] R. Hubbard, P. K. Brown, and D. Bownds, this series, Vol. 18, Part C, p. 615.

mixed by shaking. Oximes are formed and extracted by the same procedures as described above. After evaporation of the extract, oximes are dissolved in 400 μl of *n*-hexane to a final concentration of 2 pmol/μl for each oxime. The mixture can be used as a HPLC standard for a few months if it is stored in the dark under nitrogen gas in a deep-freezer.

High-Pressure Liquid Chromatography (HPLC)

HPLC is performed with normal phase-silica gel column. Good separation is obtained with spherical particle of porous silica gel with 5 or 7 μm diameter. The mobile phase we usually use is 7% diethyl ether in *n*-hexane containing 0.075% ethanol (v/v). When the column is new, often the recovery of anti-isomers of oximes from the column is very low. This problem is solved by addition of one drop of distilled water to 1 liter elution solvent. Absorbance at 360 nm is monitored with a UV detector and peak area is determined by integrating the absorbance at 360 nm with an integrator.

Figure 1A shows the chromatogram of the standard oximes of 30 pmol each. Oximes are eluted in the following order: 11-*cis*-retinaloxime, 11-*cis*-3-dehydroretinaloxime, all-*trans*-retinaloxime, and all-*trans*-3-dehydroretinaloxime. Peak areas of syn- and anti-isomers are summed for each oxime and plotted against the amount of oxime injected (Fig. 1B). Absolute quantities of retinal and 3-dehydroretinal are determined using these standard curves. A problem of the oxime method is in the syn/anti ratio which is variable as oxime formation. When visual pigment is reacted with NH$_2$OH by the extraction procedure described above, the syn/anti ratio of peak area is constant, 2.1 $\pm$ 0.1 (n = 8) both in retinal and 3-dehydroretinal. However, the ratio is about 4 when the authentic retinal and 3-dehydroretinal are reacted with NH$_2$OH in methanol/water (shown in Fig. 2C). If the oximes formed in the methanol/water are used as standards, only approximate values are obtained, because the molar extinction coefficients are different for syn- and anti-isomers.[5] The standard oximes formed in the artificial lipid membrane yield almost the same syn/anti ratio as that for visual pigment. The syn/anti ratio was 2.4 $\pm$ 0.1 in 11-*cis*-retinaloxime and 2.6 $\pm$ 0.1 in 11-*cis*-3-dehydroretinaloxime (n = 5). When we use these oximes as standards, the error of determination is reduced to 0.5% in the case of 11-*cis*-retinal. The error in determination of 11-*cis*-3-dehydroretinal may be the same order as that of 11-*cis*-retinal, although we cannot estimate it because the molar extinction coefficients of syn- and anti-isomers of 11-*cis*-3-dehydroretinal have not been determined.

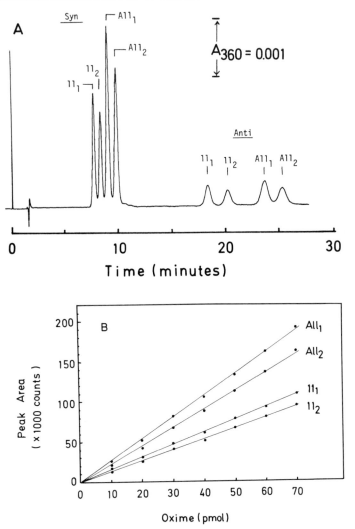

Fig. 1. Chromatogram of standard oximes and standard calibration curves. (A) HPLC chromatogram of an equimolar mixture of four standard oximes, 30 pmol each. 11_1, 11-*cis*-Retinaloxime; 11_2, 11-*cis*-3-dehydroretinaloxime; All_1, all-*trans*-retinaloxime; All_2, all-*trans*-3-dehydroretinaloxime. (B) Standard calibration curves of four oximes. Peak area is the summation of syn and anti peaks of each oxime. HPLC was performed with JASCO HPLC system equipped with a Zorbax SIL column (2.1 × 250 mm). Absorbance at 360 nm was monitored with JASCO UVIDEC 100-III in a full scale at 0.01 OD. The peak area was determined by integrating the absorbance at 360 nm with Shimadzu Chromatopac E1A. Mobile phase, 7% ether and 0.075% ethanol in *n*-hexane (v/v); flow rate, 0.6 ml/min. From Suzuki and Makino-Tasaka.[6]

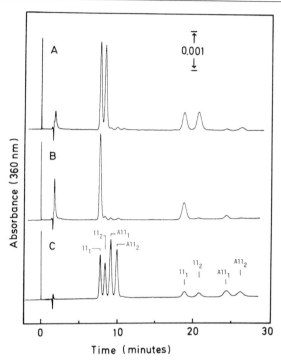

FIG. 2. Chromatograms of the extracts from bullfrog retina. (A) Dorsal half of the retina; (B) ventral half of the retina. The injection volume is 20 μl, corresponding to 1/250 of a half retina. (C) Authentic oximes formed in aqueous methanol, 15 pmol each. Analytical conditions are the same as in Fig. 1. From Suzuki and Makino-Tasaka.[6]

Comparison with a Conventional Method

The reliability of the HPLC method is evaluated by comparing the results of HPLC with the composition of the visual pigment extracted with detergent. The adult bullfrog has rhodopsin and porphyropsin in the retina and porphyropsin is localized in a dorsal half of the retina.[8] Suitable materials are obtained by separating the bullfrog retina into dorsal and ventral parts.

Figure 2 shows the chromatograms of the extracts from the dorsal and ventral retinas. From the peak areas (syn + anti), retinal and 3-dehydroretinal are quantified using the standard curves of Fig. 1B. Visual pigments are extracted from the dorsal and ventral parts of the bullfrog retina with 1.0% L1690 (Ryoto Co.). Rhodopsin and porphyropsin are

[8] T. E. Reuter, R. H. White, and G. Wald, *J. Gen. Physiol.* **58,** 351 (1971).

COMPARISON OF THE RESULTS OF HPLC WITH THE RESULTS OF BLEACHING ANALYSIS
USING THE RETINA OF BULLFROG[a]

Retina	HPLC (nmol/retina, $n = 4$)		Bleaching analysis (nmol/retina)	
	Retinal	3-Dehydroretinal	Rhodopsin	Porphyropsin
Dorsal	9.13 ± 0.42	11.13 ± 0.59 (54.9%)	8.5	10.8 (56%)
Ventral	9.39 ± 0.32	0.37 ± 0.02 (3.8%)	8.1	0.4 (5%)
Whole	18.5	11.5 (38.3%)	16.6	11.2 (40%)
Total amount	30.0		27.8	

[a] The values in parentheses represent the percentage value of 3-dehydroretinal or porphyropsin. From Suzuki and Makino-Tasaka.[6]

stable in this detergent.[9] The amounts of rhodopsin and porphyropsin in the extracts are determined by a bleaching analysis (a kinetic analysis of bleaching curve,[6] which is based on the same principle as that of the method of partial bleaching).[3] The results obtained from the two methods are shown in the table.

Comparing the results of HPLC with those of the bleaching analysis, the percentage value of 3-dehydroretinal is almost the same as that of porphyropsin in two parts of the retina. The total amount of retinal and 3-dehydroretinal determined by HPLC is slightly larger than that of visual pigments. This discrepancy is due to the green rod pigment which is present in the bullfrog retina in the amount of about 10% of total visual pigment.[10] The bleaching analysis does not estimate the green rod pigment, because this pigment is destroyed by NH_2OH in the dark. When the contribution of the green rod pigment is added to the results of HPLC, the results of the two methods are identical.

Applications

We show here three examples of applications made in our laboratory.

Determination of Chromophore of Invertebrate Visual Pigment

The HPLC method is sensitive enough that chromophore composition is easily determined even in a small eye of invertebrate. We determined

[9] S. L. Fong, A. T. C. Tsin, C. D. B. Bridges, and G. I. Liou, this series, Vol. 81, Part H, p. 113.
[10] M. Makino-Tasaka and T. Suzuki, Vision Res. 24, 309 (1984).

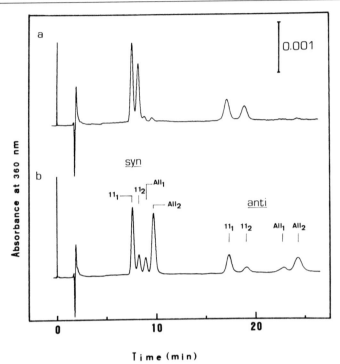

Time (min)

FIG. 3. Chromatograms of extracts obtained from eyes of a crayfish. (a) Right eye, dark-adapted; (b) left eye, irradiated with deep-red light ($\lambda > 680$ nm). Fifty microliters of 200 μl extract was injected for HPLC, corresponding to 1/4 eye. The analytical conditions are the same as in Figs. 1 and 2. Reprinted with permission from Suzuki *et al.*[11]

chromophore of visual pigment in several species of freshwater crustaceans and insects and have found that crayfish, *Procambarus clarkii*, contains retinal and 3-dehydroretinal in the eye. We show here an example of the results for the crayfish.[11]

Generally, the eye contains carotenoids which disturb the HPLC analysis. Most carotenoids are removed by washing freeze-dried material with petroleum ether. This treatment does not affect the visual pigments. In the case of the crayfish, retinal tissue is mechanically separated from the crystalline cone layer which contains carotenoids. When the isolated retina is used as a starting material the washing procedure is not necessary. The procedure of extraction of retinaloxime and 3-dehydroretinaloxime is the same as described above.

Figure 3 shows the chromatograms of the extracts obtained from a crayfish. The right eye was isolated after 2-days dark adaptation (Fig. 3a).

[11] T. Suzuki, M. Makino-Tasaka, and E. Eguchi, *Vision Res.* **24**, 783 (1984).

The major components were 11-*cis*-retinal and 11-*cis*-3-dehydroretinal. The total amount of the chromophore was 200 pmol and the proportion of 3-dehydroretinal was 48% of the total chromophore. The left eye was irradiated with deep-red light (λ>680 nm) for 4 min after the 2 days dark adaptation (Fig. 3b). The chromatogram clearly shows a selective isomerization of 11-*cis*-3-dehydroretinal to all-trans isomer. This result shows that the pigment with 11-*cis*-3-dehydroretinal absorbs the deep-red light more efficiently than the pigment with 11-*cis*-retinal. The total amount of chromophore and the proportion of 3-dehydroretinal in the irradiated left eye were the same as those in the right eye.

Invertebrate metarhodopsin (all-trans chromophore) has a long life time in the eye, and a light-adapted eye of crayfish shows four components in its chromatogram. No isomer other than 11-cis and all-trans is produced by light in the living eye. Therefore, the described analytical method can be applied also to the studies on photoreaction and regeneration of visual pigment in the living eye. Regeneration of crayfish rhodopsin was very slow and its half-time was 1 day at 20°.

Topographic Map of Visual Pigment in the Retina

Generally, visual pigment with 3-dehydroretinal (porphyropsin) is not distributed homogeneously in the retina. Spatial distributions of rhodopsin and porphyropsin in the retina are determined by the HPLC method. We divided a bullfrog retina into 38 pieces and determined retinal and 3-dehydroretinal in each piece.[12] Porphyropsin was found not only in the dorsal half but also in the peripheral regions of the ventral retina. There was a semicircular band with higher pigment densities (mainly rhodopsin) around the optic disc, which was close to the anatomically distinguished structure, area centralis. As 1/1000 part of a bullfrog retina can be analyzed by the method, more detailed maps may be made by improving the technique dividing the retinal tissue. Green rod pigments of the bullfrog retina are analyzed by the same way as red rod pigments (rhodopsin and porphyropsin).[10]

Extraocular Photoreceptor

In vertebrates, there is no extraocular photoreceptor in which chromophore of photosensitive pigment has been chemically characterized. The sensitivity of the HPLC method described here allows detection of retinal and 3-dehydroretinal in a pineal organ (*epiphysis cerebri*) of the

---

[12] M. Makino-Tasaka, T. Suzuki, K. Nagai, and S. Miyata, *Exp. Eye Res.* **40**, 767 (1985).

rainbow trout, *Salmo gairdneri.*[13] A pineal organ isolated from a dark-adapted rainbow trout contained 12.8 pmol of vitamin A aldehyde, 61% of which was 3-dehydroretinal. The proportion of 11-cis isomers was 88% of total chromophore; however, the proportion of 11-cis isomers was decreased in the pineal organ isolated from light-adapted animal. These results suggest that a paired pigment system is involved in the pineal organ also as well as in the retina of rainbow trout.

Acknowledgment

3-Dehydroretinal is the generous gift from F. Hoffman-La Roche & Co. Ltd (Basel, Switzerland).

[13] M. Tabata, T. Suzuki, and H. Niwa, *Brain Res.* **338**, 173 (1985).

[7] Enzymatic Synthesis and Separation of Retinyl Phosphate Mannose and Dolichyl Phosphate Mannose by Anion-Exchange High-Performance Liquid Chromatography

By KIM E. CREEK, DONATA RIMOLDI, MICHELE BRUGH-COLLINS, and LUIGI M. DE LUCA

Numerous *in vitro* studies have documented that membranes isolated from a variety of tissues are active in the transfer of mannose from GDP-mannose to exogenous retinyl phosphate (Ret-P) and dolichyl phosphate (Dol-P).[1–10] Although a role for dolichyl phosphate mannose (Dol-P-Man) in the elongation of the lipid-linked oligosaccharide chain Dol-P-P-

[1] L. M. De Luca, M. R. Brugh, C. S. Silverman-Jones, and Y. Shidoji, *Biochem. J.* **208**, 159 (1982).
[2] K. E. Creek, D. J. Morré, C. S. Silverman-Jones, Y. Shidoji, and L. M. De Luca, *Biochem. J.* **210**, 541 (1983).
[3] L. De Luca, N. Maestri, G. Rosso, and G. Wolf, *J. Biol. Chem.* **248**, 641 (1973).
[4] G. C. Rosso, L. De Luca, C. D. Warren, and G. Wolf, *J. Lipid Res.* **16**, 235 (1975).
[5] A. Bergman, T. Mankowski, T. Chojnacki, L. M. De Luca, E. Peterson, and G. Dallner, *Biochem. J.* **172**, 123 (1978).
[6] Y. Shidoji and L. M. De Luca, *Biochem. J.* **200**, 529 (1981).
[7] J. B. Richards and F. W. Hemming, *Biochem. J.* **130**, 77 (1972).
[8] J. S. Tkacz, A. Herscovics, C. D. Warren, and R. W. Jeanloz, *J. Biol. Chem.* **249**, 6372 (1974).
[9] C. J. Waechter, J. J. Lucas, and W. J. Lennarz, *J. Biol. Chem.* **248**, 7570 (1973).
[10] H. G. Martin and K. J. I. Thorne, *Biochem. J.* **138**, 281 (1974).

(GlcNAc)$_2$(Man)$_5$ to Dol-P-P-(GlcNAc)$_2$(Man)$_9$ seems well established, the precise molecular function if any of retinyl phosphate mannose (Ret-P-Man) awaits elucidation.[11,12] Studies on the synthesis and metabolism of Ret-P-Man have been severely hampered by its extreme lability due to its five conjugated double bonds and its allylic phosphate which is easily cleaved to yield anhydroretinol and mannose phosphate. Published procedures for the separation of Ret-P-Man and Dol-P-Man have involved solvent extraction, chromatography on columns of silicic acid, DEAE-cellulose, or DEAE-Sephacel, chromatography on thin layers of silica gel or high-pressure liquid chromatography on micro-Porasil columns.[4,13–16] These procedures are time consuming, can result in major losses of Ret-P-Man, and are not always reproducible. Here we describe detailed procedures for the enzymatic synthesis of Ret-P-Man and Dol-P-Man from exogenous Ret-P and Dol-P and two rapid and reproducible high-performance liquid chromatographic (HPLC) procedures on Mono Q anion exchange columns which result in minimal losses of Ret-P-Man. One system separates mannose, Dol-P-Man, Ret-P-Man, and Ret-P, while the other separates mannose, Ret-P-Man, mannose phosphate, and GDP-mannose. These procedures should prove valuable in studies involving Ret-P-Man and Dol-P-Man metabolism.

Preparation of Postnuclear Membranes from Liver

Liver membranes are prepared from Syrian golden hamsters or Osborne–Mendel rats which are fasted overnight. The animals are killed by bleeding from the neck under light diethyl ether anesthesia. The livers are removed and homogenized in two volumes (based on wet weight of the liver) of ice-cold 0.25 M sucrose, 50 mM Tris–HCl (pH 7.6), 5 mM MgCl$_2$, and 25 mM KCl (medium A) with approximately 7 upward and 7 downward strokes of a Potter-Elvehjem type tissue grinder. The homogenate is centrifuged at 2,500 g (5000 rpm in a Sorvall type SS-34 rotor) for 20 min at 4°. The resulting supernatant is then centrifuged at 105,000 g (40,000 rpm in a Beckman type 50 Ti rotor) for 60 min at 4°. The resulting membrane pellet is gently resuspended with either a small Potter-Elvehjem or

[11] R. Kornfeld and S. Kornfeld, *Annu. Rev. Biochem.* **54,** 631 (1985).
[12] G. Wolf, *Physiol. Rev.* **64,** 873 (1984).
[13] C. S. Silverman-Jones, J. P. Frot-Coutaz, and L. M. De Luca, *Anal. Biochem.* **75,** 664 (1976).
[14] W. Sasak, C. S. Silverman-Jones, and L. M. De Luca, *Anal. Biochem.* **97,** 298 (1979).
[15] T. Kurokawa and L. M. De Luca, *Anal. Biochem.* **119,** 428 (1982).
[16] L. Rask and P. A. Peterson, this series, Vol. 67, p. 270.

Dounce type homogenizer in a volume of medium A equal to one half the wet weight of the liver. This typically gives a protein concentration of 35–60 mg/ml when measured by the method of Lowry et al.[17] The membranes may be stored at −80° for at least 1 month without apparent loss of mannosyl transferase activity toward either exogenous Ret-P or Dol-P.

Enzymatic Synthesis of Ret-P-Man Standard from Exogenous Ret-P

All work is performed under dim yellow light (Westinghouse gold fluorescent lamps). Retinyl phosphate is chemically synthesized by the phosphorylation of all-*trans*-retinol by bisditriethylamine phosphate as described in detail by Bhat et al.[18] GDP-mannose (Sigma) or GDP-[U-^{14}C] mannose (269 mCi/mmol, New England Nuclear) is used as the mannose donor for the preparation of unlabled or [^{14}C]mannose-labeled Ret-P-Man. Unlabeled Ret-P-Man is prepared as follows: Ret-P (90 μg) in methanol is added to a glass test tube and the solvent removed under a gentle stream of N_2. The Ret-P is immediately resuspended by the addition of 180 μl of a 40 mg/ml solution of bovine serum albumin. The incubation mixture is then completed by the addition of 90 μl of 0.6 M Tris–HCl, pH 7.8, 180 μl of 0.08 M NaF, 45 μl of 0.10 M MnCl$_2$, 90 μl of 0.04 M ATP, 180 μl of 0.05 M AMP, 450 μl of 400 μM GDP-mannose, 8–10 mg protein of liver postnuclear membranes, and water to a final volume of 1.8 ml. The mixture is incubated at 37° for 30 min, stopped by cooling on ice, and immediately transferred to 12 Beckman polyallomer Airfuge tubes (150 μl/tube). The membranes are pelleted by centrifugation for 5 min in a Beckman air-driven ultracentrifuge operated at 30 psi. The supernatant is removed and the membranes resuspended and extracted in 100 μl of methanol and transferred to a glass test tube. The Airfuge tubes are rinsed with methanol (0.5 ml total) which is added to the first extract. Ret-P-Man is extracted from the membranes by vigorous mixing on a Vortex mixer. The denatured protein residue is removed from the extract by centrifugation in a clinical type centrifuge operated at full speed for 5 min. Ret-P-Man is purified from the extract by HPLC on a Mono Q HR 5/5 column as described in detail in a later section. We have determined that under these incubation conditions 80–90% of the Ret-P-Man made *in vitro* is recovered with the membrane pellet. A total of 10–15 μg of Ret-P-Man can be expected following HPLC purification. The concentration of Ret-P-Man is determined spectrophotometrically at 325 nm in 99% methanol on the

---

[17] O. H. Lowry, N. J. Rosebrough, A. L. Farr, and R. J. Randall, *J. Biol. Chem.* **193**, 265 (1951).

[18] P. V. Bhat, L. M. De Luca, and M. L. Wind, *Anal. Biochem.* **102**, 243 (1980).

basis of $E_{1\,cm}^{1\%} = 1440$. When stored at 4° in 99% methanol Ret-P-Man is typically stable for approximately 2 weeks at which time losses of Ret-P-Man become evident due to breakdown to anhydroretinol and β-mannose phosphate.

[14C]Mannose-labeled Ret-P-Man is prepared similarly to that described above for the synthesis of unlabled Ret-P-Man. GDP-[U-14C]mannose (1.0 μCi) and Ret-P (7.5 μg) are added to a Beckman polyallomer Airfuge tube and the solvent removed under a gentle stream of N_2. The Ret-P and GDP-[14C]mannose are immediately resuspended in 15 μl of 40 mg/ml bovine serum albumin. The incubation mixture is then completed by the addition of 7.5 μl of 0.6 M Tris–HCl, pH 7.8, 15 μl of 0.08 M NaF, 3.75 μl of 0.01 M MnCl$_2$, 7.5 μl of 0.04 M ATP, 15 μl of 0.05 M AMP, 1.5 mg protein of liver postnuclear membranes, and water to a final volume of 150 μl. The mixture is incubated at 37° for 30 min and the membranes pelleted by centrifugation for 5 min in a Beckman air-driven ultracentrifuge operated at 30 psi. The supernatant is removed and the [14C]mannose-labeled Ret-P-Man extracted from the membranes with up to 2 ml of methanol. The denatured protein is removed from the extract by centrifugation in a clinical type centrifuge operated at full speed for 5 min. Ret-P-[14C]Man is purified from the extract by HPLC on a Mono Q HR 5/5 column as described in detail later. Typically 30–40% of the added [14C]mannose is transferred from GDP -[14C]mannose to exogenous Ret-P under these incubation conditions.

Enzymatic Synthesis of Dol-P-Man Standard from Exogenous Dol-P

[14C]Mannose-labeled Dol-P-Man may be prepared using incubation conditions similar to those just described for the preparation of [14C]mannose-labeled Ret-P-Man. The major difference between the incubations is the inclusion of Triton X-100 which is required for the solubilization of the polyisoprenoid Dol-P. Dol-P (Grade III, Sigma, C$_{80-100}$ from porcine liver) 50 μg and GDP-[14C]mannose (1 μCi) are added to a glass test tube and the solvent removed under a gentle N_2 stream. The Dol-P is solubilized by the addition of 10 μl of a 0.5% solution of Triton X-100. The incubation mixture is completed by the addition of 20 μl of a 40 mg/ml solution of bovine serum albumin, 10 μl of 0.6 M Tris–HCl, pH 7.8, 20 μl of 0.08 M NaF, 5 μl of 0.10 M MnCl$_2$, 10 μl of 0.04 M ATP, 20 μl of 0.05 M AMP, 1.5–2.0 mg of liver postnuclear membranes, and water to a final volume of 200 μl. The incubation is thoroughly mixed and incubated at 37° for 30 min. The incubation is stopped by the addition of 1.0 ml of chloroform–methanol (2 : 1, v/v) which yields two phases, the tube is stirred and the phases separated by centrifugation in a clinical type centrifuge operated at full speed for 5 min. The lower organic phase which contains approximately 98% of the [^{14}C]mannose-labeled Dol-P-Man is removed and the solvent evaporated under a N_2 stream. The Dol-P-Man is resuspended in up to 2.0 ml of 99% methanol and is purified by HPLC on a Mono Q HR 5/ 5 column as described below. Typically 35% of the added [^{14}C]mannose is transferred from GDP-[^{14}C]mannose to the exogenous Dol-P under these incubation conditions. Dol-P-Man is stable for several weeks when stored at 4° in 99% methanol.

HPLC System for the Separation of Mannose, Dol-P-Man, Ret-P-Man, and Ret-P

This method is based on a previous observation that Dol-P-Man, Ret-P-Man, and Ret-P could be separated by anion exchange chromatography on DEAE-Sephacel columns, utilizing ammonium acetate in 99% methanol as the eluent.[14] Chromatographic separations are conducted on an Altex programmable liquid chromatography system; OD_{325} is monitored with an Hitachi variable wavelength spectrophotometer, radioactivity is monitored by a Flo-One Model HS radioactivity flow detector (Radiomatic Instrument and Chemical Co.), and the data recorded by a Linear chart recorder. The column used is the Mono Q HR 5/5 (Pharmacia Fine Chemicals). Samples are introduced to the Mono Q column in up to 2.0 ml of 99% methanol and the column is eluted with an ammonium acetate gradient in 99% methanol. The column is eluted at a flow rate of 1.5 ml/min 1 mM ammonium acetate in 99% methanol for 2 min and then the ammonium acetate concentration is increased linearly to 10 mM in 3 min. The ammonium acetate concentration is held at 10 mM for 10 min and then the ammonium acetate concentration is increased linearly to 50 mM in 20 min. The ammonium acetate concentration is then held at 50 mM for 5 min. A typical separation of standards is illustrated in Fig. 1. Mannose, Dol-P-Man, and Ret-P-Man were detected by radioactivity whereas Ret-P was monitored by its characteristic absorbance at 325 nm. As shown in Fig. 1, mannose, Dol-P-Man, Ret-P-Man, and Ret-P are eluted at 3, 13, 29, and 36 min, respectively. Under these elution conditions mannose phosphate and GDP-mannose remain bound to the column but may be eluted by washing the column with 400 mM NaCl in 70% methanol or by washing the column with 50% acetic acid. This procedure is very useful in the purification of [^{14}C]mannose-labeled Dol-P-Man and Ret-P-Man synthesized from exogenous acceptors *in vitro* since [^{14}C]mannose, [^{14}C]mannose phosphate, and GDP-[^{14}C]mannose, which can be troublesome contaminants, are easily removed by this HPLC procedure. In this system retinoic acid is eluted in the area of Dol-P-Man. We have

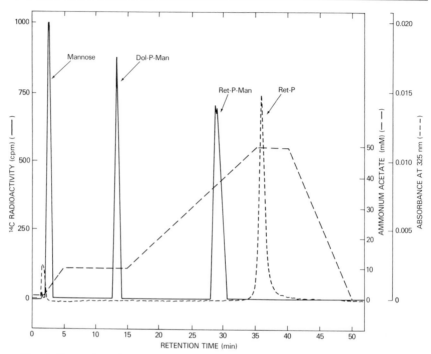

Fig. 1. Separation of mannose, Dol-P-Man, Ret-P-Man, and Ret-P by anion-exchange chromatography on a Mono Q HR 5/5 column. A mixture containing approximately 1000 cpm each of standard [^{14}C]mannose, Dol-P-[^{14}C]Man, and Ret-P-[^{14}C]Man as well as 150 ng of Ret-P was applied to the Mono Q column in 1.0 ml of 99% methanol. The column was eluted with a programmed gradient of ammonium acetate in 99% methanol as described in detail in the text. Radioactivity was monitored by a Flo-One radioactivity flow detector and OD$_{325}$ by an Hitachi spectrophotometer.

found that this procedure is highly reproducible and that column performance has remained essentially unaltered after approximately 200 analyses over a period of 1 year. Furthermore, the recovery of Ret-P-Man from the column approaches 100%.

HPLC System for the Separation of Mannose, Ret-P-Man, Mannose Phosphate, and GDP-Mannose

To carefully monitor Ret-P-[^{14}C]Man synthesis from GDP-[^{14}C]mannose and Ret-P by liver membranes *in vitro* it was necessary to develop a chromatographic procedure which separated Ret-P-[^{14}C]Man from GDP-[^{14}C]mannose and the hydrolysis products [^{14}C]mannose phosphate and [^{14}C]mannose. We found that mannose, Ret-P-Man, mannose phosphate,

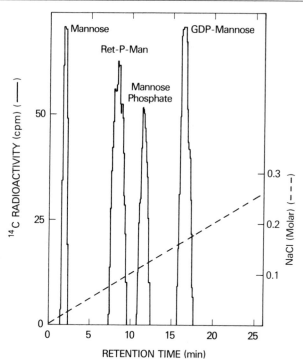

FIG. 2. Separation of mannose, Ret-P-Man, mannose phosphate, and GDP-mannose by anion-exchange chromatography on a Mono Q HR 5/5 column. A mixture containing approximately 1000 cpm each of standard [^{14}C]mannose, Ret-P-[^{14}C]Man, [^{14}C]mannose phosphate, and GDP-[^{14}C]mannose was applied to the Mono Q column in 1.0 ml of 70% methanol 30% water. The column was eluted with a programmed linear gradient of NaCl in 70% methanol–30% water as described in detail in the text. Radioactivity was monitored by a Flo-One radioactivity flow detector. Standard [U-^{14}C]mannose phosphate was generated by enzymatic cleavage of GDP-[U-^{14}C]mannose with *Crotalus atrox* phosphodiesterase I (15 mU) in 100 m*M* Tris, pH 8.8 at 37° for 30 min. Analysis of each ^{14}C-labeled standard individually determined their respective retention times.

and GDP-mannose are separated based on their difference in anionic charge on the anion exchange Mono Q HR 5/5 column. The column, pumps, and other chromatographic equipment used in this separation are identical to that described in the preceding section. Samples are introduced to the Mono Q column in up to 1.0 ml of methanol/water (70 : 30, v/ v). The column is eluted at a flow rate of 1.0 ml/min with a linear gradient of 0.4 to 300 m*M* NaCl in methanol/water (70 : 30, v/v) over a 30-min period. A typical separation of ^{14}C-labeled standards is shown in Fig. 2. Retention times for mannose, Ret-P-Man, mannose phosphate, and GDP-mannose are 2.3, 8.3, 11.3, and 16.3 min, respectively. The profile illustrated in Fig. 2 was generated by continuously monitoring the column effluent with a Flo-One radioactivity flow detector. Dol-P-Man is eluted in this system with approximately the same retention time as Ret-P-Man but in very poor yields due to poor solubility of Dol-P-Man in 70% methanol. This procedure has proved to be highly reproducible with retention times of the standards remaining essentially unaltered after numerous analyses. We recently used this procedure to demonstrate that the GDP-mannose : retinyl phosphate mannosyltransferase reaction is reversible.[19]

[19] K. E. Creek, D. Rimoldi, C. S. Silverman-Jones, and L. M. De Luca, *Biochem. J.* **227**, 695 (1985).

[8] Separation and Quantitation of Retinyl Esters and Retinol by High-Performance Liquid Chromatography[1]

By A. CATHARINE ROSS

The major form of vitamin A in liver, eye, and several other organs comprises various fatty acid esters of retinol. Before the development of HPLC, separation of some individual esters of retinol had been achieved by reverse-phase or argentation thin-layer chromatography.[2] More recent HPLC procedures for a variety of retinoid separations have achieved popularity for the obvious reasons of speed, improved resolution, and because these sensitive molecules can be kept in solvent and away from damaging oxygen and light. The HPLC technique used in our laboratory for the separation of retinyl esters[3] was originally developed in response to a need to separate completely retinyl palmitate from retinyl oleate. Whereas two additional carbon atoms in the oleate moiety of retinyl oleate confer decreased polarity, the carbon–carbon double bond has the opposite effect and, thus, separation of retinyl oleate from retinyl palmitate can be difficult. Using an octadecasilyl (C_{18})-substituted column, others found that addition of silver nitrate to the mobile phase reduced the retention of retinyl oleate sufficiently to allow its separation from retinyl palmitate.[4] Our approach has been to utilize a column with a somewhat

[1] This work was supported by United States Health Services Grant HL-25744, HD-16484, and HL-22633. Dr. Ross is recipient of a Research Career Development Award, HD-00691 from the NIH.
[2] H. S. Huang and D.S. Goodman, *J. Biol. Chem.* **240**, 2839 (1965).
[3] A. C. Ross, *Anal. Biochem.* **115**, 324 (1981).
[4] M. G. M. deRuyter, and A. P. deLeenheer, *Anal. Chem.* **51**, 43 (1979).

less hydrophobic substituent, either an octyl or phenyl group, and a simple mobile phase of acetonitrile–water. Using this procedure, we have achieved separation and quantitation of long-chain fatty acids esters of retinol in liver-derived cells, chylomicrons,[3] and plasma.[3] More recently, we have analyzed the retinyl ester pattern of milk in which a wider variety of esters including those of medium-chain fatty acids is encountered.[5] Described below is our procedure for the analysis of retinyl ester pattern and a companion procedure for assaying the total retinol content of such tissues. In each case quantitation is based on the intrinsic absorbancy of retinol at a wavelength where few other lipids interfere, and use of internal standards as a convenient means of accurate quantitation.

Instrumentation

A simple HPLC system having one pump and an absorbance monitor will suffice for either procedure. Our system includes a Waters model 6000A pump, U6K injector, Model 440 absorbance meter with a 340 nm filter, and chart recorder or electronic integrator. A variable wavelength detector set at 324–326 nm would improve sensitivity by about 25%. Separation takes place on a 5-μm Supelcosil LC-8 column, 25 cm × 4.6 mm (Supelco, Inc., Bellafonte, PA) having a hand-packed guard column of LC-8 pellicular packing. We have also used a phenyl-substituted column (10 μm μBondapak Phenyl, 30 cm × 4.6 mm, from Waters Associates, Inc.) with equal success. The LC-8 column has been in use for over 4 years with consistent results. It is washed occasionally with a solvent series of 50 ml each of acetonitrile, dichloromethane and hexane, followed again with dichloromethane and acetonitrile, and rinsed with methanol–water, 60 : 40 (v/v), before storage. A second LC-8 column from the same manufacturer produced considerably more rapid elution of esters and poorer separation. Thus, some column-to-column variation is apparent and it may be necessary to modify the proportions of solvents or test different columns to achieve complete ester separations. With improvements in column technology, this variability may have been overcome.

For quantitation of retinol, a C_{18} column, 30 cm × 3.9 mm from Waters Associates has been used with a C_{18} pellicular guard column.

Mobile Phases

Retinyl ester separation is achieved using a two-step discontinuous gradient consisting of (1) acetonitrile–water, 88 : 12 (v/v) followed by (2)

[5] A. C. Ross, M. E. Davila, and M. P. Cleary, *J. Nutr.* **115**, in press (1985).

acetonitrile–water, 98 : 2. Distilled and deionized water is filtered through a Millipore filter (HAWP, 0.45 μm and mixed with an appropriate volume of HPLC-grade acetonitrile under light vacuum for approximately 0.5 hr to remove dissolved air. For retinol analysis, a single isocratic elution is used. The mobile phase is methanol (analytical grade)–water, 90 : 10 (v/v), degassed as above.

Standards

Retinyl acetate and palmitate are commercially available but other ester standards require synthesis. We have synthesized more than 12 esters, adapting either the fatty acid anhydride procedure of Lentz et al.[5a] as described earlier,[3] or the acyl chloride procedure of Huang and Goodman[2] scaled down to use approximately 0.12 mmol of retinol. After synthesis, esters are extracted into hexane and back-washed[2]; the solvents are then completely evaporated under nitrogen or argon. Each ester is redissolved in hexane and chromatographed on an open column of aluminum oxide as previously described.[3] Retinyl esters prepared in this way have migrated as single peaks on HPLC and have had absorption maxima at 324–326 nm. Esters should be stored in the dark under nitrogen or argon. We have stored ether solutions in hexane or dried lipid for many months with little apparent deterioration. The concentration of retinyl ester standards can be determined by spectrophotometry at 324–326 nm. We showed previously that molar absorptivity is essentially equal for a number of esters and for retinol itself[3] and we have therefore applied the molar extinction coefficient for retinol [52,275 M^{-1} cm^{-1}[6]]. Our standard of all-*trans*-retinol has been prepared similarly either using crystalline all-*trans*-retinol (Sigma) or, as we prefer, after saponification of the more stable retinyl acetate. In either case, retinol is purified by chromatography on aluminum oxide[2,7] before use.

Preparation of Samples for Retinyl Ester Separation

An initial lipid extraction must be conducted and any of various procedures that provide quantitative extraction under nonhydrolyzing conditions should be satisfactory. The method is sensitive to about 75 pmol of an individual ester and sample size should be selected accordingly. With aqueous samples or cell sonicates, we have used an ethanol : hexane parti-

[5a] B. R. Lentz, Y. Barenholz, and T. E. Thompson, *Chem. Phys. Lipids* **15**, 216 (1975).
[6] J. Boldingh, H. R. Cama, F. D. Collins, R. A. Morton, N. T. Gridgeman, O. Isler, M. Kofler, R. J. Taylor, A. S. Welland, and T. Bradbury, *Nature (London)* **168**, 598 (1951).
[7] A. C. Ross, *J. Lipid Res.* **23**, 133 (1982).

tioning system in which we first add one volume of sample to four volumes of absolute ethanol in an acid-washed, ethanol-washed, or new screw-cap tube; appropriate internal standard, usually retinyl pentadecanoate (see below), is added at this time. The mixture is allowed to stand for 10–20 min, then 4 to 20 volumes of hexane are added and the mixture is shaken by hand for about 10 sec. Ten volumes of water are added, the mixture is shaken again, and then centrifuged at low speed to form a clean upper phase of hexane from which aliquots can be pipetted or weighed for further analysis. The volume of hexane used is not critical but must be sufficient to dissolve all lipid mass of which the retinyl ester content is often a very small portion (e.g., in lymph chylomicra or milk). Liver lipids may be extracted by grinding tissue with anhydrous sodium sulfate, followed by extraction with diethyl ether.[8] In the case of most lipid-rich samples, we have found it necessary to separate retinyl esters as a class from most other lipid, especially triglycerides, to reduce the lipid mass prior to solubilizing sample in acetonitrile for HPLC. We have done this using open columns of aluminum oxide,[3] as described earlier in this series.[9] The lipid extract is dried under nitrogen at 37–40° using an N-E-Vap (Organomation, Inc.) or similar evaporating device. After solubilizing the extract in hexane, the extract is chromatographed on deactivated aluminum oxide [5 ml water/100 g neutral aluminum oxide, to produce an activity grade III, 1.5 g/column in glass minicolumns (N.E.N.) plugged with glass wool or a No. 2 dental cotton ball washed with solvent]. The sample is first eluted with 12 ml hexane which will remove carotene if present and is to be discarded, then with 15 ml of 3% diethyl ether in hexane to elute retinyl esters as well as esters of cholesterol. Triglycerides, cholesterol, retinol, and more polar compounds will be retained. The eluate is concentrated to dryness with nitrogen at 37–40°, the vial rinsed down with about 1 ml of hexane and reconcentrated, and the concentrate taken to dryness just before beginning analysis by HPLC. When dry, a small portion, approximately 40–200 μl, of acetonitrile is added and the sample is warmed very briefly (3–5 sec) at 55–60° to facilitate solution in the acetonitrile. The sample is then injected onto the LC-8 HPLC column running at 3 ml/min with mobile phase 1 (acetonitrile–water, 88 : 12). After elution of retinyl oleate (about 45 min on our column), the mobile phase is switched to 2 (acetonitrile–water, 98 : 2) to more rapidly elute retinyl stearate and any other less polar esters.

Quantitation can be achieved by electronic integration, or manually after measuring peak areas by planimetry, or as the product of peak height

[8] B. D. Drujan, this series, Vol. 18, Part C, p. 565.
[9] D. S. Goodman and J. A. Olson, this series, Vol. 15, p. 462.

times width at half-height, each to be related to the known quantity of added internal standard.

Comments

Accurate quantitation requires that internal standard and endogenous retinyl esters behave identically with regard to solubility in solvents used to extract or redissolve lipids. We have found in the case of milk a variety of retinyl esters ranging from retinyl octanoate to retinyl stearate and differing quite considerably in polarity and solubility in acetonitrile. Thus, when a single internal standard, retinyl pentadecanoate, was added we found that the recovery of medium-chain esters of retinol could be overestimated under conditions where their solubility exceeded that of retinyl pentadecanoate. We have in this case added two internal standards (e.g., retinyl heptanoate and pentadecanoate) at the time of extraction and we have then quantitated retinyl esters having 8 to 12 carbon fatty acyl chains relative to retinyl heptanoate and longer chain esters relative to the longer chain retinyl ester internal standard. For samples containing only long-chain fatty acid esters of retinol, retinyl pentadecanoate alone should

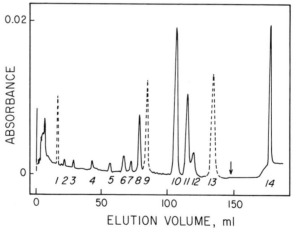

FIG. 1. HPLC chromatogram of a sample of human breast milk illustrating elution position of various fatty acid esters of retinol. Esters shown as dashed lines are synthetic esters of fatty acids having odd carbon numbers that have been used as internal standards. Identities of retinyl esters: (1) heptanoate (7:0); (2) octanoate (caprylate, 8:0); (3) decanoate (caproate, 10:0); (4) dodecanoate (laurate, 12:0); (5) γ-linolenate (18:3); (6) myristate (14:0); (7) palmitoleate (16:1); (8) linoleate (18:2); (9) pentadecanoate (15:0); (10) palmitate (16:0); (11) oleate (18:1 cis); (12) elaidate (18:1 trans); (13) heptadecanoate (17:0); (14) stearate (18:0). Vertical arrow marks change from mobile phase 1 to mobile phase 2.

suffice. This ester is desirable as an internal standard because it elutes in an open "window" between retinyl linolenate and retinyl palmitate, as shown in Fig. 1. Although retinyl esters may show better solubility in another solvent such as tetrahydrofuran and might be dissolved in such for injection, we would advise cautious testing of recoveries before adopting other solvents for there is sometimes a "sweep-through" phenomena in which a portion of esters injected in a solvent different from the mobile phase solvent elute quickly with the solvent front and thus miss appropriate detection.

Preparation of Samples for Total Vitamin A (Retinol) Analysis

Samples, either aqueous or finely minced tissues (e.g., 100–500 μl plasma or 0.05–0.2 g liver), are saponified in a freshly prepared solution of 95% ethanol/5% potassium hydroxide containing 1% pyrogallol[10,11] as antioxidant at 60° for 20 min. At least 10 volumes of solution are used per amount of tissue. After saponification, the mixture is shaken with 10–20 volumes of hexane, then 10 volumes of water. After a brief centrifugation, a measured portion or portions of the upper phase is pipetted into a vial and an appropriate amount of retinyl acetate internal standard is added, ideally 1–2 times the mass of retinol in the sample. Solvent is removed from each sample in turn just before HPLC analysis using nitrogen at 37–40°, the sides of the vial are rinsed down with hexane, and the solvent is again evaporated. A small volume, 50–100 μl, of methanol is added to dissolve the retinol and internal standard and then all or part is injected onto the C_{18} column running at 2 ml/minute with methanol-water, 90:10. Usual retention times are about 3.4 minutes for retinol and 4.6 minutes for retinyl acetate (Fig. 2).

Quantitation is based on the areas under each peak, using equal molar extinction coefficients for retinol and retinyl acetate at 326 nm. If monitoring is done at 340 nm, the molar absorbance of retinyl acetate is about

---

[10] The person preparing this reagent for the first time may be surprised by its brown and messy appearance upon addition of pyrogallol. This is normal and the upper phase of hexane to be removed after saponification and extraction will be clear. We prefer to add water (e.g., 15 ml to 5 g potassium hydroxide pellets) to dissolve the base, then add absolute ethanol to 100 ml and, last, pyrogallol.

[11] For samples with a high lipid content such as adipose or milk from some species, the tissue will not disperse adequately for complete saponification in this mixture. In these cases, we have added sample to 10–12 volumes of 95% ethanol/10% potassium hydroxide/ 1% pyrogallol, and, thereafter, 1 volume of benzene to help dissolve neutral fat. After saponification and extraction, the volume of the upper phase (hexane plus benzene) will be increased accordingly.

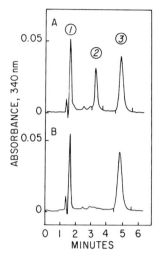

FIG. 2. HPLC analysis of retinol concentration (total vitamin A) in rat plasma (A). A 200 μl portion of plasma was saponified with 3 ml 95% ethanol/5% potassium hydroxide/1% pyrogallol for 20 min at 60°, then extracted into 4 ml of hexane. To a 3 ml portion, 350 pmol purified retinyl acetate was added. After concentration, the sample was dissolved in 100 μl methanol and approximately 80 μl was injected for HPLC. Peak 1, solvent peak; peak 2, retinol; peak 3, retinyl acetate. Vertical tick marks indicate the beginning and end of electronic integration. (B) Blank carried through same procedure.

108% of that for retinol (100%) and programming accordingly can confer a small extra degree of accuracy. This procedure is simple and sensitive to less than 50 pmol retinol. Appropriate blanks containing no sample are also saponified and run, and samples to which a known amount of retinyl acetate is added both before and after saponification should be run occasionally to check recovery throughout the procedure. Reproducibility can be excellent depending mainly on the accuracy of weighing or pipetting the sample and adding the internal standard. For fairly viscous aqueous samples such as milk or lymph, weighing is desirable; internal standard is added with a glass micropipette or calibrated gas-tight syringe. Although quantitative injection for HPLC is possible and would eliminate the need for addition of an internal standard, the internal procedure is to be preferred since evaporation and other losses of volume do not affect quantitation of retinol.

[9] Separation of Geometric Isomers of Retinol and Retinoic Acid in Nonaqueous High-Performance Liquid Chromatography

By PANGALA V. BHAT *and* ANDRE LACROIX

The development of high-performance liquid chromatography (HPLC) has aided considerably the study of the metabolism of vitamin A.[1-4] Using the HPLC system it has now been possible to demonstrate conclusively the occurrence of cis–trans isomers of retinol and retinoic acid in biological systems.[2,5,6] Many studies have reported the interconversion of all-*trans*- and 13-*cis*-retinoic acid *in vivo* in several rat tissues.[2,7] However, the methods used in these studies do not allow the separation of the entire spectrum of isomers of retinol and retinoic acid. No suitable HPLC technique has yet been described for the purification of all-*trans*-retinol and retinoic acid from their respective cis-isomers.

This chapter describes the HPLC methods developed in the author's laboratory for the separation of geometric isomers of retinol and retinoic acid. The application of the method for the purification of labeled vitamin A compounds is presented. Finally the use of HPLC and labeled retinol and retinoic acid to study the cis-trans isomerization in rat liver tissue is demonstrated.

Standard Pure Retinol and Retinoic Acid Isomers

9-*cis*-, 13-*cis*-, 11-*cis*-Retinal and all-*trans*- and 13-*cis*-retinoic acid were generously supplied by Hoffmann-LaRoche Inc. (Nutley, New Jersey); all-*trans*-retinal was obtained from Sigma Chemical Company. Retinol isomers were prepared from the corresponding retinals by reduction with sodium borohydride.[8] 11-*cis*- and 9-*cis*-Retinoic acid were synthesized by the oxidation of the respective retinals with activated MnO$_2$ as described by Bridges and Alvarez.[9] all-*trans*-[11,12-^{3}H]Retinol (specific

[1] P. V. Bhat and A. Lacroix, *Biochim. Biophys. Acta* **752**, 451 (1983).
[2] P. R. Sundaresan and P. V. Bhat, *J. Lipid Res.* **23**, 448 (1982).
[3] J. B. Williams, B. C. Pramanik, and J. L. Napoli, *J. Lipid Res.* **25**, 638 (1984).
[4] P. V. Bhat and A. Lacroix, *J. Chromatogr.* **272**, 269 (1983).
[5] M. H. Zile, R. C. Inhorn, and H. F. De Luca, *J. Biol. Chem.* **254**, 6303 (1982).
[6] C. A. Frolik, T. E. Tavela, G. L. Peck, and M. B. Sporn, *Anal. Biochem.* **86**, 743 (1978).
[7] A. M. McCormick, K. D. Kroll, and J. L. Napoli, *Biochemistry* **22**, 3933 (1983).
[8] C. D. B. Bridges, S. L. Fong, and R. A. Alvarez, *Vision Res.* **20**, 355 (1980).
[9] C. D. B Bridges, this series, Vol. 81, p. 463.

activity 51 Ci/mmol) was procured from Amersham Searly; all-*trans*-[11-^{3}H]Retinoic acid (1.6 Ci/mmol) was obtained from S.R.I. (Menlo Park, CA).

Handling and Storage of Vitamin A Isomers

The methanolic solutions of authentic vitamin A isomers were stored under N_2 at $-70°$ in screw-cap vials. Under these conditions the vitamin A isomers were stable at least for 2 weeks. To minimize the photoisomerization of vitamin A isomers, all operations were carried out under golden yellow light. Labeled vitamin A compounds were frozen in liquid nitrogen and stored at $-70°$.

Extraction of Vitamin A Isomers from Tissues

Vitamin A isomers can be extracted from the lyophilized tissues using 99% methanol as the extracting solvent in the volume ratio of 2:1 of tissue. The solution was filtered and was dried by flash evaporation, with the vacuum broken with nitrogen. The residue was immediately redissolved in 2–3 ml of absolute methanol.

Prepurification of Vitamin A Isomers Isolated from the Tissue Extract

It has been observed that the phospholipids present in the tissue extract often interfere with the separation of vitamin A isomers in nonaqueous HPLC. For this reason two simple column chromatography has been described for the initial purification of retinol and retinoic acid isomers from the tissue extract prior to injection onto HPLC.

Alumina Chromatography

Alumina chromatography can be used for the purification of retinol isomers from that of retinyl esters and polar metabolites of vitamin A. In addition, alumina column also facilitates the removal of phospholipids from the tissue extracts, which are strongly bound to alumina column.

Reagents. To 50 g of alumina (Brockmann activity 1: 80–200 Mesh from Fisher Scientific) in a glass stoppered bottle add 10% (v/w) water dropwise. The powder is shaken thoroughly and left in overnight.

Procedure. A 10 × 1.5 cm i.d. glass column was packed with water-deactivated alumina. The column was washed with 2 bed volume of hexane. The lipid extract was transferred in 100 μl of chloroform : methanol

(1 : 1). The 100 μl chloroform : methanol (1 : 1) in the column was diluted with 3 ml of hexane. The column was first eluted with 30 ml of hexane and successively eluted with 50 ml of 2% dioxane in hexane (retinyl ester fraction) and 30 ml of 15% dioxane in hexane (retinol isomer fraction).

The eluent of 15% dioxane in hexane fraction was evaporated under nitrogen to dryness and the residue was dissolved in 1 : 0 ml of absolute methanol.

A typical separation of labeled retinyl esters from a mixture of labeled retinol isomers by alumina column is shown in Fig. 1. In this experiment the lipid extract was obtained from liver tissue of a vitamin A sufficient rat injected with 10 μCi of [11,12-^{3}H]retinol. The rat was sacrificed 7 days after the injection of the label to allow the labeled vitamin A to mix with the endogenous pool of vitamin A.[1] Radioactive peak I which eluted with 2% dioxane in hexane contained retinyl esters and radioactive peak II constituted a mixture of retinol isomers.

The recovery of radioactive retinol isomers from the alumina column is of the order of 95–100% and no nonenzymatic isomerization of retinol isomers occurs in the column under these conditions.

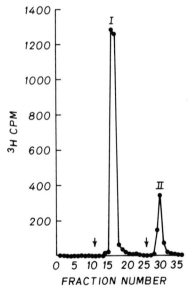

FIG. 1. Alumina chromatography of liver lipid extract from normal rat injected with all-*trans*-[11,12-^{3}H]retinol. Arrow marks indicate the change in the eluting solvent from 100% hexane to 2% dioxane in hexane (retinyl ester fraction, I) and then to 15% dioxane in hexane (retinol isomer fraction, II).

DEAE-Cellulose Chromatography

DEAE-cellulose chromatography can be used for the initial purification of retinoic acid present in the tissue extract prior to injection onto the HPLC for the separation.

Procedure. DEAE-cellulose (from Whatman) in the acetate form was prepared as described by Dankert *et al.*[10]

A 10 × 1.5 cm i.d. glass column was packed with DEAE-cellulose which has been previously washed with 99% methanol. The column was washed with 99% methanol in water (v/v). The lipid extract containing labeled retinoic acid isomer was applied on the column in methanol. The column was first eluted with 50 ml of 99% methanol in water. The mixture of retinoic acid isomers was then eluted from the column with 30 ml of 5 mM ammonium acetate in 99% methanol in water. This fraction from the column was evaporated using a flash evaporator and the residue was dissolved in absolute methanol for the separation of the isomers by HPLC.

The recovery based on the radioactivity of the retinoic acid isomers from the column is greater than 95%. Again the configuration of retinoic acid was found to be intact during the purification step.

High-Performance Liquid Chromatography Apparatus

In the authors' laboratory, a Beckman model 322 MP programmable liquid chromatography with two model 100A pumps are used for HPLC. Retinol and retinoic acid isomers are detected at 325 and 350 nm, respectively, using a UV spectrophotometer, Hitachi model 100-40 equipped with a variable wavelength between 195 and 850 nm.

Columns. Prepacked Partisil 10 ODS was obtained from Whatman (Clifton, NJ). The 5 μm Zorbax CN and 5 μm Zorbax NH$_2$ columns were purchased from Dupont Canada (Ontario, Canada).

Chemicals. HPLC grade solvents, such as acetonitrile, hexane, methanol, 2-propanol, and dichloromethane were obtained from Fisher Scientific (Pittsburgh, PA), 2-octanol was procured from Aldrich Scientific Company (Milwaukee, WI).

Separation of all-*trans*-, 9-*cis*-, 13-*cis*-, and 11-*cis*-Retinol by HPLC

We found that separation of all-*trans*, 9-*cis* pair requires opposite solvent conditions from those required for the separation of the 13-cis/11

---

[10] M. Dankert, A. Wright, W. S. Kelley, and P. W. Robbins, *Arch. Biochem. Biophys.* **116**, 425 (1966).

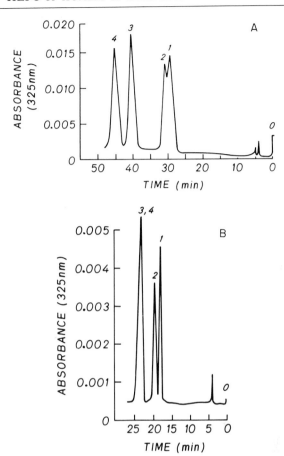

FIG. 2. (A) Chromatogram of a standard mixture of cis–trans isomers of retinol. Partisil-10-ODS (25 cm × 4.6 m i.d.) and Zorbax CN 5 μm (25 cm × 4.6 mm i.d.) columns connected in series. Mobile phase 1% of 2 octanol in hexane. Flow rate 2.0 ml/min. Sample size 400–500 ng in 20 μl of methanol. Peaks 1, 11-*cis*-retinol; 2, 13-*cis*-retinol; 3, 9-*cis*-retinol; and 4, all-*trans*-retinol. (B) Separation of standard mixture of four isomers of retinol. Conditions as in A, except the composition of mobile phase: 5% dioxane in hexane.

cis isomer pair.[11] In our laboratory we developed an HPLC system which separates the four isomers of retinol in two steps using two different solvent systems.

The Partisil 10-ODS and Zorbax CN HPLC columns are used in series. The baseline separation of 9-*cis*- from all-*trans*-retinol is achieved by eluting the column with 1% of 2-octanol in hexane (Fig. 2A). Using the

[11] P. V. Bhat, H. T. Co, and A. Lacroix, *J. Chromatogr.* **260**, 129 (1983).

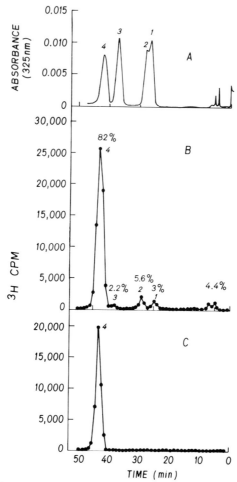

FIG. 3. Purification of labeled retinol by HPLC. Approximately 50,000 cpm of [11,12-³H]retinol was mixed with a standard mixture of retinol isomers in methanol and 50 μl was injected. (A) UV absorbing peaks of standard retinol isomers; (B) radioactive trace of the fractions collected from the colum; and (C) chromatogram of purified all-*trans*-[11,12-³H] retinol. Mobile phase: 1% of 2-octanol in hexane. Other chromatographic conditions and identification of peaks as in Fig. 2A.

same columns connected in series the baseline resolution of 13-*cis*- from 11-*cis*-retinol can be obtained using 5% of dioxine in hexane (Fig. 2B). Since the Partisil 10-ODS and the Zorbax CN columns are compatible both in normal and reverse-phase techniques it is relatively fast to equilibrate the columns with the mobile phase. When the mobile phase is changed from 1% of 2-octanol in hexane to 5% dioxane in hexane, it takes only 25 ml of the new mobile phase to equilibrate the columns.

Purification of Labeled Retinol

This HPLC method has been successfully applied to the purification of all-*trans*-[11,12-³H]retinol from its cis isomers. An example is illustrated in Fig. 3. A known amount of radioactive retinol (50,000 cpm in this case) was mixed with standard cold retinol isomers and the sample was injected onto the HPLC. Figure 3A shows the UV absorbing peak of standard retinol isomers and 3B is the tracing of the radioactive fractions. As can be seen small quantities of 11-*cis*- (3%), 13-*cis*- (2%), and 9-*cis*-(3%) retinol were detected as impurities. Figure 3B shows the pure all-*trans*-[11,12-³H]retinol, purified by HPLC. It is recommended that a standard mixture of cold retinol isomers be added to the radioactive retinol when checking for purity. This would help in identifying the all-*trans*-retinol radioactive peak. For purifying large amounts of radioactive retinol, however, one does not have to add the cold mixture of retinol isomers.

Application to Tissue Extract

An example of the application of the HPLC system to the rat liver lipid extract labeled with radioactive retinol is presented in Fig. 4. The liver lipid extract was initially purified by alumina column chromatography.

The fraction II from alumina column which contains mixture of labeled retinol isomers was injected on HPLC to separate the retinol isomers. As can be seen in Fig. 4 all-*trans*-retinol was found to be the most abundant isomer (84%), but 12.1 and 4.1% of 13-*cis*-retinol and 9-*cis*-

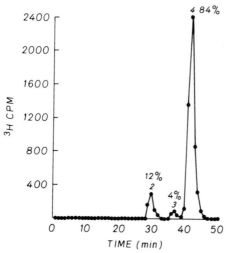

FIG. 4. HPLC of fraction II from the alumina column. Chromatographic conditions and peak identifications as in Fig. 2A.

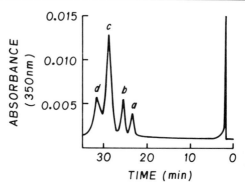

FIG. 5. Separation of a mixture of 11-*cis*-, 13-*cis*-, 9-*cis*-, and all-*trans*-retinoic acid. Column Zorbax NH$_2$, 5 μm. Mobile phase acetonitrile/dichloromethane 90/10, containing 10 m*M* acetic acid. Flow rate 2.0 ml/min. Peaks: a, 11-*cis*-retinoic acid; b, 13-*cis*-retinoic acid; c, 9-*cis*-retinoic acid; and d, all-*trans*-retinoic acid.

retinol were also found to be present. In this HPLC system the retinyl esters are eluted in the void volume of the column and the polar metabolites of retinol such as phosphate esters of retinol, retinoic acid, and 5,6-epoxyretinoic acid are retained in the column. They can be easily washed from the column with absolute methanol. Therefore precaution should be taken when one injects directly the tissue lipid extract onto HPLC. Prepurification of the tissue lipid extract by alumina would remove the retinyl esters and above-mentioned polar metabolites of retinol, thereby eliminating the step of washing the columns with absolute methanol after each injection of tissue extract.

Separation of 9-*cis*-, 11-*cis*-, 13-*cis*-, and all-*trans*-Retinoic Acid
  by HPLC

These four isomers of retinoic acid can be separated on Zorbax NH$_2$ column. To achieve the separation the column was eluted with a mixture of acetonitrile/dichloromethane (90/10) containing 10 m*M* acetic acid. Figure 5 illustrates the separation of four retinoic acid isomers. The presence of 10 m*M* acetic acid in the mobile phase facilitates the elution of retinoic acid isomers from the column.

Use of HPLC to the Purification Labeled Retinoic Acid

Approximately 50,000 cpm of all-*trans*-[11-^{3}H]retinoic acid was mixed with the mixture of standard retinoic acid isomers and separated by the HPLC system described above. One minute fractions were collected from

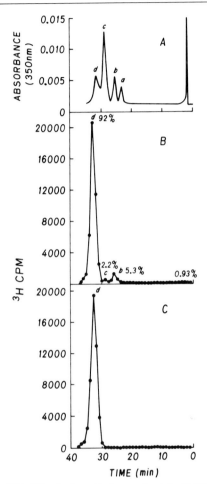

FIG. 6. Purification of [11-³H]retinoic acid. Approximately 50,000 cpm of [11-³H]retinoic acid was mixed with the mixture of standard retinoic acid isomers and injected onto HPLC. (A) UV absorbing peaks of retinoic acid isomers; (B) radioactive trace of the fractions collected from the columns; (C) chromatogram of purified all-*trans*-[11-³H]retinoic acid. Identification of peaks as in Fig. 5.

the column and counted for the radioactivity. Figure 6A shows the UV absorbing peaks of standard retinoic acid isomers and Fig. 6B is the tracing of the radioactive fractions obtained from all-*trans*-[11-³H]retinoic acid. As can be seen from the chromatogram 5.3% of 13-*cis*-retinoic acid and 2.2% of 9-*cis*-retinoic acid were present as impurities. Figure 6C is the chromatogram of purified all-*trans*-[11-³H]retinoic acid.

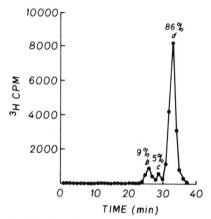

FIG. 7. Occurrence of radioactive 13-*cis*- and 9-*cis*-retinoic acid in the liver tissue of a normal rat injected with pure all-*trans*-[11-³H]retinoic acid. The rat was sacrificed 30 min after the injection of the label and the labeled isomers were isolated and purified by DEAE-cellulose prior to injection onto HPLC. Identification of peaks is as in Fig. 5.

Application of the HPLC Method to Studies *in Vivo* cis–trans Isomerization of Retinoic Acid

An example of the application of the method for the conversion of all-*trans*-[11-³H]retinoic acid to *cis* isomers in liver tissue of a vitamin A sufficient rat is shown in Fig. 7. Purified labeled retinoic acid isomers prepared from liver lipid extract by DEAE-cellulose chromatography was injected onto the column and eluted with the solvent system described above. One minute fractions from the column were collected and counted for radioactivity. As shown in Fig. 7, in addition to all-*trans*-retinoic acid (84%), 9% of 13-*cis*-retinoic acid, and 5.1% of 9-*cis*-retinoic acid were detected.

The 5 m*M* ammonium acetate fractions collected from DEAE-cellulose column, during the purification of retinoic acid isomers, also contains the oxidation products of retinoic acid, such as 4-keto-, 4-hydroxy-, and 5,6-epoxyretinoic acid; they are retained in the Zorbax NH₂ column and can be eluted from the column using methanol containing 10 m*M* acetic acid as the eluting solvent.

Precaution

We have observed that the Zorbax NH₂ column is very sensitive to change of solvent from high polarity to low polarity. Therefore when using the Zorbax NH₂ column for the purposes of separation of retinoic acid isomers, it should not be switched to other systems, especially those involving methanol/water.

Acknowledgment

This work was supported by a grant from Medical Research Council of Canada (MA 7328).

[10] Use of Enzyme-Linked Immunosorbent Assay Technique for Quantitation of Serum Retinol-Binding Protein

By NOBUO MONJI and EJAL BOSIN

We have developed an enzyme-linked immunosorbent assay (ELISA) technique for human serum retinol-binding protein (RBP).[1] The assay detects RBP via a double-antibody (rabbit anti-human RBP) sandwich technique as shown in Fig. 1. The antibody is immobilized by passive adsorption to a polystyrene tube; the assay is then carried out by successive additions containing known and unknown amounts of RBP, alkaline phosphatase linked to the same antibody, and p-nitrophenyl phosphate substrate. Colorimetric analysis of the hydrolysis of the substrate by the enzyme attached to the antigen is used for RBP quantitation.

Radial immunodiffusion, radioimmunoassay (RIA), immunonephrometric,[2] and, more recently, latex immunoassay[3] are the widely used methods for repetitive RBP analysis. The radial immunodiffusion, latex immunoassay, and immunonephrometric methods have operative extents in the submicrogram to microgram range, precluding its use when body fluids other than blood are to be analyzed without special concentrating procedures; the RIA method can be employed only at facilities where γ-radiation use is routinely allowed. The ELISA for human RBP described here is simple, sensitive, low cost, and precise; it can be carried out in any laboratory lacking radiation facilities.

[1] E. Bosin and N. Monji, *Anal. Biochem.* **133**, 283 (1983).

[2] M. T. Parviainen and P. Ylitalo, *Clin. Chem. (Winston-Salem, N.C.)* **29**, 853 (1983).

[3] A. M. Bernard, A. Vyskovil, and R. R. Lauwerys, *Clin. Chem. (Winston-Salem, N.C.)* **27**, 832 (1981).

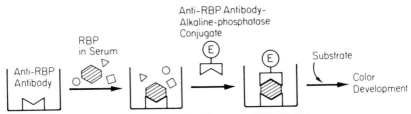

FIG. 1. Assay outline for serum RBP ELISA.

RBP ELISA

Antiserum

Monospecific Ig fraction of rabbit antisera to human RBP (145 μg/ml anti-RBP Ig) (Accurate Chemical and Scientific Co.).

Standard

Standard stabilized human serum containing 48–51 μg/ml RBP (Cal-biochem-Behring Co.).

Reagents

Carbonate–bicarbonate buffer (CBC), pH 9.6 (0.05 M containing 0.03 M NaN$_3$)
PBS-T (PBS containing 0.05% Tween 20)
10% DEA solution containing 0.02% NaN$_3$ and 0.01% MgCl$_2 \cdot$ 6H$_2$), pH 9.8

Preparation of Conjugate

Conjugation of alkaline phosphatase to the anti-RBP Ig is carried out according to the method described by Voller et al.[4] Add 1.0 mg of alkaline phosphatase to 0.6 ml of PBS containing 0.4 mg anti-RBP Ig. Dialyze the mixture against PBS for 24 hr at 4°. To the dialyzed antibody–enzyme mixture, add 2.5% glutaraldehyde to obtain a final concentration of 0.2%. After a thorough mixing, incubate the solution for 2 hr at room temperature. Dialyze the reaction mixture overnight against PBS at 4°, followed by further 24 hr dialysis against a 0.05 M Tris–HCl buffer, pH 8.0, at 4°. Centrifuge the materials for 20 min at 1000 g and 4°. Save the supernatant and dilute to 1.0 ml with 0.05 M Tris–HCl buffer, pH 8.0, containing 1.0%

---

[4] A. Voller, D. Bidwell, and A. Bartlett, in "Manual of Clinical Immunology" (N. R. Rose and H. Friedman, eds.), Chapter 69. Am. Soc. Microbiol., Washington, D.C., 1976.

BSA and 0.02% NaN$_3$. Store the prepared conjugate at 4°; under this condition the conjugate is stable for at least 6 months.

Confirmation of the Conjugation

Mix 0.1 ml of normal rabbit serum diluted 1 : 20 with 0.05 M PBS with 0.1 ml of the enzyme conjugate which is diluted 1 : 1000 with the same buffer. To the above mixture, add 0.2 ml of goat anti-rabbit antiserum and incubate for 6 hr at room temperature. Following the incubation, centrifuge the mixture for 20 min at 1000 g and 4°. Wash the pellet 3 times with 1.0 ml each of PBS by resuspension and centrifugation. More than 90% of the enzyme activity is present in the precipitate.

ELISA Procedure

1. Add 1.0 ml of anti-RBP Ig diluted 2000-fold in CBC buffer to a polystyrene tube (W. Sarstedt, Princeton, NJ).
2. Incubate the tube overnight at 4° (under these conditions, the tubes could be stored for at least 2 weeks).
3. Just prior to the assay, remove the antibody solution by aspiration and wash the tube three times with PBS.
4. Add an aliquot of 0.9 ml of either standard or unknown serum diluted 1000- to 32,000-fold or urine sample diluted 25-fold with PBS to the anti-RBP coated tube.
5. Incubate the tube for 2 hr at room temperature.
6. Remove the solution by aspiration and wash the tube 3 times with PBS.
7. Add 0.9 ml of the enzyme–antibody conjugate diluted 1 : 1000 with 0.05 M Tris–HCl buffer, pH 8.0, containing 1% BSA and 0.02% NaN$_3$ and incubate for 2 hr at room temperature.
8. Remove the excess enzyme–antibody conjugate and wash the tube 5 times with PBS-T.
9. Assay the bound enzyme by adding 0.9 ml of the substrate solution (1 mg p-nitrophenyl phosphate/ml 10% DEA solution) and by a 2 hr incubation at 37°.
10. Stop the enzyme reaction by adding 0.1 ml of a 3 N NaOH solution.
11. Measure the absorbance at 405 nm.
12. Analyze the obtained data by using the Prophet system[5] and plot the curve as the log of RBP concentration versus net absorbance (= sample absorbance − background absorbance).

[5] W. F. Raub, *Fed. Proc., Fed. Am. Soc. Exp. Biol.* **33**, 2390 (1974).

Results

RBP Assay Method

Standard curve for RBP at 3 ~ 48 ng/ml is shown in Fig. 2. Intraassay variability was examined by assaying four samples containing various RBP concentrations (2.5 ~ 51 ng/ml) in separate experiments of at least 10 samples/experiment; the results are shown in Table Ia. Seven standard curve performed on different occasions for interassay variability studies were also analyzed and the results are presented in Table Ib. The slight discrepancy between the intra- and interassay data may be due to different antigen and enzyme conjugate preparations used in those experiments. The coefficients of variation for intra- and interassay variability ranged between 4 and 7 and 9 and 12%, respectively. The presence of significant dilutional factors and the assay specificity were tested in experiments consisting of parallel dilutions (1 : 1000 to 1 : 32,000) of standard and clinical serum specimens. Results of such experiments are shown in Fig. 3. Statistical analysis did not reveal a significant difference between two curves. Experiments with urine samples showed similar behavior.

Clinical Samples

Human RBP was analyzed in sera and urine specimens of 5 patients with hepatorenal syndrome (HRS), 20 patients with liver cirrhosis and normal kidney function (NKF), 14 chronic renal failure patients, and 19 healthy subjects (Table II).

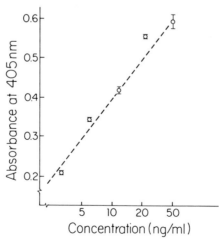

FIG. 2. Standard curve for serum RBP ELISA.

TABLE I
QUALITY CONTROL DATA OF INTRA- AND INTERASSAY VARIABILITY

RBP (ng/ml)	Absorbance[a]	Number	Coefficient of variation (%)
a. Intraassay			
2.5	0.27 ± 0.01	10	4
5	0.29 ± 0.01	10	5
25	0.44 ± 0.03	14	7
51	0.53 ± 0.03	10	7
b. Interassay[b]			
3	0.24 ± 0.03	7	12
6	0.29 ± 0.03	7	9
12	0.34 ± 0.03	7	9
24	0.40 ± 0.05	7	12
48	0.45 ± 0.04	7	9

[a] Average absorbance ± SD.
[b] Standard curves performed on different days.

Renal patients had higher serum (mean S_{RBP} 149, range 50–400 $\mu g/ml$) and urine RBP (mean U_{RBP} 8, range 1–27 $\mu g/ml$) than healthy donors (mean S_{RBP} 41, range 30–60 $\mu g/ml$; mean U_{RBP} 0.06, range 0.02–0.1 $\mu g/ml$).

Liver patients with normal renal function averaged less than half the S_{RBP} values of their normal counterparts (mean S_{RBP} 16, range 8–39 $\mu g/ml$); they had a normal RBP excretion pattern (mean U_{RBP} 0.1, range 0.03–0.2 $\mu g/ml$; compared to normal: p = NS).

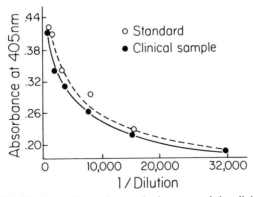

FIG. 3. Parallel dilution study on the standard serum and the clinical specimen.

TABLE II
SERUM AND URINARY RBP LEVELS IN HEALTHY, CIRRHOTIC WITH NKF, HRS, AND
RENAL FAILURE GROUPS[a]

Profile	Healthy donors	Liver cirrhosis with NKF	HRS	Renal failure
Number of subjects	19	20	5	14
Serum RBP[b]	41 ± 10[c]	16 ± 8	12 ± 4	149 ± 85
	(30–60)[d]	(8–39)	(8–17)	(50–400)
Urine RBP[b]	0.06 ± 0.03	0.1 ± 0.06	8 ± 9	8 ± 8
	(0.02–0.1)	(0.03–0.2)	(1.4–23)	(1.3–27)
Urine serum RBP × 100[e]	0.10 ± 0.09	0.8 ± 0.7	87 ± 119	5 ± 4
	(0.04–0.3)	(0.1–2)	(8–295)	(0.6–15)

[a] Biochemical profiles of urine and serum samples are available on request to the author (N.M.). NKF, Normal kidney function.
[b] μg/ml.
[c] Mean ± SD.
[d] Range.
[e] RBP index.

Patients with hepatorenal syndrome (HRS) had U_{RBP} values similar to those of the renal group (mean U_{RBP} 8, range 1–23 μg/ml); these increased values were associated with low S_{RBP} levels (mean S_{RBP} 12, range 8–17 μg/ml) notwithstanding renal failure.

Pre- and intra-HRS RBP analyses available in one patient demonstrate a 35-fold increase in U_{RBP} (from 0.04 to 1.4 μg/ml) accompanied by a decrease in S_{RBP} (from 19 to 17 μg/ml).

It is suggested that impaired hepatic release or synthesis of RBP accounts for the low serum RPB levels encountered in hepatorenal syndrome patients; a renal tubular dysfunction or injury is responsible for the increased urinary values. This particular combination of low serum and high urine RBP could help in diagnosis of this entity.

Discussion

Human RBP is a small protein of ≈21,000 Da and functions as physiologic carrier for retinol (vitamin A alcohol). As such, RBP circulates in plasma bound to an additional ≈55,000 Da protein, prealbumin (PA) and to retinol; less than 10% of plasma RBP is in a free uncomplexed form.[6,7]

[6] P. A. Peterson, F. S. Nilsson, L. Ostberg, L. Rask, and A. Vahlquist, *Vitam. Horm. (N.Y.)* **32**, 181 (1974).
[7] W. S. De Goodman, *Ann. N.Y. Acad. Sci.* **348**, 378 (1980).

RBP is synthesized and stored in the liver; it is catabolized mainly in the kidney.[6,7] Consequently, its plasma levels among other factors (e.g., vitamin A and protein deficiency) are affected by the presence or absence of liver and kidney disease. Indeed, low S_{RBP} was found in and correlated with progression of acute hepatitis and in chronic parenchymal liver disease.[8,9] Conversely, high S_{RBP} is reported in a variety of renal diseases, congential and acquired.[8,10,11]

We believe the ELISA assay reported here represents an improvement over the existing methodologies for RBP analysis. With respect to the radioimmunoassay techniques, the present assay eliminates the need for radioactive counting and disposal, for a second antibody, and for high-speed centrifugation; the time constraint imposed by the inherent radioactive decay is also eliminated. The radial immunodiffusion technique requires a minimum of 3-days incubation time and most of all requires sample concentration for expected low ranges with possible loss of antigen.

Alcohol-related liver dysfunction is the commonest form of liver disease in this country with cirrhosis as leading sequel, and renal dysfunction is common in liver disease; its occurrence estimated as high as 84% in some series.[12] The etiologic spectrum of renal dysfunction in cirrhosis includes many of the known causes of renal failure present in a general population and one entity, the hepatorenal syndrome, particular to liver disease. The diagnosis is entertained in the presence of progressive oliguria and azotemia occurring in liver disease patients without an obvious precipitating cause; a low urine Na (usually <10 mmol/liter) and high urine creatinine are characteristic though not exclusive. Its frequency has been quoted as high as 44% among hospitalized patients with severe liver cirrhosis.[13] The grim clinical outcome, the reversible nature (following liver transplant), and the paucity of renal histopathologic changes at autopsy are also well documented. Several hypotheses were postulated as to the etiology and the pathogenesis of this entity, none so far completely satisfactory.[12–16] Because renal failure is a common event in liver cirrho-

[8] F. R. Smith and W. S. De Goodman, J. Clin. Invest. 50, 2426 (1971).
[9] A. Vahlquist, A. Sjolund, A. Norden, P. A. Peterson, G. Stigmar, and B. Johansson, J. Clin. Lab. Invest. 38, 301 (1978).
[10] A. Vahlquist, P. A. Peterson, and L. Wibell, Eur. J. Clin. Invest. 3, 352 (1973).
[11] L. Scarpioni, P. Dall'Angelo, P. G. Poisetti, and G. Buzio, Clin. Chim. Acta 68, 107 (1976).
[12] L. Shear, J. Kleinerman, and G. J. Gabuzda, Am. J. Med. 39, 184 (1965).
[13] J. Rhodes, M. Bruguera, J. Teres, and J. M. Bordas, Rev. Clin. Esp. 117, 475 (1970).
[14] S. Papper, in "The Kidney in Liver Disease" (M. Epstein, ed.), p. 91. Elsevier/North Holland, New York, 1978.

sis, because HRS represents a significant percentage of such occurrence and because both organs, the liver and the kidneys, play a crucial role in RBP metabolism, we studied the effect of concomitant failure of both organs upon this protein.

Our study confirms the presence of elevated serum and urine RBP levels in renal failure patients and low RBP values in sera of cirrhotic patients. It does show that cirrhotic patients with normal kidney function have a normal excretion of RBP. Our data also show an increased urine RBP concentration in hepatorenal patients similar to that of patients with renal failure, but in contrast with the latter, associated with very low serum RBP values. The use of an RBP index takes advantage of this association to separate between the groups.

This particular combination, probably the result of opposite and concomitant derangements in RPB metabolism, could have diagnostic significance in the context of concurrent failure of both organs, the liver and the kidneys, such as in hepatorenal syndrome.

---

[15] R. Williams and S. P. Wilkinson, *in* "Liver and Biliary Tract Disease" (R. Wright, K. G. M. Alberti, and G. H. Millward-Sadler, eds.), Vol. 46, p. 1060. Saunders, Philadelphia, Pennsylvania, 1979.

[16] H. O. Conn, *in* "Disease of the Liver" (L. Schiff and E. Schiff, eds.), 5th ed., Vol. 23, p. 847. Lippincott, Philadelphia, Pennsylvania, 1982.

[11] Fourier Transform Infrared Spectroscopy of Retinoids

By NATALIE L. ROCKLEY, MARK G. ROCKLEY, BRUCE A. HALLEY, and ELDON C. NELSON

Identification of synthetic and natural retinoids requires rapid, sensitive spectroscopic methods for the identification of very small quantities of materials. It is well known that Fourier transform infrared (FTIR) spectroscopy offers sufficient instrumental sensitivity, accuracy, and speed of analysis to be the method of choice. In particular, the digital acquisition of the data combined with laser wavelength reference ensures optimum conditions for accurate band position and profile measurements. The wavelength multiplexing advantage and high energy throughput allow FTIR to be used in conjunction with low signal applications as might be the case where small quantities of material are to be analyzed or where the analysis method itself lacks efficient spectral throughput (e.g., diffuse reflectance).

There are several versatile new methods for analysis which may prove to be of immense value in future work on retinoid mixtures. In particular, attenuated reflectance cylinders allow measurement of flowing aqueous streams, such as the effluent of high-performance liquid chromatography (HPLC) systems. The ordinate performance of better quality FTIR spectrometers permits adequate blanking of the solvent concentration. Diffuse reflectance permits examination of HPLC effluent by allowing the carrier solvent to evaporate, leaving the analyte physisorbed on KCl powder. Photoacoustic spectroscopy allows examination of powders, films, or liquids without the need for sample preparation or correction for Kristiansen dispersion. All of these promise potent new alternatives for accurate and quantitative study of small quantities of retinoids.

Infrared analyses have been reported for some of the retinoids, including isomers of retinol[1-4] and retinal,[3-9] all-*trans*-dehydroretinol,[10] retinyl acetate and palmitate,[11] isomers of methyl retinoate,[12] and of retinoic acid.[13] The structures of several oxo-derivatives were resolved by various spectroscopic methods including infrared. 4-Oxoretinol and 4-oxoretinal,[14] 4-oxoretinoic acid,[15] 5,6- and 5,8-epoxy derivatives of retinol, retinyl acetate, retinal,[16] and retinoate,[17] and 11,12-epoxyretinol and

[1] C. D. Robeson, J. D. Cawley, L. Weisler, M. H. Stern, C. C. Eddinger, and A. J. Chechak, *J. Am. Chem. Soc.* **77**, 4111 (1955).

[2] H. Rosenkrantz, *Methods Biochem. Anal.* **5**, 407 (1957).

[3] K. R. Farrar, J. C. Hamlet, H. B. Henbest, and E. R. H. Jones, *J. Chem. Soc.* p. 2657 (1952).

[4] K. Chihara and W. H. Waddell, *J. Am. Chem. Soc.* **102**, 2963 (1980).

[5] C. D. Robeson, W. P. Blum, J. M. Dieterle, J. D. Cawley, and J. G. Baxter, *J. Am. Chem. Soc.* **77**, 4120 (1955).

[6] W. Oroshnik, *J. Am. Chem. Soc.* **78**, 2651 (1956).

[7] S. Saito and M. Tasumi, *J. Raman Spectrosc.* **14**, 236 (1983).

[8] B. Curry, A. Broek, J. Lugtenburg, and R. Mathies, *J. Am. Chem. Soc.* **104**, 5274 (1982).

[9] R. E. Cookingham, A. Lewis, and A. T. Lenley, *Biochemistry* **17**, 4699 (1978).

[10] C. von Planta, U. Schwieter, L. Chopard-dit-Jean, R. Ruegg, M. Kofler, and O. Isler, *Helv. Chim. Acta* **45**, 548 (1962).

[11] D. Chapman and R. J. Taylor, *Nature (London)* **174**, 1011 (1954).

[12] R. M. McKenzie, D. M. Hellwege, M. L. McGregor, N. L. Rockley, P. J. Riquetti, and E. C. Nelson, *J. Chromatogr.* **155**, 379 (1978).

[13] T. Takemura, K. Chihara, R. S. Becker, P. K. Das, and G. L. Hug, *J. Am. Chem. Soc.* **102**, 2604 (1980).

[14] H. B. Henbest, E. R. H. Jones, and T. C. Owne, *J. Chem. Soc.* p. 4909 (1957).

[15] M. S. Surekha Rao, J. John, and H. R. Cama, *Int. J. Vitam. Nutr. Res.* **42**, 368 (1972).

[16] F. B. Jungalwala and H. R. Cama, *Biochem. J.* **95**, 17 (1965).

[17] B. Morgan and J. N. Thompson, *Biochem. J.* **101**, 835 (1966).

11,12-epoxyretinal[18,19] have been prepared and identified by infrared spectroscopy. Fluorinated retinoic acids and their analogs,[20,21] sulfur[22,23] and selenium-containing retinoids,[23] several antioxidant derivatives of retinoic acid[24] and several aromatic retinoic acid derivatives[25] have been synthesized and their infrared spectral properties have been reported. Infrared spectroscopy has also been used in the identification of a metabolite of retinoic acid.[26]

Mass spectral analyses have been published on several of these retinoids,[27] and extensive investigations have been reported in recent years on the Raman and resonance Raman spectroscopy of retinal and its isomers[7–9,28–35] and retinols and retinoic acid.[28]

In the present study, the Fourier transform infrared spectra of 15 retinoids are given along with band positions which correlate with the retinoid structures.[36]

Purification Procedures and Infrared Analysis

Materials

Glass-distilled residue-free solvents should be used for high-performance liquid chromatography (HPLC) and for infrared analyses. The water used for HPLC should be deionized and redistilled in glass or of

[18] Y. Ogata, Y. Yosugi, and K. Tomizawa, *Tetrahedron* **26,** 5939 (1970).
[19] Y. Ogata, K. Tomizawa, and K. Takagi, *Tetrahedron* **29,** 47 (1973).
[20] A. J. Lovey and B. A. Pawson, *J. Med. Chem.* **25,** 71 (1982).
[21] K. Chan, A. C. Specian, and B. A. Pawson, *J. Med. Chem.* **24,** 101 (1980).
[22] M. Klaus, W. Bollag, P. Huber, and W. Kung, *Eur. J. Med. Chem.* **18,** 425 (1983).
[23] S. C. Welch and J. M. Gruber, *J. Med. Chem.* **22,** 1532 (1979).
[24] S. C. Welch, J. M. Gruber, and A. S. C. Prakasa Rao, *J. Med. Chem.* **25,** 81 (1982).
[25] M. I. Dawson, P. D. Hobbs, R. L. Chan, W. Chao, and V. A. Fung, *J. Med. Chem.* **24,** 583 (1981).
[26] K. L. Skare, H. K. Schnoes, and H. F. DeLuca, *Biochemistry* **21,** 3308 (1982).
[27] R. L. Lin, G. R. Waller, E. D. Mitchell, K. S. Yang, and E. C. Nelson, *Anal. Biochem.* **35,** 435 (1970).
[28] L. Rimai, D. Gill, and J. L. Parsons, *J. Am. Chem. Soc.* **93,** 1353 (1971).
[29] D. Gill, M. E. Heyde, and L. Rimai, *J. Am. Chem. Soc.* **93,** 6288 (1971).
[30] M. E. Heyde, D. Gill, R. G. Kilponen, and L. Rimai, *J. Am. Chem. Soc.* **93,** 6776 (1971).
[31] R. H. Callender, A. Doukas, R. Crouch, and K. Nakanishi, *Biochemistry* **15,** 1621 (1976).
[32] R. Mathies, T. B. Freedman, and L. Stryer, *J. Mol. Biol.* **109,** 367 (1977).
[33] R. Callender and B. Honig, *Annu. Rev. Biophys. Bioeng.* **6,** 33 (1977).
[34] A. Warshel, *Annu. Rev. Biophys. Bioeng.* **6,** 273 (1977).
[35] R. E. Cookingham, A. Lewis, D. W. Collins, and M. A. Marcus, *J. Am. Chem. Soc.* **98,** 2759 (1976).
[36] N. L. Rockley, B. A. Halley, M. G. Rockley, and E. C. Nelson, *Anal. Biochem.* **133,** 314 (1983). Data used with permission from Academic Press, New York.

equivalent purity. all-*trans*-Retinoic acid, 4-oxoretinoic acid, 5,6-epoxy-retinoic acid, 5,8-epoxyretinoic acid, and C_{19}-aldehyde were obtained from Hoffmann-La Roche Inc. Retinal, 9-*cis*-retinal, and 13-*cis*-retinal can be purchased. 4-Oxo-9-hydroxy-C_{19}-aldehyde, a decarboxylation product of retinoic acid, can be obtained as previously described.[37] Methyl all-*trans*-retinoate is prepared from all-*trans*-retinoic acid using diazomethane in methanol–diethyl ether solutions.[38] Isomers of methyl retinoate can be obtained by irradiating methyl all-*trans*-retinoate in dimethyl sulfoxide under fluorescent light.[39]

Purification of Retinoids by High-Performance Liquid Chromatography

All retinoids should be purified before and after the infrared analysis to monitor oxidation and isomerization. For the compounds discussed in this article, the following columns were used. Either a Partisil PXS 10/25 ODS-2 column from Whatman (Column A), or a Bondapak C_{18} column from Waters (Column B). The elution profiles for the compounds under study are as follows.

Column A. Retinal (70% methanol, 30% water, at 1 ml/min), 39 min; 9-*cis*-retinal (70% methanol, 30% water, at 1 ml/min), 20 min; 5,6-epoxyretinoic acid and 5,8-epoxyretinoic acid (80% methanol, 20% 0.01 M acetic acid, 0.5 ml/min), both at 56 min; all-*trans*-retinoic acid (85% methanol, 15% 0.01 M acetic acid, 1 ml/min), 45 min; methyl all-*trans*-retinoate and five of its isomers (85% methanol, 15% 0.01 M acetic acid, 1 ml/min) photocyclized, 13-*cis*-, 90 min; 11,13-di-*cis*-, 107 min; 13-*cis*-, 121 min; photocyclized, 13-*trans*-, 130 min; 11-*cis*-, 139 min; all-*trans*-, 172 min.

Column B. 13-*cis*-Retinal (85% methanol, 15% 0.01 M acetic acid, 0.6 ml/min), 54 min; C_{19}-aldehyde (90% methanol, 10% 0.01 M acetic acid, 0.65 ml/min), 33 min; 4-oxoretinoic acid (75% methanol, 25% 0.01 M acetic acid, 0.6 ml/min), 50 min; 4-oxo-9-hydroxy-C_{19}-aldehyde (50% methanol, 50% water, at 0.6 ml/min), 80 min.

Infrared Analysis

A variety of suitable FTIR instruments are available. We have used a Digilab FTR-20C interfaced to a Data General Nova 3/12 (Digilab Inc., Cambridge, MA) at a resolution of 4 cm^{-1}. The sample in the solvent can be applied in a liquid cell or, as we have done, on a NaCl or KBr window. The solvent (chloroform or carbon tetrachloride) can then be evaporated

[37] N. L. Rockley, B. A. Halley, and E. C. Nelson, *Biochim. Biophys. Acta* **627,** 270 (1980).
[38] H. Schlenk and J. L. Gellerman, *Anal. Chem.* **32,** 1412 (1960).
[39] B. A. Halley and E. C. Nelson, *J. Chromatogr.* **175,** 113 (1979).

to dryness before the window is inserted into the cell holder and introduced into the instrument. No solvent subtraction is necessary in this case. To get transmission data, a ratio is obtained of the background spectrum of a window without sample to that of the sample.

Summary of Absorption Bands

The structures and IR spectra of the 15 retinoids are shown in Figs. 1 and 2, respectively. Since the absorption due to the C—H stretch of methyl and methylene groups (3000–2800 cm^{-1}) remained fairly constant

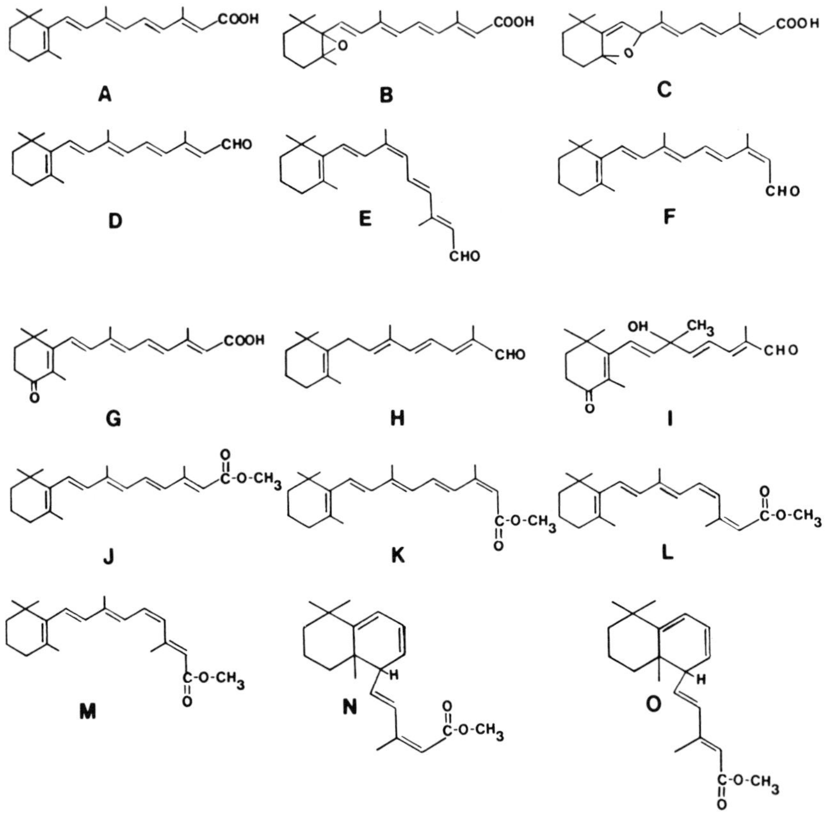

FIG. 1. Structures of 15 retinoids. (A) all-*trans*-Retinoic acid; (B) 5,6-epoxyretinoic acid; (C) 5,8-epoxyretinoic acid; (D) all-*trans*-retinal; (E) 9-*cis*-retinal; (F) 13-*cis*-retinal; (G) 4-oxoretinoic acid; (H) C$_{19}$-aldehyde; (I) 4-oxo-9-hydroxy-C$_{19}$-aldehyde; (J) methyl all-*trans*-retinoate; (K) methyl 13-*cis*-retinoate; (L) methyl 11,13-di-*cis*-retinoate; (M) methyl 11-*cis*-retinoate; (N) methyl photocyclized 13-*cis*-retinoate; (O) methyl photocyclized 13-*trans*-retinoate. Data from Rockley *et al.*[36]

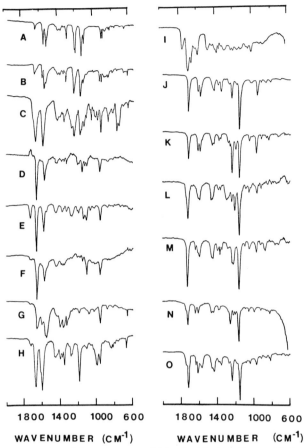

FIG. 2. Fourier transform infrared spectra of 15 retinoids. (A) all-*trans*-Retinoic acid; (B) 5,6-epoxyretinoic acid; (C) 5,8-epoxyretinoic acid; (D) all-*trans*-retinal; (E) 9-*cis*-retinal; (F) 13-*cis*-retinal; (G) 4-oxoretinoic acid; (H) C$_{19}$-aldehyde; (I) 4-oxo-9-hydroxy-C$_{19}$-aldehyde; (J) methyl all-*trans*-retinoate; (K) methyl 13-*cis*-retinoate; (L) methyl 11,13-di-*cis*-retinoate; (M) methyl 11-*cis*-retinoate; (N) methyl photocyclized 13-*cis*-retinoate; (O) methyl photocyclized 13-*trans*-retinoate. Data from Rockley *et al.*[36]

for all the samples, only the region from 2000 to 600 cm^{-1} is shown. Infrared frequencies are tabulated in Tables I and II where references to "strong," "medium," and "weak" bands are given only as approximate indications of absorption intensity.

The presence of a broad band above 3000 cm^{-1} in the spectra of all-*trans*-retinoic acid, 4-oxoretinoic acid, and 5,8-epoxyretinoic acid is due to the acid hydrogen. The hydroxyl group in 4-oxo-9-hydroxy-C$_{19}$-alde-

TABLE I
RETINOID ABSORPTION IN THE 3000 TO 1500 cm^{-1} REGION OF THE INFRARED SPECTRA[a]

Retinoid	C—H stretch	$\overset{\overset{\text{O}}{\|}}{\text{R}}$C—H stretch	C=O stretch	C=C stretch
Retinoic acid	2930–2820	—	1690m	1605s,1578s,1565m
5,6-Epoxyretinoic acid	2940–2860	—	1690m	1605m,1580s,1565m
5,8-Epoxyretinoic acid	2960–2860	—	1680s	1650m,1600s
Retinal	2950–2850	2720w	1660s	1578s,1560m
9-cis-Retinal	2960–2860	2720w	1730m,1664s	1585s,1555m
13-cis-Retinal	2950–2850	2720w	1730w,1660s	1580s,1560m
4-Oxoretinoic acid	2940–2850	—	1735w,1660s	1605m,1560s,1555m
C$_{19}$-Aldehyde	2945–2860	2720m	1725w,1670s	1620m,1660s
4-Oxo-9-hydroxy-C$_{19}$-aldehyde	2960–2855	2720w	1730m,1672s	1640s,1600w,1565m
Methyl all-trans-retinoate	2950–2860	—	1712s	1609w,1583m,1570w
Methyl 13-cis-retinoate	2950–2850	—	1711s	1606m,1581m,1560w
Methyl 11,13-di-cis-retinoate	2940–2850	—	1715s	1600m,1585m
Methyl 11-cis-retinoate	2920–2850	—	1712s	1630w,1600w,1582m
Methyl photocyclized 13-trans-retinoate	2950–2850	—	1735w,1719s	1635w,1612m,1570m
Methyl photocyclized 13-cis-retinoate	2960–2850	—	1735w,1718s	1635m,1602m

[a] Data from Rockley et al.[36] Absorption: s, strong; m, medium; w, weak.

hyde absorbs in the same region. The weak band which appears at 2720 cm^{-1} in the spectra of the aldehydes reflects the presence of an aldehyde C—H stretch.[40] Strong bands between 1735 and 1660 cm^{-1} are attributed to the C=O stretch of ketones, aldehydes, carboxylic acids, and esters. Two carbonyl bands appear in some of the spectra (Fig. 2E–I), a first band around 1660 cm^{-1} and a second band around 1730 cm^{-1}. This 1730 cm^{-1} band is weaker and is thought to be due to non-hydrogen-bonded mono-mers, while the first band near 1660 cm^{-1} reflects the presence of dimers with hydrogen bonding involving the carbonyl or the carboxyl groups.[4,13] In the case of retinoic acid and its isomers (Fig. 2A–C) only the associated C=O stretch band at 1680 cm^{-1} is observed when the infrared spectra are recorded from evaporated films.[4] The spectra of the methyl retinoates all have bands between 1735 and 1711 cm^{-1} reflecting the presence of a conjugated ester carbonyl.

[40] R. M. Silverstein and G. C. Bassler, in "Spectrometric Identification of Organic Compounds," 2nd ed., p. 64. Wiley, New York, 1967.

TABLE II

RETINOID ABSORPTIONS IN THE 1500 TO 950 cm^{-1} REGION OF THE INFRARED SPECTRA[a]

Retinoid	H—C—H bending (nonsymmetrical)	C-1 gem-Me$_2$ and bending H—C—H (symmetrical)	1300–1100 cm	H bending of H H / RC=CR' trans, unsubstituted
Retinoic acid	1445w	1350m	1260s,1255s,1190s,1160m	970m,955m
5,6-Epoxyretinoic acid	1445w	1345m	1258s,1182s,1158m,1120w	975m,955m
5,8-Epoxyretinoic acid	1445m	1365w	1280m,1252s,1190s,1145m,1120m	960m
Retinal	1450w	1335w	1212w,1200w,1162m,1134w,1110w	968m
9-cis-Retinal	1450m	1380w,1335w	1270w,1215w,1200w,1145m,1110m	962m
13-cis-Retinal	1450w	1378w,1360w	1280w,1220w,1190w,1160w,1110m	964m
4-Oxoretinoic acid	1450w	1370m	1280w,1252w,1230w,1200w	968m
C$_{19}$-Aldehyde	1450w	1360m	1290m,1270m,1190s,1115w	962m
4-Oxo-9-hydroxy-C$_{19}$-aldehyde	1455m	1355w	1270w,1200w,1180w,1130w	970m
Methyl all-trans-retinoate	1433m	1354m	1267w,1240m,1189w,1152s	968m
Methyl 13-cis-retinoate	1450m,1430m	1378w,1358w	1275w,1258s,1236s,1194m,1155s	972m
Methyl 11,13-di-cis-retinoate	1458m,1445m	1379w,1358w	1290m,1270w,1236m,1203m,1156s	965m
Methyl 11-cis-retinoate	1445m,1432m	1385w,1358m	1275w,1260w,1225m,1212m,1150s	965m
Methyl photocyclized 13-trans-retinoate	1459w,1437w	1358w	1273w,1240m,1156s	975w
Methyl photocyclized 13-cis-retinoate	1458w,1434w	1380w,1370w	1250m,1218w,1199w,1159s	990w

[a] Data from Rockley et al.[36] Absorption: s, strong; m, medium; w, weak.

The carbon–carbon double bond stretch frequency (1650 to 1555 cm^{-1}) is affected by unsaturation.[40-42] Conjugation with another $C=C$ or $C=O$ leads to a shift to a lower frequency. The highest frequencies are found for the 4-oxo-9-hydroxy-C_{19}-aldehyde (Fig. 2I) and two photocyclized methyl retinoate isomers which have two $C=C$ bonds in conjugation with each carbonyl (Figs. 2N and O). The C_{19}-aldehyde (Fig. 2H) and 5,8-epoxyretinoic acid (Fig. 2C) have three $C=C$ groups conjugated with the carbonyl. The remaining compounds all have five or six conjugated double bonds and therefore the $C=C$ stretch band has a lower frequency.

Methyl deformations are found between 1500 and 1300 cm^{-1}. The region between 1459 and 1430 cm^{-1} is similar in the spectrum of each of the 15 retinoids and is due to the asymmetric HCH bending of the C-9 and C-13 methyl groups[9,28,29] and the HCH bending (scissoring) of methylene groups.[42,43] The absorptions between 1380 and 1335 cm^{-1} have been attributed to *gem*-dimethyl symmetric deformation and to C-9 and C-13 symmetric methyl deformations in the spectra of retinal and its isomers.[8,9]

The 1300 to 1100 cm^{-1} region, also known as the fingerprint region, is sensitive to conformation and to the identity of the end group, particularly in the raman spectrum.[28,31] It has several bands due to a large number of vibrations which contribute to it. These are primarily C—C single bond stretches and C—C—H bends,[33,41,44] but also include the C—O stretch of a carboxylic acid at 1320 to 1210 cm^{-1}.[40] The presence of a carboxylic ester can readily be detected by the strong absorption at 1150 to 1159 cm^{-1} as in the spectra of the six methyl retinoate isomers (Fig. 2J–O). The band is due to a symmetrical C—O—C stretching.

The strong absorbances at 1258 cm^{-1} in the spectrum of 5,6-epoxyretinoic acid (Fig. 2B) and 1252 cm^{-1} in that of 5,8-epoxyretinoic acid (Fig. 2C) are attributable to symmetrical stretching of the epoxide ring.[16] The presence of bands at 1065, 1083, and 1175 cm^{-1} confirms the furan structure in the 5,8-epoxyretinoic acid.[14]

A trans-ethylenic out-of-plane C—H bending leads to the absorbances at 990 to 955 cm^{-1} present in all the spectra.[28,43-46] Theoretical calculations

[41] D. J. Pasto and C. R. Johnson, *in* "Organic Structure Determination," p. 109. Prentice-Hall, Englewood Cliffs, New Jersey, 1969.

[42] E. R. Blout, M. Fields, and R. Karplus, *J. Am. Chem. Soc.* **70**, 194 (1948).

[43] R. S. Rasmussen and R. Brattain, *J. Chem. Phys.* **15**, 120 (1947).

[44] W. Krauss, P. Ruckert, F. Scheidel, and C. Wagner-Bartak, *Ber. Bunsenges. Phys. Chem.* **72**, 415 (1968).

[45] N. Sheppard and G. B. B. M. Sutherland, *Proc. R. Soc. London, Ser. A* **196**, 195 (1949).

[46] J. E. Jackson, R. F. Paschke, W. Tolberg, H. M. Boyd, and D. H. Wheeler, *J. Am. Oil Chem. Soc.* **29**, 229 (1952).

suggest that these bands also involve significant contributions from C—CH$_3$ stretching.[47]

Lines below 900 cm^{-1} have been assigned to out-of-plane modes and torsional deformations.[47] The only strong bands observed in these studies appear at 880 cm^{-1} (m), 792 cm^{-1} (s), and 760 cm^{-1} (m) in the spectrum of 5,8-epoxyretinoic acid (Fig. 2C). They reflect the presence of an unsymmetrical stretching of an epoxide ring in which a C—C bond is stretching during contraction of the adjacent C—O bond.[16,40] These bands also appear in the spectrum of 5,6-epoxyretinoic acid (Fig. 2B) at 880, 840, and 780 cm^{-1}, but are much weaker.[15]

These spectra compare very well with those which have already been published. For a few of these compounds no other data are available in the literature. Resonance Raman spectroscopy has proved to be a powerful tool for the understanding of the effects of conformation on the vibrational spectra of various isomers of retinal and many of the observed bands, particularly in the fingerprint region, have been assigned to specific vibrational modes for these few compounds.[7-9,33] The vibrational assignments for all-*trans*-retinal and its isomers have also been presented based on a theoretical approach using normal mode analysis with intensity calculations.[34,47] This opens up the possibility for a more exacting analysis of the vibrations responsible for the various bands in other retinoids. This should provide considerable predictive capability in the analysis of new analogs.

In summary, retinoids which have C=O groups (aldehydes, acids, or 4-oxo compounds) will have a band between 1730 and 1660 cm^{-1}. Compounds with two to three conjugated unsaturations will have a C=C stretch between 1650 and 1620 cm^{-1}, while more unsaturations will lead to bands between 1610 and 1555 cm^{-1}. An epoxy ring leads to bands around 1250, 880, and 790 cm^{-1}. A furan structure is confirmed by bands at 1065, 1083, and 1175 cm^{-1}. Unsubstituted trans C=C bonds, present in most retinoids, are detectable by absorbances at 990–955 cm^{-1}.

It is expected that Fourier transform infrared spectroscopy, which permits analysis of small amounts (nanogram) of sample in short measurement times, will prove to be a useful tool in the future identification of synthetic and natural retinoids.

Acknowledgments

This research was supported in part by Public Health Service Research Grant AM-09191 and by the Oklahoma Agricultural Experiment Station.

[47] A. Warshel and M. Karplus, *J. Am. Chem. Soc.* **96,** 5677 (1974).

[12] Purification of Interstitial Retinol-Binding Protein from the Eye

By S.-L. FONG, G. I. LIOU, and C. D. B. BRIDGES

The occurrence of specific proteins that may be involved in the intracellular transport and utilization of retinoids has been known for a number of years. To date, four have been purified and characterized (CRAlBP,[1] CRBP, CRBP II, CRABP.[1a-4] Until recently, the only known example of a protein that carried retinoid extracellularly was plasma retinol-binding protein (RBP).[5] Evidence for the extracellular transport of retinoids through the interstitial spaces between the cells in organs and tissues has been accumulating in recent years, however. Blomhoff et al.[6,7] have shown that when [^{3}H]retinol is fed to rats it is initially taken up by the hepatocytes, but then it appears to be transferred to the stellate cells for storage. It has not been established whether a special binding protein is involved in this process. With respect to the testis, Carson et al.[8] showed that an established cell line derived from murine Sertoli cells secretes a retinol-binding protein that under physiological conditions could be involved in the transport of retinol into the lumen of the seminiferous tubules or to the developing germ cells.

In the vertebrate eye, the transport of retinol between cells of the neural retina and the retinal pigment epithelium has been established and appears to be an essential component of the visual cycle of rhodopsin bleaching and regeneration.[9-12] The interstitial space between these tissue

[1] CRAlBP, cellular retinal-binding protein; CRBP, cellular retinol-binding protein; CRBP II, cellular retinol-binding protein II; CRABP, cellular retinoic acid-binding protein.

[1a] G. W. Stubbs, J. C. Saari, and S. Futterman, J. Biol. Chem. **254**, 8529 (1979).

[2] J. C. Saari, this series, Vol. 81, p. 819.

[3] F. Chytil, and D. E. Ong, in "The Retinoids" (M. B. Sporn, A. B. Roberts, and D. S. Goodman, eds.), Vol. 2, p. 89. Academic Press, New York, 1984.

[4] D. E. Ong, J. Biol. Chem. **259**, 1476 (1984).

[5] D. S. Goodman, in "The Retinoids" (M. B. Sporn, A. B. Roberts, and D. S. Goodman, eds.), Vol. 2, p. 41. Academic Press, New York, 1984.

[6] R. Blomhoff, P. Helgerud, M. Rasmussen, T. Berg, and K. R. Norum, Proc. Natl. Acad. Sci. U.S.A. **79**, 7326 (1982).

[7] R. Blomhoff, K. Holte, L. Naess, and T. Berg, Exp. Cell Res. **150**, 186 (1984).

[8] D. D. Carson, L. I. Rosenberg, W. S. Blaner, M. Kato, and W. J. Lennarz, J. Biol. Chem. **259**, 3117 (1984).

[9] J. E. Dowling, Nature (London) **188**, 114 (1960).

[10] C. D. B. Bridges, Exp. Eye Res. **22**, 435 (1976).

layers is filled with interphotoreceptor matrix (IPM), which has been shown to contain a high M_r glycoprotein that carries endogenous all-*trans*- and 11-*cis*-reʋnol.[13,14] This protein is referred to as interphotoreceptor matrix or interstitial retinol-binding protein (IRBP).[13,14] Its postulated role as a physiological carrier for retinol is supported by the finding that the amount of all-*trans*-retinol bound to IRBP increases when eyes are exposed to light[14,15] and by the observation that its loss in rats with hereditary retinal dystrophy is associated with malfunctions of retinol transport and rhodopsin regeneration in the eye.[16]

IRBP is synthesized and secreted by cells of the neural retina,[17–20] which may therefore have an important function in regulating its amount in the subretinal space. The cells in question may be the photoreceptors.[16,21] The estimated concentration of IRBP in bovine IPM is 30–100 μM, or 429–475 μg for an eye with a neural retina surface of 25 cm².

Purified bovine IRBP binds approximately two molecules of all-*trans*-retinol and contains 8.4% by weight of carbohydrate that consists of sialic acid, neutral hexoses (mannose, fucose, galactose), and glucosamine in the molar ratio of approximately 1 : 3 : 2. It has about four isoelectric forms with p*I* ranging from 4.4 to 4.8. Bovine, rat, and mouse IRBPs all have the same apparent M_r of 144K on SDS–polyacrylamide gels. Human and monkey IRBPs have lower apparent M_r of 135K in the same system. IRBP appears to be widely distributed throughout the vertebrates, as might be expected if it played a critical role in the visual cycle. Polypeptides that are immunologically cross-reactive with anti-bovine IRBP se-

[11] C. D. B. Bridges, *in* "The Retinoids" (M. B. Sporn, A. B. Roberts, and D. S. Goodman, eds.), Vol. 2, p. 125. Academic Press, New York, 1984.

[12] C. D. B. Bridges, R. A. Alvarez, S.-L. Fong, F. Gonzalez-Fernandez, D. M. K. Lam, and G. I. Liou, *Vision Res.* **24**, 1581 (1984).

[13] G. I. Liou, C. D. B. Bridges, and S.-L. Fong, *Invest. Ophthalmol. Visual Sci.* **22**, 65 (1982).

[14] G. I. Liou, C. D. B. Bridges, and S.-L. Fong, *Vision Res.* **22**, 1457 (1982).

[15] A. J. Adler, and K. J. Martin, *Biochem. Biophys. Res. Commun.* **108**, 1601 (1982).

[16] F. Gonzalez-Fernandez, R. A. Landers, P. A. Glazebrook, S.-L. Fong, G. I. Liou, D. M. K. Lam, and C. D. B. Bridges, *J. Cell Biol.* **99**, 2092 (1984).

[17] C. D. B. Bridges, S.-L. Fong, G. I. Liou, R. A. Alvarez, and R. A. Landers, *Prog. Retinal Res.* **2**, 137 (1983).

[18] S.-L. Fong, G. I. Liou, R. A. Landers, R. A. Alvarez, F. González-Fernández, P. A. Glazebrook, D. M. K. Lam, and C. D. B. Bridges, *J. Neurochem.* **42**, 1667 (1984).

[19] S.-L. Fong, G. I. Liou, R. A. Alvarez, and C. D. B. Bridges, *J. Biol. Chem.* **259**, 6534 (1984).

[20] B. Wiggert, L. Lee, P. J. O'Brien, and G. J. Chader, *Biochem. Biophys. Res. Commun.* **118**, 789 (1984).

[21] F. Gonzalez-Fernandez, R. A. Landers, S.-L. Fong, G. I. Liou, P. Glazebrook, D. M. K. Lam, and C. D. B. Bridges, *Neurochem. Int.* **7**, 533 (1985).

rum have also been identified in the eyes of amphibians, fishes, birds, and reptiles. In the teleosts, the M_r is half that in the other vertebrates. All IRBPs characteristically have much higher apparent M_r on gel-filtration columns than on SDS gels. Bovine IRBP, for example, has an apparent M_r at 249,000 while human is at 205,000. The difference is believed to be due to shape factors,[22,23] and not to dimerization as originally suggested by Liou et al.[14]

A 1.5-kb cDNA encoding three consecutive bovine tryptic peptides has been recently cloned and sequenced by Liou et al.

Small-Scale Purification of IRBP

The first reported purification of IRBP to homogeneity was by high-performance size-exclusion chromatography of crude bovine IPM,[14] as described below. Sufficient IRBP was obtained by this method to demonstrate that it consisted of a single concanavalin A (Con A)-binding polypeptide with apparent M_r of 140–145K by SDS–polyacrylamide gel electrophoresis. A similar purification of human IRBP was reported by Fong et al.[19] The following technique is based on Liou et al.[14]

High-Performance Size-Exclusion Liquid Chromatography[14]

Aliquots of IPM (see below for details of preparation: the amount used in this experiment is roughly equivalent to one-fifth of an eye) are freed of particulate matter by centrifugal filtration (Rainin 0.2-μm cellulosic membranes) and injected (Waters U6K injector) in 0.15–1.0 ml volumes into an HPLC system equipped with two 30 × 0.75 cm Bio-Rad TSK-400 and TSK-250 columns connected in series. The operable range of the first column is 5–1000K and of the second column, 1–300K. The mobile phase consists of 0.1 M Na_2SO_4, 0.02 M Na phosphate (pH 6.8), 0.05% NaN_3 at a flow rate of 0.9 ml/min (Beckman model 100A pump).

The effluent is continuously monitored at 280 nm with an Altex-Hitachi absorbance detector, then with a Gilson Spectra/Glo filter fluorometer (excitation at 330–400 nm; emission at 430–600 nm). A typical chromatogram is illustrated in Fig. 1. Fractions (0.9 ml) are collected, dialyzed against water (Spectrapor membrane, 6000–8000 MW cutoff), and lyophilized for analysis by SDS–PAGE.

The lyophilized samples are dissolved in dissociation buffer (1.25% SDS, 40 mM dithiothreitol, 0.5 mM EDTA, 5% sucrose, 10 μg/ml pyronin Y, 10 mM Tris–HCl, pH 8.0) and heated in a boiling water bath for 3 min. Electrophoresis is performed on 1.5-mm-thick 5–20% polyacrylamide continuous gradient slab gels, with 3% stacking gels, based on the system

[22] J. C. Saari, D. C. Teller, J. W. Crabb, and L. Bredberg, *J. Biol. Chem.* **260,** 195 (1985).
[23] A. J. Adler, C. D. Evans, and W. F. Stafford III, *J. Biol. Chem.* **260,** 4850 (1985).

of Laemmli.[24] The current is initially set at 12.5 mA, then increased to 25 mA after the pyronin Y tracking dye has passed from the stacking gel into the resolving gel. The gels are stained with Coomassie Brilliant Blue and destained in 10% acetic acid containing a small amount of Whatman DE-23 fibrous anion exchanger. The stained gel is also illustrated in Fig. 1.

The glycoprotein nature of IRBP can be demonstrated by incubating the gel slabs in ^{125}I-labeled Con A followed by autoradiography.[25] The use of other ^{125}I-labeled lectins such as *Lens culinaris* hemagglutinin, *Ricinus communis* agglutinin I, and wheat germ agglutinin to probe the carbohydrate structure of IRBP has been described by Fong *et al.*[18,19]

Preparative-Scale Purification of IRBP

Bridges *et al.*[17] and Fong *et al.*[19] used anion-exchange chromatography on DEAE-cellulose (DE-52) followed by gel-filtration chromatography on Sepharose CL-4B to obtain purified preparations of bovine and human IRBPs so permitting their partial characterization. For rat IRBP, Gonzalez-Fernandez *et al.*[21] used Con A-Sepharose followed by anion-exchange chromatography on DEAE-cellulose. Adler and Evans[26] described a rapid isolation technique for bovine IRBP that also employed an immobilized Con A column. The method described below for the preparation of bovine IRBP in milligram quantities is essentially that of Fong *et al.*[19]

Step 1. Preparation of Interphotoreceptor Matrix (IPM)

Bovine eyes are placed on ice as soon as possible and used between 2 and 5 hr after the animal has been killed. If the following procedure is carried out under room lighting, the bleaching of rhodopsin leads to the accumulation of all-*trans*-retinol bound to IRBP and permits the column effluents to be monitored for endogenous fluorescence. The maximum degree of saturation achieved by this method ranges up to 29% of the full binding capacity, which is 2 mol of retinol per mol of IRBP.

All dissections are carried out on ice. The anterior one-third of the eye is cut away with a sharp razor blade and the vitreous is removed. Care should be taken to ensure that the retina is not disturbed by this procedure. the vitreous surface of the retina is rinsed with about 3 ml of ice-cold buffer A (0.15 M NaCl, 0.01 M sodium phosphate, 0.1 mM phenylmethylsulfonyl fluoride). The neural retina is then peeled away from the retinal pigment epithelium and placed in buffer A. One hundred neural retinas are prepared in this way, incubated for 1 hr on ice in 100 ml of buffer A,

[24] U. K. Laemmli, *Nature (London)* **227**, 680 (1970).
[25] C. D. B. Bridges and S.-L. Fong, *Neurochemistry Int.* **1**, 255 (1980).
[26] A. J. Adler and C. D. Evans, *Biochim. Biophys. Acta* **761**, 217 (1983).

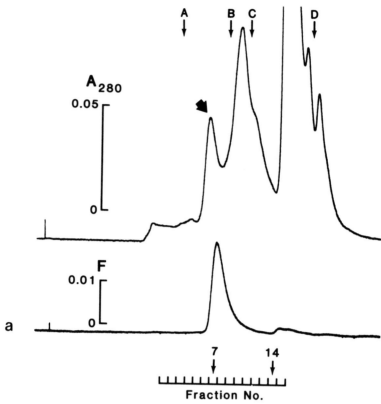

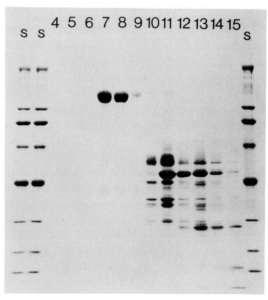

a

b

and centrifuged at 1000 g for 10 min. The supernatant, which contains most of the IPM, is then centrifuged at 100,000 g for 1 hr and dialyzed twice at 4° against 4 liters of buffer B (10 mM Tris-acetate, pH 7.5, 0.1 mM phenylmethylsulfonyl fluoride). An SDS–polyacrylamide gel pattern for this crude preparation is illustrated in Fig. 2, track 1. For comparison, a similar IPM preparation from human eyes[18] was electrophoresed in track 2. The preparation of IPM from monkey eyes has also been described.[27]

If desired, the yield of IPM can be improved by soaking the retinas for a second time in buffer A, and also by rinsing the surface of the retinal pigment epithelium cells after the neural retina has been removed.

Step 2. Anion-Exchange Chromatography

The dialyzed preparation of IPM is loaded onto a DEAE-cellulose (Whatman DE-52, 2.5 × 30 cm). The column is washed with 1 liter of buffer B, then eluted with a 2-liter linear gradient, 0–0.05 M NaCl in buffer B. Ten milliliter fractions are collected and monitored by their A_{280} and fluorescence emission at 470 nm when irradiated at 330 nm. A typical chromatographic profile is shown in Fig. 3. The first fluorescent peak is centered at fraction 48 and has an apparent M_r of 15,000 if it is examined by gel-filtration chromatography on Sephadex G-75. This fraction contains CRBP and CRABP. The second fluorescent peak is centered at fraction 76 (0.23 M NaCl). It has an apparent M_r of 250,000–260,000 by gel-filtration chromatography on Sepharose CL-4B. When 40-μl aliquots from every tenth tube are analyzed by SDS–polyacrylamide gel electrophoresis, as illustrated in Fig. 3A, it can be established that the second fluorescent peak is associated with a single polypeptide of M_r 144,000, attributable to IRBP.

[27] T. M. Redmond, B. Wiggert, F. A. Robey, N. Y. Nguyen, M. S. Lewis, L. Lee, and G. J. Chader, *Biochemistry* **24**, 787 (1985).

FIG. 1. (a) High-performance size-exclusion chromatography of bovine IPM on Bio-Rad TSK columns. The effluent was continuously monitored by its absorbance at 280 nm (A_{280}, upper trace) and its fluorescence (F, lower trace). Because fluorescence was monitored after absorbance, the fluorescence peak is slightly displaced with respect to the corresponding absorbance peak (arrowed). Molecular weights: A, 669K (thyroglobulin); B, 150K (IgG); C, 43K (ovalbumin); D, included volume (cyanocobalamin). Fractions were collected, numbered as indicated, and analyzed by SDS–PAGE. (b) Analysis by SDS–PAGE of eluted fractions from the high-performance size-exclusion columns. Coomassie Blue stained 5–20% polyacrylamide gradient gel; track numbers correspond to the fraction numbers. Tracks labeled "s" contain molecular weight standards. From the top of the gel, they are myosin (210K); β-galactosidase (130K); phosphorylase B (94K); bovine serum albumin (68K); ovalbumin (43K); carbonate dehydratase (31K); soybean trypsin inhibitor (21K); lysozyme (14.3K). After Liou *et al.*[14]

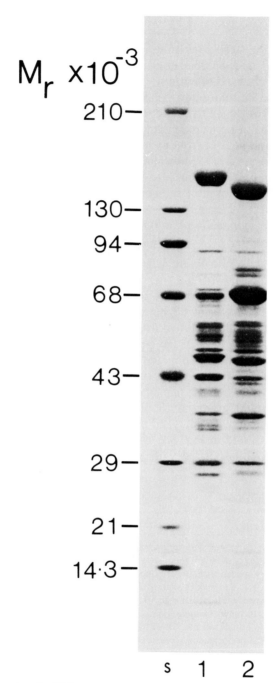

$M_r \times 10^{-3}$

210—

130—

94—

68—

43—

29—

21—

14·3—

s   1   2

FIG. 2. Comparison by SDS–PAGE of crude bovine (track 1) and human IPM (track 2). The gel is the same as in Fig. 1. Track labeled "s" contains molecular weight standards as in Fig. 1. The prominent protein at 144K (bovine) and 135K (human) corresponds to IRBP.

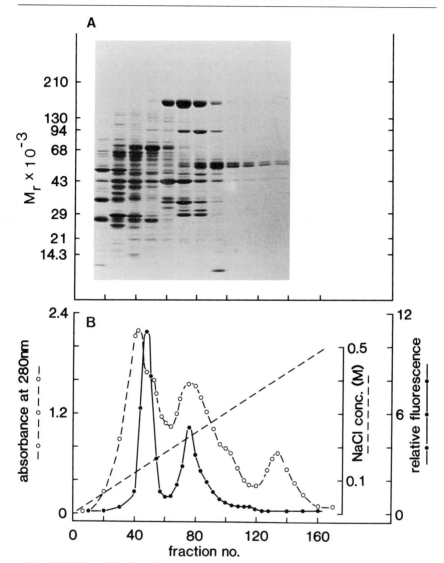

FIG. 3. Initial purification of interstitial retinol-binding protein by ion-exchange chromatography of crude bovine interphotoreceptor matrix. (B) fluorescence (●) and A_{280} (○) elution profiles of DEAE-cellulose column. (A) SDS–PAGE of eluted fractions (abscissa corresponds to B). The fluorescent peak centered at fraction 76 and associated with the M_r 144,000 protein band on SDS–PAGE was used for further purification by concanavalin A affinity chromatography (Fig. 4). Reproduced from Fong et al.[19]

Step 3. Affinity Chromatography on Concanavalin A-Sepharose

The fluorescent material from the peak centered at fraction 76 in Fig. 3 is mixed with calcium chloride and manganese chloride to give final concentrations of 1 mM and is then loaded on a 1.5 × 15 cm column packed with Con A-Sepharose (Pharmacia). The column is first eluted with buffer C (50 mM Tris–HCl, 0.15 M NaCl, 1 mM calcium chloride, 1 mM manganese chloride, pH 7.5). When the A_{280} of effluent 5-ml fractions has dropped to baseline levels (Fig. 4), the column is eluted with buffer C containing 0.2 M methyl- α-D-mannopyranoside (Calbiochem). One hundred-microliter aliquots from every fourth tube may be examined by SDS–polyacrylamide gel electrophoresis, as shown in Fig. 4A.

Step 4. Gel-Filtration Chromatography

This step is sometimes necessary to remove a minor protein of M_r 29,000 that is present if the original preparation of interphotoreceptor matrix is appreciably contaminated with proteins of intracellular origin. Gel-filtration chromatography also serves to remove the large amount of methyl-α-D-mannopyranoside. Fractions 44–78 from the Con A-Sepharose column are pooled and concentrated to 5–10 ml in an Amicon concentration cell (PM10 membrane) and applied to a 1 × 150 cm column packed with Sepharose CL-4B (Pharmacia). The column is eluted with buffer B. One milliliter fractions are monitored for A_{280} and fluorescence as in Fig. 3. Additionally, aliquots can be removed and examined by SDS–polyacrylamide gel electrophoresis.

Yield

Using the above technique, the overall yield of protein from 100 bovine neural retinas is about 14 mg. As determined by rocket immunoelectrophoresis[12,19] the quantity of IRBP in a crude IPM preparation from 100 eyes is 43–48 mg. The yield from the ion-exchange and Con A columns is usually 30–45 mg. A major loss occurs during the concentration step prior to gel-filtration. This is because IRBP tends to precipitate (perhaps because of its high proportion of hydrophobic amino acids), particularly in concentrated solutions. This can be prevented by the addition of octyl-β-D-glucopyranoside (Calbiochem) to a concentration of 1% (w/v).[19] After concentration, the detergent-containing solution may be subjected to gel-filtration chromatography as described above. Under these circumstances, the yield may be increased to 20–23 mg. However, the protein may be partially denatured by this procedure and may lose some of its retinol-binding capacity.

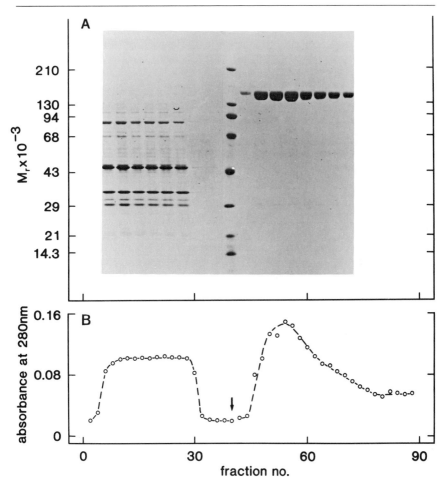

FIG. 4. Con A-Sepharose affinity chromatography. The column was first eluted with buffer C, then 0.2 M methyl-α-D-mannopyranoside (arrow). (B) A_{280} elution profile. (A) SDS–PAGE of eluted fractions (abscissa corresponds to B); the center lane corresponds to protein standards (from top of gel: myosin, β-galactosidase, phosphorylase B, bovine serum albumin, ovalbumin, carbonate dehydratase, soybean trypsin inhibitor, lysozyme). Reproduced from Fong et al.[19]

Acknowledgments

The investigation was supported by the Retina Research Foundation of Houston, National Institutes of Health Grants EY 02489, EY02502, EY03829, and RR-05425, and an unrestricted departmental grant from Research to Prevent Blindness.

[13] Quantification of Physiological Levels of Retinoic Acid

By Joseph L. Napoli

Retinoic acid is a physiological metabolite of retinol with appreciable tissue concentrations[1] that is generally more potent than retinol in promoting the differentiation of cells and tissues in culture.[2] Retinoic acid is likely the retinoid that directly supports the differentiation function of vitamin A.[3] The need for sensitive and specific methods of assaying retinoic acid is motivated by its role in epithelial differentiation, the growing body of evidence that suggests an inverse relationship exists between tissue retinoid concentrations and the incidence of multiple neoplastic and dermatological diseases,[4] and a desire to study the physiological mechanisms that must underly the pharmacological efficacy of its analogs.

Potential Problems and Controls

The charged nature of retinoic acid, its sensitivity to light and oxygen, and the numerous charged retinoids that occur in biological samples[5] present potential problems in assay design. Careful resolution of retinoic acid from other retinoids is necessary to ensure specificity. This is currently best achieved by high-performance liquid chromatography (HPLC). Sephadex LH-20 chromatography is contraindicated because it affords only poor resolution, and its use is likely to result in overestimations, particularly if it is coupled with a nonspecific quantification step.[6] The potential presence of geometric isomers in samples, or isomerization during handling, can and should be addressed in choosing controls and chromatographic steps. Work should be done under yellow or red light in so far as possible, and samples should be stored under an inert atmosphere (nitrogen, or better argon) at low temperatures ($-80°$) to minimize degradation. Irradiation with ultraviolet light is an effective method of destroying retinoic acid.[7] Blanks can be prepared by irradiating samples exposed to the atmosphere with fluorescent light for 24–36 hr or with ultraviolet light overnight.

[1] A. M. McCormick and J. L. Napoli, *J. Biol. Chem.* **257,** 1730 (1982).
[2] R. Lotan, *Biochem. Biophys. Acta* **605,** 33 (1980).
[3] J. B. Williams and J. L. Napoli, *Proc. Natl. Acad. Sci. U.S.A.* **82,** 4658 (1985).
[4] D. E. Ong and F. Chytil, *Vitam. Horm. (N.Y)* **40,** 105 (1983).
[5] J. L. Napoli and A. M. McCormick, *Biochim. Biophys. Acta* **666,** 165 (1981).
[6] Y. Shidoji and N. Hosoya, *Anal. Biochem.* **104,** 457 (1980).
[7] B. S. Sherman, *Clin. Chem. (Winston-Salem, N.C.)* **13,** 1039 (1967).

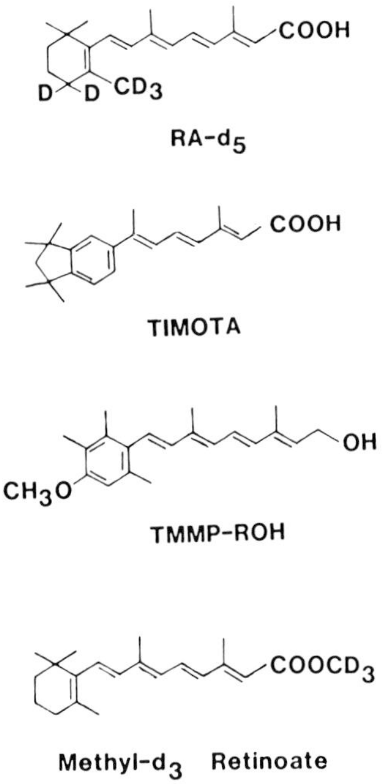

FIG. 1. Structures of the internal standards used during the assay of retinoic acid and retinol.

Internal Standards

Mass Spectrometry. An excellent internal standard for use during mass spectrometric quantification is all-*trans*-4,4,18,18,18-pentadeuterated retinoic acid (RA-d$_5$) (Fig. 1), which was synthesized by M. I. Dawson and colleagues.[8] This analog is chromatographically indistinguishable from retinoic acid and is chemically equivalent, i.e., methylated at the same rate and equally sensitive to light. Moreover it produces signals that are five mass units higher than those of the intended analyte, i.e., isotope peaks produced from retinoic acid are unlikely to interfere with the internal standard. A reasonable substitute is methyl-d$_3$ retinoate, which is readily prepared from retinoic acid (see below). Its use, however, does not control for methylation efficiency, and its extraction efficiency may differ from that of retinoic acid.

[8] M. I. Dawson, P. D. Hobbs, R. L.-S. Chan, and W.-R. Chao, *J. Med. Chem.* **24**, 1214 (1981).

HPLC. For quantification by HPLC, several of the synthetic retinoids are possible internal standards, such as all-*trans*-7-(1,1,3,3-*t*etramethyl-5-*i*ndanyl)-3-*m*ethyl-*o*cta-2,4,6-*t*rienoic *a*cid (TIMOTA),[9] or perhaps all-*trans*-13-demethylretinoic acid. A good internal standard to use when measuring retinol is all-*trans*-9-(4-methoxy-2,3,6-trimethylphenyl)-3,7-dimethyl-2,4,6,8-nonatetraen-1-ol, i.e., the TMMP analog of retinol (TMMP-ROH). All three compounds are available from Hoffmann-La Roche. Although retinyl acetate has been used as an internal standard in retinol assays, it is potentially troublesome. Retinyl acetate is susceptible to hydrolysis, and when this occurs not only is the internal standard concentration diminished, but the analyte (retinol) concentration is increased.

Extraction Procedures

Methanolic HCl/Hexane. With gas chromatography/mass spectrometry as the quantification method, retinoic acid and internal standard are recovered from plasma, serum, or cultured cells and their media (we have used up to a total of 6 ml of the latter, but foresee no problems with reasonably larger amounts), by adding an equal volume of 0.2 N HCl in methanol to the sample, mixing well with a Vortex mixer, and extracting with 1 volume of hexane (cells and media), or with three 0.5 ml portions of hexane (serum or plasma).[10] The procedures are conducted in disposable glass culture tubes (13 × 100 mm, for serum or plasma) or in glass screw cap culture tubes (16 × 150 mm, for cells and media). Centrifugation is used to achieve clean separation between the phases. The hexane phase is transfered to a clean tube, and the solvent is evaporated under a stream of nitrogen. Recovery is 83% ± 2 (±SD, $n = 10$). This procedure can be used to recover retinol if acid is not used in the methanol.

Phosphate Buffer/Ethanol/Hexane. An equally effective method involves modification of a procedure that uses phosphate buffer.[6] To the sample is added 4 volumes of 0.1 M potassium phosphate buffer, pH 5.4, and 2 volumes of ethanol. The resulting suspension is mixed well and is extracted 3 times with 7 volumes of hexane each time. This method is gentle; retinol can be recovered. But it is unwieldy and more time consuming with samples of 1.5 ml or greater, because larger volumes mandate the use of separatory funnels and solvent evaporation by rotary evaporator.

Alkaline Ethanol/Hexane: Acidic Ethanol/Hexane. An extraction

---

[9] P. Loeliger, W. Bollag, and H. Mayer, *Eur. J. Med. Chem.* **15,** 9 (1980).
[10] J. L. Napoli, B. C. Pramanik, J. B. Williams, M. I. Dawson, and P. D. Hobbs, *J. Lipid Res.* **26,** 387 (1985).

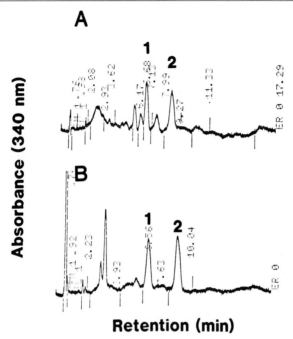

Retention (min)

FIG. 2. The effect of the extraction procedure on retinoid degradation and the facility of quantifying retinoids by HPLC. To 6 ml of cell culture medium was added 180 nmol of retinol (final concentration: 30 μM), and 100 pmol each of all-*trans*-retinoic acid, **1**, and TIMOTA, **2**. The sample was extracted with the alkaline ethanol/hexane : acidic ethanol/ hexane procedure: (A) 0.2 N KOH was used; (B) 0.025 N KOH was used. The samples were eluted as their free acids from a normal phase HPLC column (0.62 × 8 cm) with 0.2% acetic acid in 7% dichloromethane/hexane at a flow rate of 3 ml/min. Note the peaks in the vicinity of retinoic acid resulting from breakdown of retinoic acid and the internal standard when the higher concentration of base was used. Contrast the appearance of these chromatograms with those obtained using the methanolic HCl/hexane extraction procedure (Fig. 4C).

procedure that eliminates neutral lipids may be required prior to quantifi- cation by HPLC if UV absorbing impurities interfere with accurate mea- surements. This is achieved by adding to a 6 ml sample in a screw-cap culture tube (16 × 150 mm) 1.5 g of sodium chloride, 1 volume of 0.025 N potassium hydroxide in ethanol, and extracting the alkaline solution twice with 6 ml of hexane. To the aqueous phase is added 0.5 ml of 4 N HCl. Retinoic acid is recovered from the resulting acid solution by extracting it into 6 ml of hexane. Recovery is 82% ± 13 (±SD, n = 14). The use of stronger base (0.2 N NaOH or KOH), as reported in similar procedures,[11] results in partial breakdown of retinoids (Fig. 2).

[11] A. P. De Leenheer, W. E. Lambert, and I. Claeys, *J. Lipid Res.* **23**, 1362 (1982).

Methylation

Diazomethane. The hexane extract is dissolved in methanol (50 μl) and placed on ice. A solution of diazomethane in diethyl ether (approx. 1 drop from a Pasteur pipette) is added. After standing for 5 min, the solvents and excess diazomethane are removed under a stream of nitrogen. A solution of diazomethane in ether can be generated from *N*-methyl-*N*-nitroso-*p*-toluenesulfonamide or *N*-methyl-*N'*-nitro-*N*-nitrosoguanidine. Diazomethane is a highly toxic and potentially explosive gas. It must be generated, handled, and stored in glassware free from scratches, sharp edges, and ground glass joints behind a safety shield in a well-ventilated hood. This procedure should be conducted in specialized glassware, used solely for the production of diazomethane. Such glassware is available commercially.

Dialkyl Acetals. The use of diazomethane is an efficient, rapid, and clean way of converting retinoic acid into its methyl ester. A less hazardous method is to allow retinoic acid to react with a dimethylformamide dialkyl acetal. For example, heating retinoic acid in dimethylformamide hexadeuterodimethyl acetal (Tri-Deutero-8, Pierce Chemical Co.) at 65° for 1 hr produces methyl-d_3 retinoate.[10] Other alkyl derivatives are commercially available, so that a variety of esters can be prepared with these reagents. Care must be taken, however, to effect complete reaction when attempting to derivatize picomolar quantities of retinoic acid by this method.

High-Performance Liquid Chromatography (HPLC)

HPLC is performed with normal phase columns packed with 5–10 μm silica particles. Normal phase chromatography is preferred because it affords better resolution than reverse phase chromatography, and because evaporation of the mobile phase during sample recovery is easier. Two column dimensions are used: the standard analytical column, 0.46 × 25 cm or the 0.62 × 8 cm configuration. The former is usually eluted at a flow rate of 1–2 ml/min; the latter at 3–5 ml/min. Thus, the latter configuration allows higher flow rates at lower back pressures without loss in resolution, and also offers a higher capacity. Samples are monitored with an ultraviolet absorbance detector at 340 nm. If only retinol is measured, somewhat greater sensitivity can be attained at 313 nm, with a fixed wavelength detector.

Methyl Esters. A mobile phase of 2% *t*-butyl methyl ether in hexane (mobile phase 1) elutes methyl 13-*cis*-retinoate in 7 ml (3.5 min) and methyl all-*trans*-retinoate in 8.5 ml (4.3 min) as sharp peaks (0.46 × 25 cm column). Thus, this system can be used to obtain total methyl retinoates (Fig. 3). In contrast, a mobile phase of 45% toluene in hexane (mobile phase 2) with the same column elutes methyl 13-*cis*-retinoate in 11.5 ml (5.8 min) and methyl all-*trans*-retinoate in 18.5 ml (9.3 min). A similar result can be achieved with 35% dichloromethane (or 1,2-dichloroethane) in hexane (mobile phase 3) which elutes methyl 13-*cis*-retinoate in 18 ml (4–6 min, depending on the flow rate) and methyl all-*trans*-retinoate in 27 ml (5.4–9 min) (0.62 × 8 cm column). The latter two systems provide the opportunity to quantify all-*trans*-retinoic acid itself, as opposed to only total retinoic acids. A drawback of these very nonpolar systems is the tendency of the retention times to shift during use as a consequence of moisture accumulating in the mobile phase or on the column. We have found this to be less of a problem with dichloromethane (or dichloro-ethane)/hexane in contrast to toluene/hexane.

Free Acids. The free acids of retinoids also can be analyzed on normal phase columns. For example, a mobile phase of 0.2% acetic acid in 5% dichloromethane/hexane (mobile phase 4) elutes 13-*cis*-retinoic acid in 22.5 ml (4.5–7.5 min, depending on the flow rate), all-*trans*-retinoic acid in 26 ml (5.2–8.7 min), and TIMOTA in 34 ml (6.8–11.3 min) (0.62 × 8 cm column) (Fig. 4). An alternative mobile phase, 0.1% acetic acid in 25%

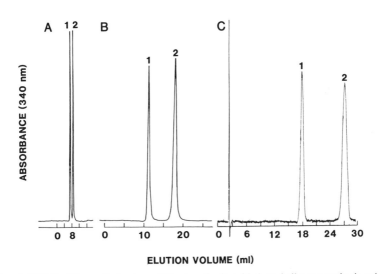

ELUTION VOLUME (ml)

FIG. 3. HPLC of the methyl esters of 13-*cis*-retinoic acid, **1**, and all-*trans*-retinoic acid, **2**. Samples were eluted from the 0.46 × 25 cm column at 2 ml/min (A, B) or from the 0.62 × 8 cm column at 3 ml/min (C). The mobile phases used were (A) mobile phase 1 (2% *t*-butyl methyl ether in hexane); (B) mobile phase 2 (45% toluene in hexane); (C) mobile phase 3 (35% dichloromethane in hexane).

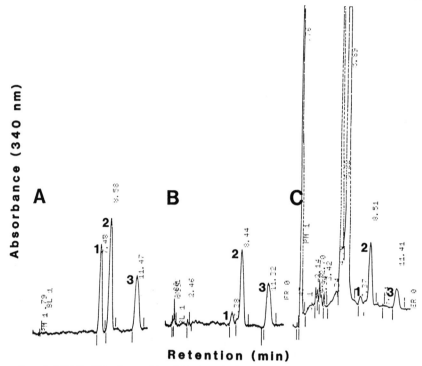

Retention (min)

Fig. 4. HPLC of the free acids of 13-*cis*-retinoic acid, **1**, all-*trans*-retinoic acid, **2**, and TIMOTA, **3**. Samples were eluted from the 0.62 × 8 cm column at 3 ml/min with mobile phase 4 (0.2% acetic acid in 5% dichloromethane in hexane). The three compounds were applied directly to HPLC (A); only all-*trans*-retinoic acid and TIMOTA were added to the medium (6 ml) of pig kidney cells (LLC-PK$_1$) in culture and were extracted by the methanolic HCl/hexane procedure (B); or 10 μM retinol was incubated with LLC-PK$_1$ cells for 1 hr, TIMOTA was added, and the sample was extracted by the methanolic HCl procedure (C). The cells are synthesizing all-*trans*-retinoic acid from retinol; in the absence of cells, a peak much smaller than that of the internal standard was observed in the retinoic acid area; in the absence of retinol, the cells did not produce peaks migrating with retinoic acid. Note the amount and intensity of retinol-derived peaks other than retinoic acid that are observed with the methanolic HCl/hexane extraction procedure. This is in contrast to the results observed with the alkaline ethanol/hexane : acidic ethanol/hexane extraction procedure performed on a sample that had a 3-fold higher concentration of retinol (Fig. 2B).

1,2-dichloroethane/hexane (mobile phase 5), elutes 13-*cis*-retinoic acid in 14.4 ml (2.9–4.8 min), all-*trans*-retinoic acid in 19.5 ml (3.9–6.5 min), and TIMOTA in 22.5 ml (4.5–7.5 min). Small increases in the acid produce large decreases in retention time. Therefore, the retention times or capacity factors (k) are more conveniently controlled by changes in the chlorinated hydrocarbon.

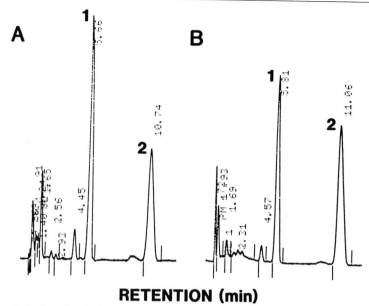

RETENTION (min)

Fig. 5. HPLC of retinol. Retinol, **1** (10 μM) was added to LLC-PK$_1$ cells in culture and incubated for 0 hr (A) or for 8 hr (B). TMMP-ROH, **2**, was added as internal standard; the samples were extracted with methanol/hexane and were analyzed on normal phase HPLC (0.62 × 8 cm column) eluted with mobile phase 6 (5% acetone in hexane) at a flow rate of 3 ml/min. After 8 hr of incubation, there was 38% conversion of retinol to metabolites.

Retinol. Retinol is also analyzed on a normal phase system (0.62 × 8 cm column). It elutes with 5% acetone in hexane (mobile phase 6) in 17 ml (3.4–5.7 min), whereas TMMP-ROH elutes in 33 ml (6.6–11 min) (Fig. 5). The potential differences between normal and reverse phase HPLC in their ability to resolve retinoids is illustrated well with retinol. Normal phase systems, like the one described above, readily resolved commercially available high specific activity (40–60 Ci/mmol) [11,12-^{3}H]retinol into a number of radioactive peaks. We have found, by analyzing three different batches, that all-*trans*-[11-^{3}H]retinol accounted for only 48, 57, and 83% of the radioactivity. Analyses of the same materials on reverse phase systems (ODS; 85–100% methanol in water or 60–100% acetonitrile in water) failed to resolve the impurities and *gave the appearance* of >90% purity.

Gas Chromatography (GC)

Retinoic acid and its isomers are usually chromatographed over GC columns as their esters. A variety of stationary phases and conditions have been used. Examples include: 3% SP-2100-DOH on Supelcoport (2 mm × 3 ft column) at 230° and a methane flow rate of 20 ml/min (methyl 13-*cis*-retinoate and methyl all-*trans*-retinoate coelute in 3.1 min)[10]; 3% SE-30 on Chromosorb W HP (2 mm × 6 ft) at 230° and a helium flow rate of 50 ml/min (methyl 13-*cis*-retinoate elutes in 8 min and methyl all-*trans*-retinoate elutes in 8.5 min).[12] The first system offers the advantage of short analysis time, whereas the second system seems to marginally resolve methyl 13-*cis* and all-*trans*-retinoates. A GC system that partially resolves the geometric isomers of methyl retinoates could present complications during routine analyses of total methyl retinoate, especially with deuterated methyl all-*trans*-retinoate as internal standard. Variations in resolution would change the results in an unpredictable manner, depending on the degree of separation of the other isomers from methyl all-*trans*-retinoate.

Mass Spectrometry (MS)

Electron Impact (EI). Compounds that produce spectra with a few intense peaks of known origin, including the molecular ion (M^+), provide the best opportunity for quantification by MS. Unfortunately, the electron impact mass spectra of methyl retinoate and the other naturally occurring retinoids, produced at 70 eV, show extensive fragmentation (Fig. 6) and only a few of the peaks are unique to or characteristic of individual compounds.[13] This situation can be simplified somewhat by using a lower ionizing potential (e.g., 32 eV), but this usually diminishes the proportion of sample ionized, i.e., the sensitivity. Besides the extensive fragmentation and the relatively low proportion of sample ionized, at least one other factor diminishes the sensitivity of GC/EI/MS: the need for a jet separator to remove the carrier gas before the sample enters the mass spectrometer source. This process of separating the GC carrier gas from the sample also results in sample loss.

Chemical Ionization (CI). An improvement in sensitivity is afforded by chemical ionization MS. A larger amount of the sample is ionized in the CI mode, relative to EI. Moreover, depending upon the compound type and the reagent gas, useful negative, as well as positive, ions are produced. Thus two monitoring options are available: positive ion chemical ionization (PCI) and negative ion chemical ionization (NCI). Frequently, negative ion spectra are simpler and more intense than positive

[12] T.-C. Chiang, *J. Chromatogr.* **182**, 335 (1980).
[13] R. L. Lin, G. R. Waller, E. D. Mitchell, K. S. Yang, and E. C. Nelson, *Anal. Biochem.* **35**, 435 (1970).

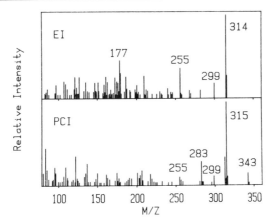

FIG. 6. A comparison of the electron impact (EI, 70 eV) and positive ion chemical ionization (PCI) mass spectra of methyl all-*trans*-retinoate. Both modes produce extensive fragmentation and only a few peaks that are unequivocally assignable: EI, m/z 314 (M^+), 299 ($M^+ - CH_3$), 255 ($M^+ - COOCH_3$); PCI, m/z 343 ($M^+ + C_2H_7$), 315 ($M^+ + H$), 314 (M^+), 313 ($M^+ - H$), 299 ($M^+ - CH_3$), 283 ($M^+ - OCH_3$), 255 ($M^+ - COOCH_3$). Lowering the voltage from 70 to 32 eV produces only a marginal change in the EI spectrum. These results are in contrast with those obtained in the CI mode by monitoring negative ions (Fig. 7). Spectra were obtained with a Finnigan 4021 EI/CI/GC MS served by an INCOS 2000 Data System. Samples were introduced through the GC (3% SP21000-DOH on Supelcoport). In the EI mode the column was eluted with helium; in the CI mode methane (10^{-5} Torr) was used as GC carrier gas and as CI reagent gas.

ion spectra. Such is the case with methyl retinoate when methane is used as the CI reagent gas. The PCI mode is somewhat of an improvement over EI (Fig. 6), but the NCI mode is clearly more advantageous (Fig. 7). Not only is the proportion of sample ionized about 6-fold greater with NCI than with PCI (data not shown), but the response is concentrated in a single intense peak, which represents the molecular ion. Therefore, GC/ MS assays reported here are based on selected ion monitoring (SIM) of peaks produced in NCI MS.

Assay Examples

General. Internal standards are added to samples in 10–25 μl of ethanol; the solutions are mixed well and allowed to stand for 10–15 min before extraction. The amounts of internal standards added are such that the area of their peaks will be within an order of magnitude of the areas of the sample peaks. Therefore, the amounts of internal standards used in the examples described below should be considered as guidelines only.

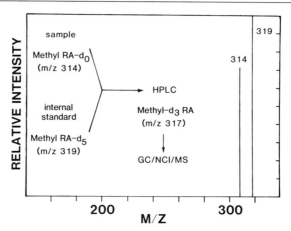

Fig. 7. Quantification of retinoic acid by GC/NCI/MS. One negative ion peak each, representing the molecular ions (M^+), is produced from methyl retinoate and methyl RA-d_5 in the CI mode with methane as reagent gas. Thus methyl retinoate provides a peak at m/z 314 and the internal standard provides a peak at m/z 319. The HPLC standard methyl-d_3 retinoate (methyl-d_3 RA) provides a peak at m/z 317; therefore contamination resulting from this chromatographic marker will not affect the accuracy of the assay. Since only two peaks are important to quantification, these are the only peaks monitored; i.e., the spectrometer is not set to scan the mass range, but is set only to detect m/z 314 and m/z 319 (selected ion monitoring). The results from a mixture of 0.56 pmol of retinoic acid and 0.78 pmol of internal standard are shown.

Hexane is evaporated under a gentle stream of nitrogen. Larger volumes (>10 ml) may be removed with the aid of a rotary evaporator. Blood to prepare plasma is collected in an EDTA containing Vacutainer. Serum or plasma is obtained by centrifuged blood at 7800 g_{ave} for 15 min. For best results, neither the gravitational force nor the spin time should be decreased. Clean glassware, rinsed with high-purity organic solvents and dried in a oven at 120° (or higher), should be used.

Determination of Retinoic Acid by GC/MS. To 40–100 µl of serum plasma is added 0.83 pmol of all-*trans*-RA-d_5 as internal standard or to cells and their culture media is added 17–67 pmol of all-*trans*-RA-d_5. Endogenous retinoic acid and the internal standard are extracted with the methanolic HCl/hexane procedure and are methylated with diazomethane. The methylated residues from extracts of 100 µl or less of serum or plasma can be injected directly into GC/MS in 5 µl of methanol. The mass range need not be scanned; only the peaks at m/z 314 and m/z 319 need to be monitored (selected ion monitoring, SIM). In the NCI mode, the relative response factor of retinoic acid with all-*trans*-RA-d_5 as internal standard is 1; the response is linear from sample to internal standard ratios of

0.1–20, and is probably linear well beyond these points.[10] The lower limit of detection during GC/MS with a packed column (2 mm × 3 ft, SP2100-DOH) is 120 fmol with direct injection of the standards. The lower limit of detection during assays will depend upon the sample recovery, but should be no less than 240 fmol.

For larger amounts of serum or plasma or for extracts of cells and their culture media, an HPLC purification step may be necessary before GC/ MS. This can be accomplished with normal phase HPLC eluted with mobile phase 1 (see HPLC section). To determine the contribution of all-*trans*-retinoic acid to the total retinoic acid pool, normal phase HPLC eluted with mobile phase 2 or 3 can be used to isolate all-*trans*-retinoic acid prior to GC/MS (Fig. 3). Care must be taken not to contaminate samples as they are being processed through HPLC by using either the analyte or the internal standard as chromatographic markers. Contamination can be avoided by using methyl-d₃ 13-*cis*-retinoate and methyl-d₃ all-*trans*-retinoate as HPLC standards. These two compounds produce molecular ions at m/z 317 in NCl MS and therefore do not interfere with measurements (Fig. 7).

Determination of Retinoic Acid by HPLC. To cells and their media are added 0.1–10 nmol of TIMOTA as internal standard. The samples are extracted with either the methanolic HCl/hexane method (Fig. 4) or with the alkaline ethanol/hexane : acid ethanol/hexane procedure (Fig. 2), if larger amounts of impurities are present. (This occurs during the synthesis of retinoic acid from higher concentrations of retinol, ~30 μM or greater.) The retinoic acid and internal standard recovered are applied to normal phase HPLC (0.62 × 8 cm column) as their free acids in 100 μl of mobile phase 4 or 5 and are eluted with mobile phase 4 or 5. The area of the sample and internal standard peaks produced by the ultraviolet absorbance detector are recorded with an integrator. The lower limit of detection, after extraction and handling losses, is 20 pmol. This HPLC assay is less sensitive than the GC/MS assay, but demands less sophisticated equipment, fewer steps, and is more rapid.

Determination of Retinol by HPLC. To cells and their media incubated with retinol (60 nmol) are added 14 nmol of TMMP-ROH. The samples are extracted with the methanol/hexane procedure or with the phosphate buffer/ethanol/hexane procedure. The samples are applied to normal phase HPLC in 100 μl of mobile phase 5 and are eluted with mobile phase 6 (Fig. 5).

Determination of Retinoic Acid and Retinol in the Same Sample. Internal standards for both retinoic acid and retinol are added to the sample. The retinoids are extracted with either of the two methods compatable with retinol. Retinol is separated from retinoic acid by applying the sample in methanol to DEAE Sephadex A-25 (either in the form of a small column of batchwise in a small disposable culture tube). Retinol is recovered in the methanol phase. After the methanol phase is removed, retinoic acid is recovered by treating the resin with 2% formic acid in methanol. Only 0.2 g of resin is sufficient for separating retinol from retinoic acid in 100 μl serum or plasma extracts. The sample is applied in 0.5 ml of methanol, retinol is washed off in a further 5.5 ml, and retinoic acid is recovered in 4 ml of formic acid/methanol. The recovered retinoids are then quantified by the appropriate procedures. Alternatively, the alkaline ethanol/hexane : acidic ethanol/hexane extraction procedure described above can be used. Retinol is recovered from the hexane extract of the alkaline phase; retinoic acid is recovered from the hexane extract of the acidic phase.

Summary

The methods discussed here are versatile procedures that have been effective for the quantification of retinoic acid and retinol in plasma or serum, cells in culture, and animal tissues. They are capable of measuring a wide range of concentrations: the GC/MS assay for retinoic acid is especially useful when low amounts of retinoic acid are to be accurately quantified and is unrivaled for sensitivity and specificity; the HPLC assays are less sensitive, but are more convenient, more accessible, and preclude the need for the expensive and demanding technology of GC/MS.

With these procedures, retinoic acid has been measured in human plasma (~4.9 ng/ml), rat serum (~2 ng/ml), fetal bovine serum (~2 mg/ml), and rat tissues (0.05–1.0 μM). In the first two cases, all-*trans*-retinoic acid accounts for approximately 75% of the total. The proportion of all-*trans*-retinoic acid in other cases has not yet been determined.

Acknowledgments

The development of these methods has been supported in part by NIH Grants CA 42092 and AM 36870. The assistance of Cheryl W. Cummings, Charles O. Shields, and John B. Williams in the development of these methods is gratefully acknowledged.

Section III

Vitamin D Group

[14] Isolation and Identification of Vitamin D Metabolites

By Joseph L. Napoli, Nick J. Koszewski, and Ronald L. Horst

In the last 4–5 years the work of several research groups has shown that far from being simple and well defined, the metabolism of vitamin D and its metabolites is complex; its regulation and its relationship to vitamin D function are not yet completely understood.[1-4] During the same period a more sophisticated realization of the functions of vitamin D has emerged. 1,25-Dihydroxyvitamin D_3 [1,25-$(OH)_2D_3$] promotes cellular differentiation[5,6] in addition to its classical actions in the maintenance of calcium homeostasis[7]; these actions may be directed by *in situ* metabolic modification of the hormone. In fact, extensive metabolism of 25-OHD_3 and 1,25-$(OH)_2D_3$ occurs in target tissues and is accelerated by 1,25-$(OH)_2D_3$ through enzyme induction.[8,9] That is, 1,25-$(OH)_2D_3$ tissue concentrations are dependent not only on feedback inhibition of its synthesis, but also on its metabolism in tissues. The induction process appears to depend on new protein synthesis and the 1,25-$(OH)_2D_3$ receptor[10]; therefore, it provides an opportunity to approach the nuclear actions of the hormone. Other topics to be addressed by isolation and identification techniques are the study of the extrarenal conversion of 25-OHD_3 into 1,25-$(OH)_2D_3$, the species-specific metabolic differences between vitamins D_2 and D_3,[11] and the lesser tendency of vitamin D_2 to cause hypercalcemia relative to vitamin D_3.

[1] J. L. Napoli and R. L. Horst, *in* "Vitamin D: Basic and Clinical Aspects" (R. Kumar, ed.), p. 91. Nijhoff, The Hague, 1984.

[2] J. L. Napoli and R. L. Horst, *Biochemistry* **22**, 5848 (1983).

[3] S. Yamada, M. Ohmori, H. Takayama, Y. Takasaki, and T. Suda, *J. Biol. Chem.* **258**, 457 (1983).

[4] G. Jones, M. Kung, and K. Kano, *J. Biol. Chem.* **258**, 12920 (1983).

[5] E. Abe, C. Kiyura, H. Sakagami, M. Takeda, K. Konno, T. Yamazaki, S. Yoshiki, and T. Suda, *Proc. Natl. Acad. Sci. U.S.A.* **78**, 4990 (1981).

[6] D. J. Mangeldorf, H. P. Koeffler, C. A. Donaldson, J. W. Pike, and M. R. Haussler, *J. Cell Biol.* **98**, 391 (1984).

[7] J. A. Kanis, D. F. Guilland-Cumming, and R. G. G. Russell, *in* "Endocrinology of Calcium Metabolism" (J. A. Parsons, ed.), p. 321. Raven, New York, 1982.

[8] K. Colston and D. Feldman, *J. Biol. Chem.* **257**, 2504 (1982).

[9] J. L. Napoli and C. A. Martin, *Biochem. J.* **219**, 713 (1984).

[10] J. S. Chandler, S. K. Chandler, J. W. Pike, and M. R. Haussler, *J. Biol. Chem.* **259**, 2214 (1984).

[11] R. Horst, J. L. Napoli, and E. T. Littledike, *Biochem. J.* **204**, 185 (1982).

Experimental Approaches

A successful isolation and identification of a vitamin D metabolite usually consists of extraction, prepurification (optional), several high-performance liquid chromatography (HPLC) procedures, ultraviolet (UV) absorbance spectroscopy, mass spectrometry (MS), chemical derivatization, and may include proton nuclear magnetic resonance (NMR) spectrometry.

Large samples (e.g., 6–16 liters of plasma[12,13]) can be extracted by the procedure of Bligh and Dyer.[14] Smaller samples (incubation of tissue homogenates, cultured cells and their media, small volumes of serum or plasma) can be extracted by the diethyl ether/dichloromethane procedure of Horst et al.,[15] or by the addition of 1–3 volumes of methanol/dichloromethane (2 : 1).[16,17] For investigations with cultured cells,[9] prepurification of the extracts prior to HPLC may not be necessary. For most studies, however, a large amount of contaminating lipid must be removed before attempting HPLC. There are several methods available. Thin-layer chromatography is useful, but may not always provide optimal recovery.[18] Filtration through Sep-Paks (i.e., commercially available mini silica gel or ODS columns) with polar solvents (e.g., ethyl acetate) removes materials capable of irreversibly binding to HPLC columns and increases column longevity. Such procedures can also be used to isolate particular groups of metabolites, if used discriminantly.[19–21] Larger amounts of lipids can be removed by column chromatography with Sephadex LH-20[12,15,16] or with mixtures of Celite and silica gel.[22,23] Although these procedures

[12] J. L. Napoli, B. C. Pramanik, J. J. Partridge, M. R. Uskokovic, and R. L. Horst, *J. Biol. Chem.* **257,** 9634 (1982).

[13] T. A. Reinhardt, J. L. Napoli, D. C. Beitz, E. T. Littledike, and R. L. Horst, *Arch. Biochem. Biophys.* **213,** 163 (1982).

[14] E. G. Bligh and W. J. Dyer, *Can. J. Biochem. Physiol.* **37,** 911 (1959).

[15] R. L. Horst, E. T. Littledike, J. L. Riley, and J. L. Napoli, *Anal. Biochem.* **116,** 189 (1981).

[16] J. L. Napoli, B. C. Pramanik, P. M. Royal, T. A. Reinhardt, and R. L. Horst, *J. Biol. Chem.* **258,** 9100 (1983).

[17] R. L. Horst, P. M. Wovkulich, E. G. Baggiolini, M. R. Uskokovic, G. W. Engstrom, and J. L. Napoli, *Biochemistry* **23,** 3973 (1984).

[18] M. J. Thierry-Palmer and T. K. Gray, *J. Chromatogr.* **262,** 460 (1983).

[19] J. S. Adams, T. L. Clemens, and M. F. Holick, *J. Chromatogr.* **226,** 198 (1981).

[20] A. A. Redhwi, D. C. Anderson, and G. N. Smith, *Steroids* **39,** 149 (1982).

[21] T. A. Reinhardt, R. L. Horst, J. W. Orf, and B. W. Hollis, *J. Clin. Endocrinol. Metab.* **58,** 91 (1984).

[22] J. L. Napoli and R. L. Horst, *Biochem. J.* **206,** 173 (1982).

[23] J. L. Napoli, J. L. Sommerfeld, B. C. Pramanik, R. Gardner, A. D. Sherry, J. J. Partridge, M. R. Uskokovic, and R. L. Horst, *Biochemistry* **22,** 3636 (1983).

are useful for the preparation of samples for more rigorous purification and analysis, they do not provide high resolution and therefore do not demonstrate cospecificity. *Comigration on Sephadex LH-20 of a metabolite with a standard should not be construed as compelling evidence for coidentity.*

Compelling evidence for the identity of a known metabolite from a new source consists of comigration in at least two unique HPLC systems and spectral characterization. Preferably the more powerful normal phase HPLC should be used. Changing the amount of polar modifier, i.e., 15–10% 2-propanol in hexane, does not constitute a new system. The identity of new metabolites can be determined after rigorous purification by HPLC through their spectral characteristics, by mass spectral analysis of chemical derivatives, and by independent testing of the validity of the structure assigned, e.g., by determining biological precursor–product relationships.[2,22,24] Failure to achieve sufficient purity, misinterpretation of mass spectral data, and relying on inconclusive mass spectral data have been the underlying causes of invalid structural assignments.

High-Performance Liquid Chromatography

HPLC is the best method known for the resolution and purification of vitamin D metabolites. It has much higher resolving power than conventional methods, yet it is not foolproof. A number of dissimilar vitamin D derivatives comigrate in a single HPLC system. This emphasizes the need for more than one system in determining coidentity or for comparing metabolites from biological sources to those obtained *in vitro* from known precursors. For instance, as the concentration of $1,25-(OH)_2D_2$ in an incubation mixture with isolated intestinal cells increases, the amount of product that migrates with $24-oxo-1,23,25-(OH)_3D_3$ on a hexane-based HPLC system increases. Reanalysis of the recovered "$24-oxo-1,23,25-(OH)_3D_3$" on a second HPLC system, with a dichloromethane-based mobile phase, showed that the original peak was heterogeneous; and in fact, as the substrate concentration increased, not only did the production of authentic $24-oxo-1,23,25-(OH)_3D_3$ increase, but other products began to be formed.[2] Conclusions based only on the first analysis would have been invalid. This situation also reemphasizes the inherent limitations in using chromatographic techniques with much less resolving power than HPLC, such as Sephadex LH-20, as evidence of identity.

[24] R. L. Horst, T. A. Reinhardt, B. C. Pramanik, and J. L. Napoli, *Biochemistry* **22,** 245 (1983).

The elution positions on HPLC with several mobile phases of two groups of vitamin D derivatives are compared: 25-OHD and derivatives (Table I) and 1,25-(OH)$_2$D and derivatives (Table II) [19-nor-10-oxo-25-OHD$_3$ is also discussed with the 1,25-(OH)$_2$D derivatives, since it migrates close to 1,25-(OH)$_2$D$_3$ on hexane-based normal phase systems]. Exact elution volumes depend on the specific circumstances (column brand and condition, precise mobile phase composition, quality of solvents, degree of moisture in solvents). Therefore the data in the tables are reported as α values, i.e., the k (capacity factor) of the test compound is divided by the k of either 25-OHD$_3$ or 1,25-(OH)$_2$D$_3$. *Generally the elution order, i.e., relative α values or degree of resolution, does not change*

TABLE I

ELUTION POSITIONS OF 25-OH-D DERIVATIVES RELATIVE TO 25-OH-D$_3$ IN HPLC[a]

	HPLC System[b]				
	Silica			Cyano	ODS
Substance	I	II	III	IV	V
			α^c		
25-OHD$_2$	0.84	0.73	0.76	0.80	1.09
25-OHD$_3$[d]	1.00	1.00	1.00	1.00	1.00
23S,25-(OH)$_2$D$_3$	1.92	3.77	3.51	2.52	0.24
24R,25-(OH)$_2$D$_2$	2.31	2.23	2.18	2.12	0.31
Lactone	2.65	1.00	1.00	4.27	0.13
24R,25-(OH)$_2$D$_3$	2.65	3.06	2.90	2.60	0.23
X^e	2.65	3.47	3.30	3.17	0.28
25,26-(OH)$_2$D$_2$ (I)^f	3.10	2.68	2.50	2.60	0.37
25,26-(OH)$_2$D$_2$ (II)^f	3.14	3.39	3.07	2.86	0.37
25,26-(OH)$_2$D$_3$ (25R and 25S)	4.80	4.11	4.06	3.90	0.31
19-nor-10-oxo-25-OHD$_3$ (I)^g	5.28	3.34	3.68	3.96	0.24
19-nor-10-oxo-25-OHD$_3$ (II)^g	6.01	4.62	4.62	4.50	0.23
1,25-(OH)$_2$D$_2$	7.31	6.02	6.41	5.00	0.39
1,25-(OH)$_2$D$_3$	8.33	8.00	8.60	6.40	0.32

[a] Columns used were Dupont Zorbax (0.46 × 25 cm). The mobile phase volumes (V_m) were 3.5 ml.

[b] Mobile phases used were I, methanol/2-propanol/hexane (1 : 3 : 96); II, methanol/acetonitrile/dichloromethane (1 : 20 : 180); III, methanol/1,2-dichloroethane (1 : 79); IV, acetonitrile/ethanol/dichloromethane/hexane (1 : 1.5 : 6 : 191.5); V, water/methanol (1/3).

[c] $\alpha = k$ (substance)/k (25-OHD$_3$); $k = V_r - V_m$.

[d] Elution volumes for 25-OHD$_3$ range from I, 17–20 ml; II, 10–13 ml; III, 15–18 ml; IV, 11–14 ml; V, 155–165 ml.

[e] This 25-OHD$_2$ metabolite has not yet been positively identified.

[f] C-25 stereochemistry unknown.

[g] C-5 stereochemistry unknown.

TABLE II
ELUTION POSITIONS OF 1,25-(OH)₂D DERIVATIVES RELATIVE
TO 1,25-(OH)₂D₃ IN NORMAL PHASE HPLC[a]

Substance	Mobile Phase[b]		
	VI	VII	VIII
	α		
1,25-(OH)$_2$D$_2$	0.90	0.83	0.90
19-nor-10-oxo-25-OH-D$_3$	0.96	0.43	—
1,25-(OH)$_2$D$_3$	1.00	1.00	1.00
24-oxo-1,25-(OH)$_2$D$_3$	1.20	0.71	—
23-oxo-1,25-(OH)$_2$D$_3$	1.53	0.83	—
24-oxo-1,23,25-(OH)$_3$D$_3$	1.57	1.17	1.06
1,23,25-(OH)$_3$D$_3$	1.60	2.25	1.42
1,24,25-(OH)$_3$D$_2$	1.88	1.81	1.67
1,24,25-(OH)$_3$D$_3$	2.07[c]	2.60	1.85
1α-OH-lactone	2.29[c]	1.06–1.16[c]	1.68
1,25,28-(OH)$_3$D$_2$	2.41	3.60	2.20
1,25,26-(OH)$_3$D$_2$	2.47	2.60	2.23
1,25,26-(OH)$_3$D$_3$	3.10	4.25	2.70

[a] Data were obtained with a Dupont Zorbax-Sil column (0.46 × 25 cm). The mobile phase volume (V_m) was 3.5 ml. Formulae used were $\alpha = k$ substance/k 1,25-(OH)$_2$D$_3$; $k = V_r - V_m$. Differences of 0.1 indicate noticeable separation, if not complete resolution.

[b] Mobile phases used were VI, 10% 2-propanol/hexane; VII, 5–7% 2-propanol/dichloromethane; VIII, 12–15% 2-propanol in hexane/dichloromethane (8:1). Elution volumes for 1,25-(OH)$_2$D$_3$ were VI, 25–30 ml; VII, 19–22 ml; VIII, 13–17 ml.

[c] These values depend on the individual column used; the relative elution positions in hexane of 1α-OH-lactone and 1,24,25-(OH)$_3$D$_3$ can be the reverse of that shown here.

significantly from column to column within a given system; but this occurs occasionally. Note that the relative elution orders of lactone (25-OH-D₃-26,23-lactone) and 25-OHD₃ and of 1α-OH-lactone [1,25-(OH)₂D₃-26,23-lactone] and 1,24,25-(OH)₃D₃ in hexane-based normal phase HPLC can interchange depending on the column used and its condition. Moreover, the elution of 1α-OH-lactone relative to 1,25-(OH)₂D₃ is also column dependent: it varies from near comigration to baseline resolution in dichloromethane-based normal phase systems.

25-Hydroxylated Vitamin D Derivatives. The elution positions of 14 25-OHD metabolites are compared to 25-OHD₃ on five unique HPLC systems (Table I). Generally, the metabolites that comigrate in one system can be separated easily by using a different system based on the silica acid column. For example, lactone, 25,28-$(OH)_2D_2$, and 24,25-$(OH)_2D_3$ comigrate in system I. However, adequate resolution of these three compounds can be achieved with dichloromethane or with 1,2-dichloroethane-based systems (systems II and III). The reverse-phase column (system V) provides the least selectivity among the columns tested. Both system III and system V have the disadvantage of creating higher back pressures relative to the other systems.

1α-Hydroxylated Vitamin D Derivatives. The elution positions of 12 vitamin D metabolites or derivatives relative to 1,25-$(OH)_2D_3$ were obtained on a normal phase column with three different mobile phases (Table II). No single system is capable of resolving all of the derivatives tested; on the other hand, no two compounds comigrate through all three systems. The use of the first two systems (VI and VII) sequentially will resolve all of the compounds. For example, 19-nor-10-oxo-25-OHD$_3$ elutes just prior to 1,25-$(OH)_2D_3$ in the hexane-based system (system VI), close enough to be mistaken for each other; but the two are well separated on the dichloromethane-based system (system VII). A similar situation exists for 1,25-$(OH)_2D_3$ and 24-oxo-1,25-$(OH)_2D_3$. Nor are 1,23,25-$(OH)_3D_3$, 23-oxo-1,25-$(OH)_2D_3$, and 24-oxo-1,23,25-$(OH)_3D_3$ well resolved by system VI, whereas resolution is complete with system VII. The same is true for 1,25,28-$(OH)_3D_2$ and 1,25,26-$(OH)_3D_2$. On the other hand, 1,25-$(OH)_2D_2$ and 23-oxo-1,25-$(OH)_2D_3$ are not separated by system VII, but are by system VI. The same is true for 1,24,25-$(OH)_3D_3$ and 1,25,26-$(OH)_3D_2$. As a final example, 1α-OH-lactone migrates close to 1,24,25-$(OH)_3D_3$ in system VI and close to 1,25-$(OH)_2D_3$ in system VII, but is well resolved from both with a mobile phase consisting of dichloromethane and hexane (system VIII).

Ultraviolet Absorbance Spectroscopy

The UV spectra of vitamin D and its metabolites having the 5,6-cis (5Z,7E,10[19]-triene, **1**, Fig. 1) geometry obtained in ethanol are smooth curves with one absorbance maximum at 265 nm and a molar absorbtivity, ε, of 18,200 for D$_3$, and 19,200 for D$_2$. Isomerization of the 5,6-cis system to a 5,6-trans system (5E,7E,10[19]-triene, **2**, Fig. 1) results in a bathochromic shift to an absorbance maximum at 273 nm and an increase in absorbance intensity ($\varepsilon = 24,300$). Rearrangement of the double bonds, as in tachysterol [5(10),6E,8-triene, **3**, Fig. 1] has a significant effect on the absorbance characteristics ($\lambda_{max} = 280$ nm; $\varepsilon = 24,600$). Several other derivatives, e.g., isovitamin D (1[10],5E,7E-triene) and isotachysterol

FιG. 1. Structures of vitamin D and its derivatives. The triene systems shown are **1**, 5,6-*cis*-vitamin D [5*Z*,7*E*,10[19]-triene]; **2**, 5,6-*trans*-vitamin D [5*E*,7*E*,10[19]-triene]; **3**, tachysterol [5(10),6*E*,8-triene]; **4**, (5*E*)19-nor-10-oxo-vitamin D.

(5[10]6*E*,8[14]-triene), produce spectra like that of tachysterol, i.e., three peaks with a molar absorptivity of the maximum greater than that of vitamin D. One isomer, 6,7-*cis*-isotachysterol (5[10],6*Z*,8[14]-triene), has a lower absorbance maximum (253 nm), and a lesser molar absorbtivity (ε = 13,000). Replacement of the 10(19)-methylene group with an oxygen atom to give 19-nor-10-oxo-vitamin D (**4**, Fig. 1) results in an absorbance maximum at 308 nm for the 5*Z* isomer and 312 nm for the 5*E* isomer.[23]

Mass Spectrometry

Electron Impact. The electron impact (EI) mass spectra of vitamin D metabolites are usually dominated by peaks that result from rearrangement of the 5,7,10(19)-triene system, followed by cleavage between C-7 and C-8 to give a fragment composed of the steroidal A ring and carbons 6 and 7.[25] The 5*E* and 5*Z* isomers produce nearly identical spectra, which are characterized by a base peak at either m/z 136 or 118 (m/z 134—H$_2$O), depending on the compound and the circumstances under which the spectrum is obtained. 1α-Hydroxylation does not interfere with the rearrangement and fragmentation, but the peaks are observed at m/z 152 and 134 (m/z 152—H$_2$O). The peak at m/z 134 is always more intense than the m/z 152 peak and is usually the base peak. As occurs with the UV spectra, rearrangement or substitution in the triene system produces mass spectra that do not have peaks at m/z 118 and 136 or at m/z 152 and 134. This is illustrated by the mass spectra of 23*S*,25-(OH)$_2$D$_3$[12] and 19-nor-10-oxo-25-OHD$_3$[23] (Fig. 2).

The parent peak (molecular ion) of vitamin D metabolites is intense when the spectrum is obtained on magnetic instruments. It is present,

[25] W. H. Okumura, M. L. Hammond, H. J. C. Jacobs, and J. van Thuijl, *Tetrahedron Lett.* **52**, 4807 (1976).

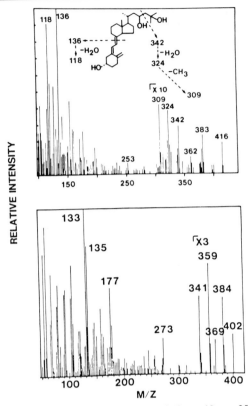

FIG. 2. EI mass spectra of 23,25-(OH)₂D₃ (top) and 19-nor-10-oxo-25-OHD₃ (bottom).

albeit much less intense, when the spectrum is obtained on a quadrupole instrument. Many metabolites also display fragmentation patterns in the high mass region that are characteristic of their particular side chains. Note the peaks in the spectrum of 23S,25-(OH)₂D₃ (Fig. 2) which result after cleavage between C-23 and C-24; similar cleavage is observed in the spectrum of 1,23S,25-(OH)₃D₃.[26] Cleavage between C-25 and C-26, particularly in the persilyl derivatives, is characteristic of 25,26-dihydroxylated metabolites.[27,28] The 23-oxo-25-hydroxy side chain can be readily distinguished from the 24-oxo-25-hydroxy side chain by the high mass fragmentation pattern, in both the 1α-hydroxylated compounds and those lacking

[26] J. L. Napoli and R. L. Horst, Biochem. J. 214, 261 (1984).
[27] J. L. Napoli, R. T. Okita, B. S. Masters, and R. L. Horst, Biochemistry 20, 5865 (1981).
[28] T. A. Reinhardt, J. L. Napoli, B. Pramanik, E. T. Littledike, D. C. Beitz, J. J. Partridge, M. R. Uskokovic, and R. L. Horst, Biochemistry 21, 6230 (1981).

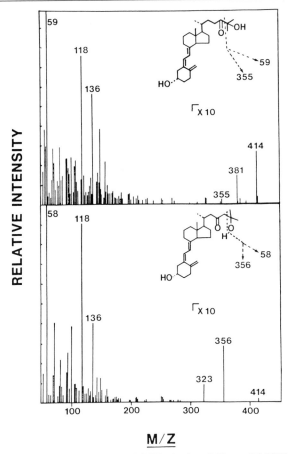

FIG. 3. EI mass spectra of 24-oxo-25-OHD$_3$ (top) and 23-oxo-25-OHD$_3$ (bottom).

a 1α-hydroxy group[16,24,29] (Fig. 3). Note that these two side chains are also distinguished by base peaks at m/z 58 and m/z 59, respectively, rather than at m/z 118. The more complicated side chain of 24-oxo-1,23,25-(OH)$_3$D$_3$ produces a fragmentation pattern in the high mass region that could be interpreted in a number of ways in the absence of supporting data.[2] The mass spectrum of the silylated metabolite, however, firmly establishes the loci of the side chain oxo and hydroxyl groups (Fig. 4).

A double bond at C-22 in combination with a 25-hydroxyl group, such as in 25-OHD$_2$ and its metabolites, can also produce characteristic frag-

[29] R. L. Horst, T. A. Reinhardt, and J. L. Napoli, *Biochem. Biophys. Res. Commun.* **107,** 1319 (1982).

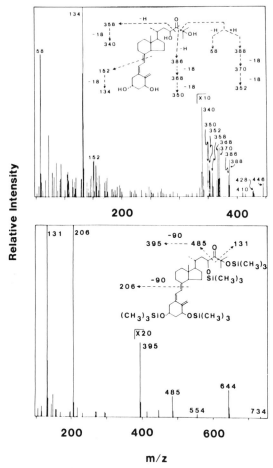

FIG. 4. EI mass specra of 24-oxo-1,23,25-(OH)$_3$D$_3$ (top) and its silylated derivative (bottom).

mentation. Accordingly, 25-OHD$_2$, 24,25-(OH)$_2$D$_2$, and 1,25-(OH)$_2$D$_2$ produce peaks at $M^+ - 58$, and 25,26-(OH)$_2$D$_2$ produces a peak at $M^+ - 74$ [$M^+ - (58 + 16)$], due to cleavage between C-24 and C-25. Thus, a peak at $M^+ - 58$ can be produced by 25-hydroxylated vitamin D$_2$ compounds, as well as by 23-oxo-25-OHD$_3$ side chains.

Chemical Ionization. Vitamin D metabolites produce strong positive ions in the chemical ionization mode. The nature of the positive ion chemical ionization (PCI) spectra is dependent on the reagent gas. Methane and isobutane provide intense high mass peaks representing the molecular ion and ions resulting from the losses of neutral fragments from the molecular ion, particularly water. The molecular ion is more intense with isobutane. These several peaks, M^+, $M^+ - H_2O$, $M^+ - 2 \times H_2O$, given their intensity, provide the opportunity to quantify vitamin D metabolites by selected ion monitoring. Both reagent gasses provide only weak peaks from the steroidal A ring plus C-6 and C-7 fragment, i.e., the peak at m/z 136 in the EI spectrum which is often the base peak [or at m/z 134 (152—H_2O) in the case of 1α-hydroxylated metabolites]. Quite a different spectrum results when nitrogen is used as reagent gas. The PCI spectrum of 1,25-$(OH)_2D_3$ obtained with nitrogen appears similar to its EI spectrum, with the notable difference that the high mass peaks are more intense.[30]

Chloride ion addition, negative ion, chemical ionization (Cl-NCI) MS is a very useful technique for the unequivocal establishment of the molecular weight of metabolites, particularly those that are prone to extensive fragmentation and therefore give weak molecular ions.[2,12,16,24] This technique relies on a mixture of methane and dichloromethane as reagent gas, which produces negatively charged chloride ion adducts of the metabolite. The adducts are easily distinguished from other peaks because chlorine is a mixture of two isomers, ^{35}Cl and ^{37}Cl. Therefore, each adduct provides two peaks. Compounds that fragment extensively in the EI or PCI mode show little or no fragmentation in Cl-NCI. The base peak is produced by chloride addition to the molecular ion. Compare the EI mass spectrum of 24-oxo-1,23,25-$(OH)_3D_3$ (Fig. 4) to its Cl-NCI mass spectrum (Fig. 5). This example is somewhat atypical since 24-oxo-1,23,25-$(OH)_3D_3$, unlike the eight other derivatives tested, does show carbon–carbon bond cleavage because of the labile bond between C-24 and C-25. In most cases only minor fragmentation is noted (0–20% relative intensity), usually produced by loss of water from the parent peak.

Chemical Modification

Silylation. Formation of this type of derivative is helpful for three reasons: the number of silylation sites (usually hydroxyl groups) can be determined, the mass spectra of silylated compounds often have diagnostic fragmentation patterns that are not clear in the mass spectra of the underivatized materials,[2,24,28] and the elution positions on HPLC of silylated derivatives shift relative to underivatized metabolites and impurities, thus affording material of greater purity. To the metabolite (100–400 ng) in a dry mini reaction vial is added N-methyl-N-trimethylsilyltrifluoroacetamide (50 μl). The reaction is heated at 90° for 1–2 hr to ensure

---

[30] J. L. Napoli, unpublished results.

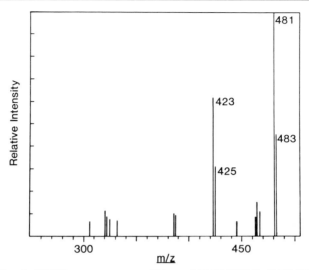

FIG. 5. Cl-NCI mass spectrum of 24-oxo-1,23,25-(OH)₃D₃ (MW 446).

silylation of tertiary hydroxyl groups such as the 25-hydroxyl group. The reaction is cooled and excess reagent is removed under a stream of nitrogen. The product is purified by HPLC and analyzed by MS. We have been successful with a Whatman Partisil PXS 10/25 ODS-3 column (0.42 × 25 cm) eluted with 0.1% methanol in dichloromethane.[2,12,16,22,23,24,26,31]

Periodate Cleavage. This reaction tests for the presence of vicinal alcohol functions or an alcohol group vicinal to a ketone. To the metabolite (100–900 ng) in methanol (20–50 μl) is added 5% aqueous sodium metaperiodate (50 μl). After 20–120 min at ambient temperature, the product is extracted with chloroform or the solvent is removed under a stream of nitrogen. Purification is achieved by HPLC[13,27,28] and analysis is by MS.

Reduction. Reduction of a ketone to an alcohol is a good way to verify its presence. Moreover, if the alcohol product is a known metabolite, very reliable evidence for the structure of the ketone is obtained. If the alcohol product is not a known metabolite, certain structures can be confidently ruled out. To metabolite (200–900 ng) in methanol (30–40 μl) is added a molar excess of sodium borohydride. After 20 min, acetone is added to decompose excess reagent. The solvent is removed under a stream of nitrogen and the product is purified by HPLC.[16,23,24]

[31] G. E. Lester, R. L. Horst, J. L. Napoli, *Biochem. Biophys. Res. Commun.* **120,** 919 (1984).

TABLE III

NMR Signals of Vitamin D and Its Metabolites[23,32,33]

Metabolite	6(d)[b]	7(d)	22,23(sm)	19Z(s)[c]	19E(s)[c]	1β(m)	3α(m)	28(d)	21(d)	26,26(d)[d]	18(s)
Vitamin D$_3$	6.2	6.0	—	5.0	4.8	—	3.9	—	0.92	0.87	0.54
Vitamin D$_2$	6.2	6.0	5.2	5.0	4.8	—	3.9	0.90	1.0	0.83, 0.81	0.54
1,25-(OH)$_2$D$_3$	6.4	6.0	—	5.0	5.3	4.4	4.2	—	0.94	1.2^c	0.53
19-nor-10-oxo-D$_3$	7.6	5.9	—	—	—	—	4.2	—	0.92	0.86	0.54
24-OHD$_2$	6.2	6.0	5.4	5.0	4.8	—	4.0	1.2	1.0	0.88, 0.87	0.56
25-OHD$_2$	6.2	6.0	5.3	5.0	4.8	—	3.9	0.99	1.0	1.16, 1.12^f	0.55
24,25-(OH)$_2$D$_2$	6.2	6.0	5.6	5.0	4.8	—	3.9	1.27	1.0	1.21, 1.18^f	0.56
(Δ22)1,25-(OH)$_2$D$_3$^{g}	6.4	6.0	5.4	5.0	5.3	4.4	4.2	—	1.0	1.19^c	0.56

Assignment (δ)^a

^{a} Chemical shifts (δ) are given in ppm downfield from tetramethylsilane as internal standard. Spectra were taken in CDCl$_3$.

^{b} Letters in parentheses: d, doublet; sm, sharp multiplet; s, singlet; m, multiplet.

^{c} These signals appear as singlets in low field spectra; they actually are multiplets ($J = 1$–2 Hz).

^{d} Each signal is a doublet. The two sets of doublets in D$_2$, versus a single set in D$_3$, result from the magnetic nonequivalency of the C-26 and C-27 methyl groups due to the presence of a C-28 methyl group. This would be observed only in high field NMR.

^{e,f} A singlet or two sets of singlets; further splitting into doublets is precluded by the presence of the 25-hydroxyl group.

^{g} We are grateful to Drs. Enrico G. Baggiolini and Milan R. Uskokovic for sharing this compound with us.

Nuclear Magnetic Resonance Spectroscopy

Nucler magnetic resonance (NMR) spectroscopy can be a powerful aid in identifying new metabolites but is not used frequently because it requires larger amounts of sample (10–500 μg) than MS (25–200 ng) and because the mass spectra of metabolites and their derivatives usually provide conclusive structural evidence. Nevertheless, the NMR spectra of vitamin D compounds have distinctive signals that are diagnostic of the alkene protons, the side chain substitution, and the A-ring substitution.[23,32,33] Table III provides detailed NMR information about vitamins D_2 and D_3 and a variety of their metabolites. Note that in the vitamin D_2 series, the protons of the C-26 and C-27 methyl groups are not magnetically equivalent due to the proximity of the chiral center at C-24, and therefore, generate two distinct signals in the (300-MHz proton) NMR. *Of special interest, is our assignment for the C-21 and C-28 methyl group resonances in vitamin D_2.* Based on the consistency of the data collected from the listed vitamin D_2 metabolites, as well as from (Δ22)1,25-(OH)$_2$D$_3$, it appears that the doublet at 1.0 ppm downfield from TMS should actually be assigned to the C-21 methyl group of vitamin D_2 and the doublet at 0.90 ppm should be assigned to the C-28 methyl group. In the past, these assignments have been reversed.[32] Past studies, however, have not had the benefit of comparison of (Δ22)1,25-(OH)$_2$D$_3$ and the several vitamin D_2 metabolites. These revised assignments should alleviate future confusion, as the use of NMR becomes more widespread in the elucidation of other vitamin D_2 metabolites.

Acknowledgments

We are grateful to Cathleen A. Martin and Ricardo Aspiroz for help in preparing this manuscript.

[32] R. M. Wing, W. H. Okamura, A. Rego, M. R. Pirio, and A. W. Norman, *J. Am. Chem. Soc.* **97**, 4980 (1975).

[33] E. Berman, Z. Luz, Y. Mazur, and M. Sheves, *J. Org. Chem.* **42**, 3325 (1977).

[15] A New Pathway of 25-Hydroxyvitamin D₃ Metabolism

By GLENVILLE JONES

Using the isolated perfused rat kidney we[1] are able to demonstrate the *in vitro* formation of several metabolites of 25-hydroxyvitamin D_3 (25-OHD_3), which can be logically placed on a single metabolic pathway:

$$25\text{-}OHD_3 \rightarrow 24,25\text{-}(OH)_2D_3 \rightarrow 24\text{-}oxo\text{-}25\text{-}OHD_3$$
$$\rightarrow 24\text{-}oxo\text{-}23,25\text{-}(OH)_2D_3 \rightarrow\rightarrow 24,25,26,27\text{-}tetranor\text{-}23\text{-}OHD_3$$

Each metabolite can be identified by a combination of HPLC, UV spectrophotometric, and mass spectrometric techniques. The enzymes involved in the pathway are present only in the vitamin D repleted animal, are stimulated by vitamin D intoxication, and probably play some role in excretion of the potentially toxic molecule, 25-OHD_3.

Methods

Isolated Perfused Rat Kidney[2,3]

Perfusion Apparatus. Perfusate from the venus reservoir (Fig. 1) is passed through a filter (1.2-μm pore size, Millipore) and delivered to the multibulb glass oxygenator (Johns Scientific, Toronto, Ontario) by a roller pump (Masterflex, Cole Parmer, Chicago); a mixture of 95% O_2 and 5% CO_2 is delivered into the oxygenator through an inlet in its lower end and escapes through an outlet at its top. The perfusate, entering the top of the oxygenator, forms a thin layer as it flows down the sides of the bulbs of the oxygenator. The P_{O_2} and P_{CO_2} values of the perfusate should approximate 400 and 30 mm Hg, respectively. The oxygenated perfusate enters through a glass arterial reservoir and is then delivered to the renal artery of the isolated kidney by a second roller pump. The perfusate flow rate is measured with a flowmeter (F-1200, Gilmont Instruments, Great Neck, NY) inserted between the pump and arterial reservoir. The perfusion pressure is monitored with an aneroid manometer connected to the top of the arterial reservoir. The pressure is kept constant at 100 mm Hg through the experiment by adjusting the flow rate.

[1] Supported by grants from the Canadian Medical Research Council MT-5777 and DG304.
[2] A. M. Rosenthal, G. Jones, S. W. Kooh, and D. Fraser, *Am. J. Phyiol.* **239**, E12 (1980).
[3] G. S. Reddy, G. Jones, S. W. Kooh, and D. Fraser, *Am. J. Physiol.* **243**, E265 (1982).

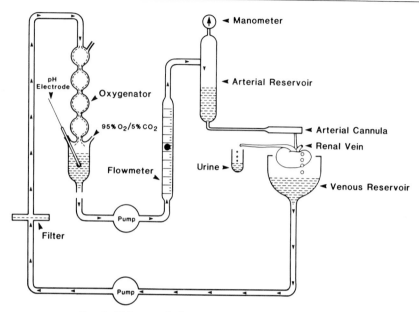

FIG. 1. Kidney perfusion apparatus. From Reddy *et al.*[3]

Glass equipment and tubing should be used throughout the apparatus except for the compressible tubing within the pumps (Tygon, i.d. $\frac{1}{8}$ × o.d. $\frac{1}{4}$ in.) and very short lengths of narrow-gauge (Tygon, i.d. $\frac{3}{32}$ × o.d. $\frac{5}{32}$ in.) connecting tubing to provide flexible connections elsewhere in the system. A plastic three-way stopcock is inserted between the second pump and the oxygenator to permit sampling of perfusate from the system at various times.

The perfusion apparatus, except for the two pumps, flowmeter, and filter, is enclosed in a Plexiglas chamber maintained at 37° by a heater. After each perfusion the glassware should be washed with distilled water and detergent and boiled in distilled water for 4–6 hr before reusing. New Tygon tubing should be used in each perfusion.

Perfusate Composition. The perfusion medium consists of a salt solution containing (in mM) Na$^+$, 140; K$^+$, 5.9; Mg^{2+}, 1.0; SO$_4^{2-}$, 1.0; Cl$^-$, 105; HCO$_3^-$, 25; glucose, 5; lactate, 5, to which is added bovine serum albumin (BSA, fraction V, Miles Laboratories, Elkhart, IN) 6 g/100 ml. Perfusate contains 2.6 mM calcium and 2 mM inorganic phosphate. The perfusate should be filtered (0.45 μm pore size, Millipore, Bedford, MA) before use in each experiment. We normally make up 150 ml of such perfusate prior to each perfusion.

Surgical Procedure and Perfusion Technique. Rats (250–350 g body weight) are reared on a standard rodent diet containing nutritionally adequate amounts of vitamin D, calcium, and phosphorus. The right kidney is used for perfusion, and the surgical technique is similar to that of Nishitsuttsuji-Uwo *et al.*[4] To reduce the time required to isolate the kidney from the animal, we cannulate only the renal artery; we let the venus effluent flow freely from the cut uncannulated renal vein into the venous reservoir. The rat is anesthetized with an intraperitoneal injection of Inactin (100 mg/kg body weight) or sodium pentobarbital (50 mg/kg body weight). The right renal and superior mesenteric arteries are separated by blunt dissection. The adrenal branch of the right renal artery is ligated. Loose ligatures are placed around the right renal and superior mesenteric arteries. Sodium heparin (1500 U in 0.5 ml) is injected through a 26-gauge needle into the caudal segment of the inferior vena cava. The superior mesenteric artery is ligated distally, and a spring clamp is applied to the artery at its origin from the aorta.

The arterial cannula consists of a 2-cm-long 20-gauge thin-walled stainless-steel needle attached at right angles to another 2-cm-long stainless-steel tubing (i.d. 0.23 cm) (Fig. 1). The tip of the cannula is beveled slightly to facilitate insertion. The arterial cannula, connected to the arterial reservoir with a short Tygon tube, is filled to its tip with the perfusate in readiness for cannulation. A small opening is made at the distal end of the superior mesenteric artery using fine scissors. The tip of the cannula is passed through the opening into the superior mesenteric artery until it reaches the spring clamp. The spring clamp is then released and the cannula is quickly advanced across the lumen of the aorta into the right renal artery. The moment that the cannula enters the renal artery, the flow of oxygenated perfusate is started and the cannula tied securely into the right renal artery with a ligature. Fifty milliliters of oxygenated perfusate is pumped through the kidney to wash out blood. The kidney is transferred to the apparatus. The volume of the perfusate within the apparatus is then adjusted to about 100 ml. The kidney, supported by the arterial cannula, is positioned within the perfusion box directly above the venous reservoir that holds the perfusate effluent. The entire operative procedure should take less than 15 min. A further period of 15 min of perfusion should be allowed for stabilization before metabolic studies are started.

Study of 25-OHD₃ Metabolism. This can be studied in two ways, either by using analytical techniques involving radioactive 25-OHD₃ substrate, HPLC, and liquid-scintillation counting or by preparative techniques involving using large concentrations of 25-OHD₃ and HPLC cou-

---

[4] J. M. Nishitsuttsuji-Uwo, B. D. Ross, and H. A. Krebs, *Biochem. J.* **103**, 852 (1967).

pled with diode array-spectrophotometry. In either case, the radioactive 25-OHD$_3$ ([26,27-^{3}H]25-OHD$_3$ or [23,24-^{3}H]25-OHD$_3$, 1 μCi, specific activity 20 Ci/mmol; Amersham, Arlington Heights, IL) or nonradioactive 25-OHD$_3$ (crystalline, 50 μg to achieve a final concentration 500 ng/ml perfusate; Hoffmann-LaRoche, Nutley, NJ) must be dissolved in 200 μl ethanol, mixed with 5–10 ml perfusate, and added directly to the venous reservoir.

Samples of perfusate (2–4 ml) are removed every hour after perfusion is begun and perfusion periods can last up to 6–8 hr. At the termination of perfusion 80 ml of perfusate normally remains.

HPLC Analysis of Perfusate Extracts

Lipid extracts of perfusate can be analyzed directly on HPLC without prior chromatographic purification on minicolumns of Sephadex or Sep-Pak cartridges. Lipid extraction is based upon the methanol–chloroform procedure of Bligh and Dyer[5] except that we substitute dichloromethane for chloroform. The extract is dichloromethane is evaporated to dryness under nitrogen and redissolved in hexane–isopropanol–methanol (94:5:1 or 91:7:2). Analytical samples from 2–4 ml perfusate are redissolved in 200 μl 94:5:1 solvent whereas preparative samples from 80 ml perfusate are dissolved in 2 ml of 91:7:2 solvent. Reactivials (5 ml; Pierce, Rockford, IL) can be used during manipulation of the samples so as to minimize evaporation of samples and permit centrifugation of any particulate matter insoluble in hexane–isopropanol–methanol mixtures.

Analytical samples dissolved in 200 μl of hexane–isopropanol–methanol (94:5:1) are run on a 4.6 mm × 25 cm Zorbax-CN column eluted with the same solvent. Effluent from the column is collected in 0.5 min/0.65 ml fractions evaporated to dryness and radioactivity measured by liquid scintillation counting. Figure 2 depicts the type of results obtained with this technique. It is evident that several radioactive peaks are formed from [^{3}H]25-OHD$_3$ and that these contain different amounts of radioactivity depending upon the position of radioactive label used. For instance, the 24-oxo-23,25-(OH)$_2$D$_3$ peak is devoid of radioactivity if we start with 23,24-^{3}H label, whereas it is a major radioactive peak if we use 26,27-^{3}H label.

The study of the pathway of nonradioactive 25-OHD$_3$ metabolism in preparative sized samples requires a different approach. The scheme shown in Fig. 3 illustrates our approach to purification of metabolites of 25-OHD$_3$ by HPLC.[6] Three steps are involved from bulk extract to pure

[5] E. G. Bligh and W. J. Dyer, *Can. J. Biochem. Physiol.* **37**, 911 (1959).
[6] G. Jones, M. Kung, and K. Kano, *J. Biol. Chem.* **258**, 12920 (1983).

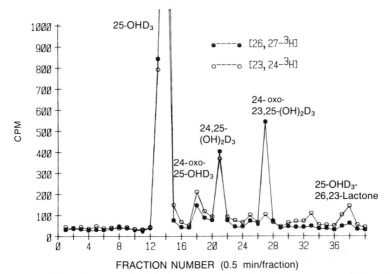

FIG. 2. High-performance liquid chromatography of perfusate extracts of vitamin D-repleted rat kidneys using different labeled substrates. (●) 25-OH[26,27-³H]D₃; (○) 25-OH[23,24-³H]D₃. The perfusate extracts represent samples taken at 6 hr from kidneys perfused with 1 μCi of 25-OH[³H]D₃ and 25 nM 25-OHD₃. Chromatography, Zorbax-CN (4.6 mm × 25 cm); solvent, hexane/isopropanol/methanol (94 : 5 : 1); flow rate, 1.3 ml/min; pressure, 500 psi. From Jones et al.[6]

material suitable for mass spectrometry: a preparative HPLC step on a 9.4 mm × 25 cm Zorbax-SIL column eluted with hexane–isopropanol–methanol (91 : 7 : 2) at a flow rate of 4 ml/min[7]; a second step on a 4.6 mm × 25 cm Zorbax-CN column eluted with hexane–isopropanol–methanol (94 : 5 : 1) at a flow rate of 1.5 ml/min[8]; and a third step back on Zorbax-SIL with the initial solvent to confirm purity. Such a scheme separates most of the metabolites of 25-OHD₃ based upon their hydroxyl content whilst avoiding the column overloading problems encountered with large lipid contents of bulk perfusate extracts. Further fractionation on Zorbax-CN of the "24,25-(OH)₂D₃" region into 24,25-(OH)₂D₃, 24-oxo-23,25-(OH)₂D₃, and 25-OHD₃-26,23-lactone is based upon the chemical selectivity of the column packing. Interactions between the C=O groups of some vitamin D metabolites and the C≡N packing cause major changes in retention over comparable Zorbax-SIL elution patterns.[8]

[7] G. Jones, J. Chromatogr. **221**, 27 (1980).
[8] G. Jones, J. Chromatogr. **276**, 69 (1983).

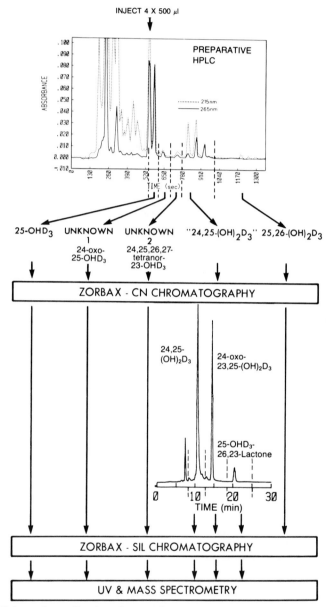

FIG. 3. Scheme for purification of metabolites of 25-OHD$_3$ from kidney perfusate. From Jones *et al.*[6]

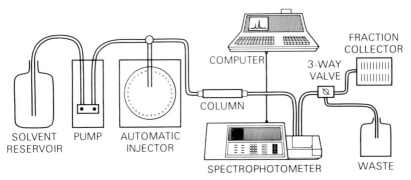

FIG. 4. HPLC configuration for diode-array UV spectrophotometry.

Diode-Array Spectrophotometric Detection of Metabolites of 25-OHD₃

We use diode-array spectrophotometry to search for new metabolites of vitamin D and in the final HPLC step to establish purity of isolated metabolites. 25-OHD₃ and its metabolites all possess the same characteristic ultraviolet spectrum (ε_{265} = 18,300; λ_{max} = 265 nm; λ_{min} = 228 nm; $\lambda_{max}/\lambda_{min}$ = 1.8–2). Use of high substrate concentrations (1250 nM; 500 ng/ml perfusate) permits use of the diode-array spectrophotometer as a detection device in HPLC. The technique possesses good sensitivity (~10 ng/peak) as well as affording improved selectivity for vitamin D compounds. Though several specialized diode-array detectors are now commercially available (e.g., Model 1040A, Hewlett Packard, Palo Alto, CA; Model 2140, LKB, Bromma, Sweden) we use a modified spectrophotometer (Model 8450A, Hewlett Packard, Palo Alto, CA) coupled to a computer (Model 9816S, Hewlett Packard, Palo Alto, CA) (Fig. 4). Using software we have developed ourselves,[9] it is possible to collect complete UV spectra (200–400 nm) every 3 sec during a 20 min HPLC run. The resulting data can be displayed in several ways.

A few wavelengths (e.g., 265 or 228 nm) can be selected to give the conventional absorbance vs run time profile (Fig. 5A). Alternatively, a three-dimensional plot of absorbance vs wavelength vs run time can be constructed from the total data (Fig. 5B). The example given in Fig. 5 represents the chromatographic profile that is observed when a kidney is perfused with 1250 nM 24,25-(OH)₂D₃ as substrate and a portion of the bulk extract from 80 ml perfusate is processed as in the scheme in Fig. 3. The single wavelength plot (Fig. 5A) allows one to recognize peaks which are not evident in the original perfusate and which increase with perfusion

---

[9] This program, written in Basic, is available on request from the author.

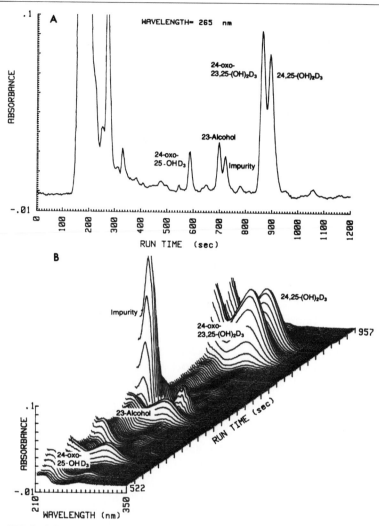

Fig. 5. HPLC of the extract of perfusate from kidneys perfused in the presence of 1250 nM 24(R),25-(OH)$_2$D$_3$: (A) absorbance at 265 nm vs run time; (B) absorbance vs wavelength vs run time. A and B represent the same chromatographic run. Note in A that the peak marked as an impurity appears minor because the wavelength used for monitoring is 265 nm. The impurity peak absorbs strongly at 225 nm and in the 275–280 nm region. Note also in B that all vitamin D metabolites have λ_{max} = 265 nm and λ_{min} = 228 nm. Chromatographic conditions were as follows: Zorbax-SIL 25 cm × 6.2 mm column; solvent of hexane–2-propanol–methanol (94 : 5 : 1, v/v), 1.5 ml/min. Reprinted with permission from Jones *et al.* *Biochemistry* **23**, 3749. Copyright 1984 American Chemical Society.[10]

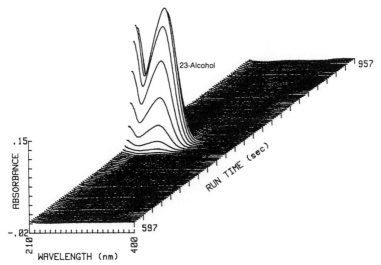

FIG. 6. Final HPLC of putative 24,25,26,27-tetranor-23-OHD₃ from kidney perfusate. Chromatographic conditions were as follows: Zorbax-SIL 25 cm × 6.2 mm column; solvent of hexane–2-propanol–methanol (94 : 5 : 1, v/v), 1.5 ml/min. Material from this column was used for mass spectrometry. Reprinted with permission from Jones *et al. Biochemistry* **23,** 3749. Copyright 1984 American Chemical Society.[10]

time. The three-dimensional plot, however, permits us to recognize vitamin D-like compounds and distinguish impurities on the basis of their UV-absorptive characteristics. Thus this technique allows us to recognize that at least three compounds are formed from 24,25-$(OH)_2D_3$ in the vitamin D-replete kidney: 24-oxo-25-OHD₃, 24-oxo-23,25-$(OH)_2D_3$, and 24,25,26,27-tetranor-23-OHD₃. Note that use of radioactive 25-OHD₃ with ³H at 26,27 carbon of the side chain does not allow us to recognize the formation of 24,25,26,27-tetranor-23-OHD₃ because the 26,27-³H is cleaved from the molecule during its formation.

In addition to providing a screening procedure in the search for metabolites with the vitamin D-like UV spectrum, diode-array spectrophotometry also establishes purity of HPLC peaks prior to mass spectrometry. Metabolites are purified through the first two HPLC steps of the scheme shown in Fig. 3. The final HPLC step generates a three-dimensional plot which can be examined for evidence of UV contamination. Purification of 24,25,26,27-tetranor-23-OHD₃ leads to a chromatographic profile of the type illustrated in Fig. 6.[10] It is devoid of other chromatographic peaks

[10] G. Jones, K. Kano, S. Yamada, T. Furusawa, H. Takayama, and T. Suda, *Biochemistry* **23,** 3749 (1984).

or shoulders and in this example was followed by successful mass spectrometry.

24-Oxo-23,25-$(OH)_2D_3$ generated from 25-OHD$_3$ and purified through the scheme shown in Fig. 3 shows an HPLC peak similarly devoid of contamination (Fig. 7). Here the data are depicted as sequential UV scans, represented by increasing number, overlaid so as to compare all salient features of the spectrum. Commercially available detectors can also normalize such scans so as to attempt to superimpose similar spectra.

Mass Spectrometry

Purified renal metabolites are identified by subjecting them to electron-impact mass spectrometry with a direct insertion probe (Model HP598S, Hewlett Packard, Palo Alto, CA). Ionization voltage is 70 eV and the source temperature should be programmed in the range 25–400°. As little as 250 ng of metabolite will yield a definitive spectrum. Peaks purified through the scheme depicted in Fig. 3 give mass spectra similar to those shown in Fig. 8.

Identification of Metabolites of Pathway

24-Oxo-25-OHD$_3$. The perfusate peak emerging at 590 sec on Zorbax-SIL (Fig. 5) and 9.8 min on Zorbax-CN shows the UV absorption spectrum of vitamin D and comigrates with synthetic 24-oxo-25-OHD$_3$. On mass spectral analysis the perfusate peak had a molecular ion at m/z 414 and a fragmentation pattern identical with that published for 24-oxo-25-OHD$_3$ (Fig. 8B). Particularly characteristic is the fragment m/z 355 which corresponds to C-24–C-25 cleavage also found in the spectra of 24,25-$(OH)_2D_3$, 24-oxo-23,25-$(OH)_2D_3$ (Fig. 8A), and other 24-substituted vitamin D compounds. The increased fragility of the C-24–C-25 bond indicates that the additional substitution of the perfusate metabolite is in the C-24 position and the addition of 14 mass units over 25-OHD$_3$ probably represents a 24-oxo group. The structure is thus 24-oxo-25-OHD$_3$.

24-Oxo-23,25-$(OH)_2D_3$. The perfusate peak emerging at 870 sec on Zorbax-SIL (Fig. 5) and 14.8 min on Zorbax-CN (Fig. 3) and gives the UV spectrum shown in Fig. 7. On mass spectrometry this compound shows a molecular ion at m/z 430 suggesting the introduction of two oxygen functions and one degree of unsaturation into the vitamin D molecule. The fragmentation pattern (Fig. 8A) shows fragments at m/z 397, 372, 339, 271, 253, 136, 118. The strong retention of this metabolite on Zorbax-CN is indicative of a C=O containing compound and the molecular ion of the mass spectrum is consistent with this. The additional oxygen functions

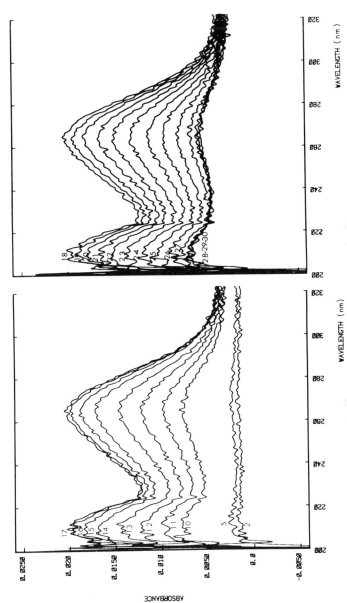

FIG. 7. Ultraviolet spectrum of 24-oxo-23,25-(OH)$_2$D$_3$ from the renal metabolism of 25-OHD$_3$. Sequential UV scans (2 sec apart) of the emerging chromatographic peak of putative 24-oxo-23,25-(OH)$_2$D$_3$ in a scanning spectrophotometer. Scan numbers 2–17 represent scans from the front side of the chromatographic peak. Scan numbers 18–30 represent scans from the tail side of the peak. From Jones et al.[6]

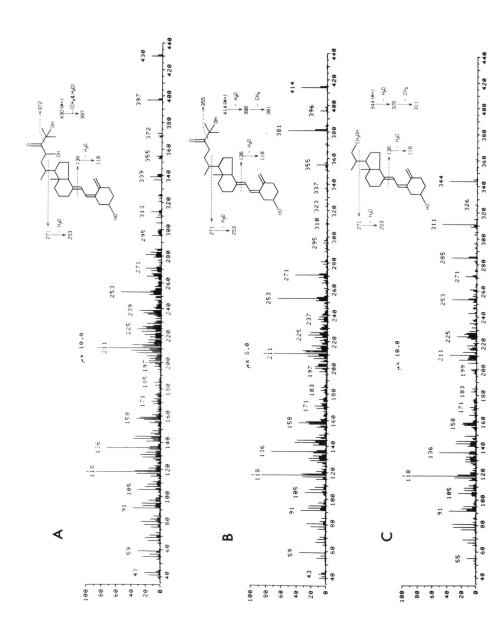

must be in the side chain since the fragments at m/z 271, 253, 136, 118 indicate no ring substitution.

Periodate treatment of 26,27-^{3}H-labeled metabolite gives a complete loss of tritium from the molecule and generates a product eluting slightly ahead of the 24,25-$(OH)_2D_3$ cleavage product on Zorbax-CN chromatography. The periodate cleavage product of the metabolite shows a molecular ion m/z 342 on mass spectrometry, indicative of a C_{23} aldehyde. These data suggest that the metabolite is a side chain-substituted derivative of 25-OHD_3 with one additional hydroxyl and one additional oxo group on carbon positions C-23 and C-24. The exact structure can be established by trimethylsilylation of the molecule.

Trimethylsilylation of 24-oxo-23,25-$(OH)_2D_3$ gives two products, one the tristrimethylsilylated derivative with molecular ion m/z 646 and major fragments at m/z 556, 487 (C-23–C-24 cleavage), 397, 307, 208, 131, and 118, and the other the ditrimethylsilylated compound with molecular ion m/z 574 and fragments at 487, 397, 307. The C-23–C-24 fragility shown by these compounds is indicative of 23-hydroxylation and is observed in other 23-hydroxylated vitamin D metabolites. The structure is thus 24-oxo-23,25-$(OH)_2D_3$.

24,25,26,27-Tetranor-23-OHD$_3$. The perfusate peak emerging at 700 sec on Zorbax-SIL and 9.5 min on Zorbax-CN has UV spectral characteristics of the *cis*-triene of vitamin D_3 on scanning spectrophotometry (Fig. 6). The unknown peak is totally devoid of radioactivity if we start with 25-OH[23,24-^{3}H]D_3 or 25-OH[26,27-^{3}H]D_3. The purified peak subjected to mass spectrometry (Fig. 8C) shows a molecular ion m/z 344 indicative of a vitamin D molecule with a side chain cleaved between carbons C-23 and C-24. Fragments are present at m/z 311, 285, 271, and 253. This pattern is identical to that of chemically synthesized material.[10] The mass spectrum of the di-TMS derivative features a molecular ion at m/z 488, with major fragments at m/z 473, 398, 383, 308, 293, 253, 208, and 118. The structure is thus 24,25,26,27-tetranor-23-OHD_3.

Evidence for a Renal Pathway

The metabolite 24,25-$(OH)_2D_3$ has been known for some years as an *in vivo* and *in vitro* product of 25-OHD_3 metabolism.[11] Using 24,25-$(OH)_2D_3$

---

[11] H. F. De Luca, "Vitamin D: Metabolism and Function," p. 1. Springer Verlag, Berlin and New York, 1979.

FIG. 8. Mass spectra of metabolites of 25-OHD_3 purified from renal perfusates. (A) 24-Oxo-23,25-$(OH)_2D_3$; (B) 24-oxo-25-OHD_3; (C) 24,25,26,27-tetranor-23-OHD_3. From Jones *et al.*[6]

as a substrate for the perfused rat kidney it can be shown that the three metabolites described here appear in the perfusate within 2–4 hr. If a vitamin D-depleted donor rat is used, no such metabolites are formed from 24,25-$(OH)_2D_3$ until 4–6 hr has elapsed. It appears that the enzyme(s) of the pathway must be induced over the 0- to 4-hr period by the vitamin D metabolite. Use of 24-oxo-25-OHD_3 as substrate in the vitamin D-repleted kidney results in synthesis of 24-oxo-23,25-$(OH)_2D_3$ and 24,25,26,27-tetranor-23-OHD_3 but not 24,25-$(OH)_2D_3$. Use of 23(S),25-$(OH)_2D_3$ gives only 25-OHD_3-26,23-lactone. Use of 24-oxo-23,25-$(OH)_2D_3$ as substrate in the vitamin D replete kidney results in synthesis of only 24,25,26,27-tetranor-23-OHD_3. From examination of the time course of appearance of each metabolite it is evident that 24-oxo-25-OHD_3 reaches its maximum concentration before 24-oxo-23,25-$(OH)_2D_3$ and this in turn before tetranor-23-OHD_3. Consideration of the structure of the side chains make it likely that the tetranor-23-OHD_3 is derived from one of the other metabolites, probably 24-oxo-23,25-$(OH)_2D_3$, by a one or two step cleavage. Use of other radioactive labels should help establish the exact pathway of metabolism.

The production of the metabolites of this pathway is stimulated by prior treatment with vitamin D_3.[10] This and the fact that the enzymes of the pathway appear to be stimulated in the perfused kidney along with the 24-hydroxylase[10] argues for some degradative role for the pathway though this is by no means certain. The poor binding of 24-oxo-23,25-$(OH)_2D_3$ to plasma vitamin D binding globulin may explain why this metabolite has never been observed in plasma *in vivo*.

Acknowledgments

The author would like to acknowledge the important contributions made by Drs. Donald Fraser, S. W. Kooh, G. S. Reddy, A. Rosenthal, K. Kano, and S. Yamada.

[16] Enzyme-Linked Immunoabsorption Assay for Vitamin D-Induced Calcium-Binding Protein

By BARBARA E. MILLER and ANTHONY W. NORMAN

One of the principal mediators of vitamin D activity is the metabolite 1,25-dihydroxyvitamin D [1,25-$(OH)_2D$]. This metabolite has been shown to function as a steroid hormone through a receptor-mediated mechanism of action.[1] Among the known responses to 1,25-$(OH)_2D_3$ is the induction of a vitamin D-dependent calcium-binding protein (CaBP).[2,3] There are four known classes of vitamin D-dependent CaBPs based on molecular weight. One class with a molecular weight of 28,000 has been shown to have a high affinity for calcium and is present in the chick intestine. The species and tissue distribution of this protein as well as the other three classes of vitamin D-dependent CaBPs are given in Ref. 2. The appearance of CaBP is a good physiological indicator of vitamin D activity. A high correlation exists between the presence of receptors for 1,25-$(OH)_2D$ and the presence of CaBP in tissues.[2] Also the amount of CaBP present in the intestine has been shown to be directly proportional to the intestinal localization of 1,25-$(OH)_2D$.[4] Therefore assay techniques which measure the vitamin D-dependent CaBP are a valuable tool in studies of the vitamin D endocrine system.

The most sensitive assay technique available for the detection and quantitation of the chick intestinal 28K CaBP has been a radioimmunoassay (RIA).[5,6] This RIA has a sensitivity of 1 ng which has enabled the detection of CaBP in a wide number of tissues.[5] Thus this assay has proven to be extremely useful in vitamin D research. However, the RIA is not without certain disadvantages, most of which result from radioactive ^{125}I that is coupled to CaBP. To avoid the problems associated with handling radioactive material, we have developed an enzyme-linked immunoabsorbent assay (ELISA) for the chick intestinal CaBP. This ELISA is as sensitive as the RIA for measuring CaBP.

Principle

The ELISA is based on the competition for antibody binding between CaBP that has been immobilized on polystyrene wells and nonimmobilized CaBP free in solution. Antisera (prepared in the rabbit against CaBP[5,6]) that binds the free CaBP is washed away while that bound to the immobilized CaBP remains in the well. The addition of alkaline phosphatase conjugated goat anti-rabbit γ-globulin, followed by the substrate p-nitrophenyl phosphate, quantitates the amount of antibody that has inter-

[1] A. W. Norman, "Vitamin D: The Calcium Homeostatic Steriod Hormone." Academic Press, New York, 1979.

[2] A. W. Norman, J. Roth, and L. Orci, *Endocr. Rev.* **3**, 331 (1982).

[3] P. Siebert, W. Hunziker, and A. W. Norman, *Arch. Biochem. Biophys.* **219**, 286 (1982).

[4] E. J. Friedlander, H. Henry, and A. W. Norman, *J. Biol. Chem.* **252**, 8677 (1977).

[5] S. Christakos, E. J. Friedlander, B. R. Frandsen, and A. W. Norman, *Endocrinology* **104**, 1495 (1979).

[6] B. E. Miller and A. W. Norman, this series, Vol. 102, p. 291.

acted with the bound CaBP. The amount of CaBP adsorbed to the wells and the amount of antisera added are both fixed and limiting so that the intensity of the yellow color produced during the reaction is inversely proportional to the amount of free CaBP present. The amount of CaBP present in a tissue sample can be determined by comparing the absorbance at 405 nm to a standard curve prepared using reference CaBP of known concentration.

Materials

Coating buffer: 1.59 g of Na_2CO_3, 2.93 g of $NaHCO_3$, and 0.20 g of NaN_3 are dissolved in distilled water and diluted to a final volume of 1 liter. The buffer which is pH 9.6 is stored at 4°. This buffer should be freshly prepared every 2 weeks.

PBS–Tween–BSA buffer: 8.0 g of NaCl, 0.2 g of KH_2PO_4, 1.15 g of anhydrous $NaHPO_4$, 0.2 g of KCl, 0.2 g of NaN_3, 0.5 ml of polyoxyethylene sorbitan monolaurate (Tween 20), and 5 g of bovine serum albumin are dissolved in distilled water for a final volume of 1 liter. The buffer, which is pH is 7.4, is stored at 4°.

Diethanolamine buffer: to 800 ml of distilled water is added 97 ml of diethanolamine, 0.10 mg of $MgCl_2 \cdot 6H_2O$, and 0.20 g of NaN_3. The pH is adjusted to 9.8 with 1 N HCl and the volume is adjusted to 1 liter with distilled water. This buffer is stored in the dark at 4°.

Substrate solution: p-nitrophenyl phosphate (available from Sigma, St. Louis, MO, as 5 mg tablets) is dissolved at room temperature in the diethanolamine buffer to produce a final concentration of 1 mg/ml. The substrate solution is prepared just prior to use.

Alkaline phosphatase conjugated second antibody: goat anti-rabbit γ-globulin to which alkaline phosphatase has been conjugated can be obtained from Sigma, St. Louis, MO or can be prepared by the method of Voller et al.[7]

Terminator of alkaline phosphatase activity: 1.5 N NaOH.

Carrier surface: disposable polystyrene enzyme immunoassay cuvettes (Gilford Instruments, Palo Alto, CA) are used.

Vitamin D-dependent 28K calcium-binding protein (CaBP): the CaBP protein which is employed as a standard in the assay can be isolated and purified by the procedures of this laboratory.[6,8] This purified CaBP may be used to generate a high titer antiserum.

[7] A. Voller, D. Bidwell, and A. Bartlett, in "Manual of Clinical Immunology" (N. Rose and H. Friedman, eds.), p. 507. Am. Soc. Microbiol., Washington, D.C., 1976.

[8] E. J. Friedlander and A. W. Norman, in "Vitamin D: Biochemical, Chemical and Clinical Aspects Related to Calcium Metabolism" (A. W. Norman, K. Schaefer, H. G. Grigoleit, and D. von Herrath, eds.), p. 241. de Gruyter, Berlin, 1977.

Procedure

Chick intestinal CaBP is adsorbed (immobilized) to EIA wells by incubating each well with 5 ng of purified CaBP in 300 μl of coating buffer. Following overnight incubation at 4° the coating buffer is removed and the wells are washed three times with PBS–Tween–BSA buffer. These coated wells can be covered with parafilm and stored at 4 or −20° for at least 6 weeks. Thus coated EIA wells can be kept readily available for future assays.

Assay: in the ELISA for CaBP the antisera dilution and the amount of coating CaBP were selected so that wells containing only immobilized CaBP, i.e., no reference or sample CaBP, yield an optical density reading of 1.0 at 405 nm. With our antisera this occurred at 1 : 80,000 dilution of the antisera and CaBP coating concentration of 5 ng per well. The standard curve for the ELISA is generated using reference CaBP of known concentration ranging from 0.8 to 40 ng/100 μl. Dilutons of the antisera, the tissue sample to be measured, and the reference CaBP are made using the PBS–Tween–BSA buffer.

The ELISA involves incubating for 2 hr at room temperature (or overnight at 4°) 100 μl of sample or reference CaBP with 300 μl of a 1 : 80,000

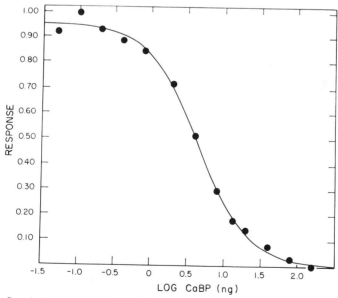

FIG. 1. Standard curve in the ELISA for the vitamin D_3-dependent chick intestinal 28K CaBP. This curve was computer generated.[9] Response is defined as the ratio of the optical density reading obtained for the sample to the difference between the optical density readings obtained at low and high antigen concentrations. The correlation coefficient is 0.997.

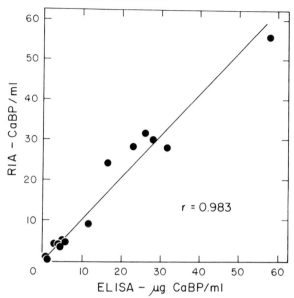

FIG. 2. Correlation of values for chick intestinal CaBP obtained by the RIA and ELISA. CaBP concentrations were simultaneously determined by RIA and ELISA in renal homogenates from four vitamin D_3-repleted chicks and in duodenal homogenates of vitamin D_3-deficient chicks which were administered 13 nmol of 1,25-$(OH)_2D_3$ and sacrificed 0–24 hr following the injection. The correlation coefficient is 0.983.

dilution of rabbit antisera in EIA wells coated with 5 ng CaBP. The wells are then washed three times with PBS–Tween–BSA buffer, and 300 μl alkaline phosphatase conjugated goat anti-rabbit immunoglobulin (diluted 1 : 1000 with PBS–Tween–BSA) is added. Following a 2-hr incubation at room temperature (or overnight at 4°) the wells are again washed (3×) with PBS–Tween–BSA, and 300 μl of the p-nitrophenyl phosphate substrate solution is added. Alkaline phosphatase activity is terminated after 30 min by the addition of 100 μl of 1.5 N NaOH. The absorbance at 405 nm is then measured. Once the reaction has been terminated the color developed during the enzymatic reaction is stable for several hours.

The standard curve is formed by plotting the optical density at 405 nm versus the log of the reference CaBP concentration. Alternatively the curve can be computer generated from a statistical package designed to fit sigmoidal standard curves for competition ELISA using a four-parameter logistic function.[9] A typical standard curve is shown in Fig. 1. The stan-

[9] P. F. Canellas and A. E. Karu, J. Immunol. Methods 47, 375 (1981).

dard curve for the ELISA for chick intestinal CaBP is linear from 0.8 to 20 ng; the sensitivity of this assay is 0.8 ng.

Discussion

The ELISA has been gaining popularity as an alternative to the RIA. This assay involves the same principles as the RIA except that the label used is an enzyme instead of a radioisotope. Therefore the health hazards as well as the disposal and handling problems associated with radioisotopes are eliminated. Also the enzymes chosen for the ELISA have long shelf-lives, usually up to 1 year or longer, and are readily available; thus they can easily be maintained in the laboratory. The ELISA described here has the same sensitivity as the RIA for chick intestinal 28K CaBP (0.8 versus 1 ng), and measurements of calcium binding protein obtained by the ELISA correlate highly with those obtained by the RIA (Fig. 2). An additional advantage of the ELISA over the RIA for CaBP is a shortened assay time; the RIA requires 24–48 hr, the ELISA can be completed in 5 hr. Thus the ELISA for CaBP should prove extremely valuable in future vitamin D research.

[17] Measurement of 25-Hydroxyvitamin D 1α-Hydroxylase Activity in Mammalian Kidney

By Bruce Lobaugh, Janice R. Almond, and Marc K. Drezner

The conversion of 25-hydroxyvitamin D to its hormonal form, 1,25-dihydroxyvitamin D, is a regulated process catalyzed by the renal mitochondrial enzyme, 25-hydroxyvitamin D 1α-hydroxylase (1α-hydroxylase). Until recently assays of 1α-hydroxylase have been performed exclusively in avian kidneys because of the presence in mammalian renal homogenates of an "inhibitory factor(s)" which precludes demonstration of the enzyme activity. The inhibition is likely due to contamination of the preparations with a circulating vitamin D binding protein which binds the majority of available substrate. In the past several years, a number of investigators have developed methods for measurement of mammalian 1α-hydroxylase activity employing extensively washed kidney mitochondria,[1] isolated renal tubules,[2] kidney cortical slices,[3] or renal homogenates prepared from the thoroughly perfused organs of exsanguinated animals.[4] While these techniques permit assay of enzyme activity with the successful removal of binding protein, their routine use is hindered by several obstacles. First, the efforts to remove binding protein are time consuming and not always completely effective. Second, assay sensitivity is limited, a factor which prevents accurate estimation of the relatively low level of enzyme activity maintained in the kidneys of normal animals raised under basal conditions. Finally, use of 25-hydroxy[³H]vitamin D_3 as substrate for the reaction in each of these techniques introduces inaccuracy and adds complexity to the measurement of the 1α-hydroxylase. The inaccuracy is due largely to the reliance upon recovery of total radioactivity as an estimate of reaction product recovery. The added complexity is due to the need for extensive chromatographic purification to separate 1,25-dihydroxy[³H]vitamin D_3 from 19-nor-10-oxo-25-hydroxy[³H]vitamin D_3, a product formed during the enzymatic biosynthesis of calcitriol. This contaminant coelutes with 1,25-dihydroxy[³H]vitamin D_3 during Sephadex LH-20 chromatography and normal phase HPLC employing various mobile phase solvent systems commonly in use. The method described here increases the sensitivity of the assay, permits accurate estimation of reaction product recovery, and limits the purification procedures necessary to adequately prepare the 1,25-dihydroxyvitamin D_3 for quantification. The technique represents a collection of procedures adapted from several sources[5,6] and integrated in a unique fashion to overcome the major problems previously plaguing assay of mammalian renal 1α-hydroxylase activity.

Properties of 25-Hydroxyvitamin D 1α-Hydroxylase

The renal 25-hydroxyvitamin D 1α-hydroxylase enzyme is a mitochondrial mixed-function oxidase consisting of three proteins. Two of these, a nonheme iron protein, ferredoxin, and ferredoxin reductase function in the transfer of electrons from NADPH to the terminal component,

[1] R. Vieth and D. Fraser, *J. Biol. Chem.* **254,** 12455 (1979).

[2] R. T. Turner, B. L. Bottemuller, G. A. Howard, and D. J. Baylink, *Proc. Natl. Acad. Sci. U.S.A.* **77,** 1537 (1980).

[3] H. J. Armbrecht, T. V. Zenser, and B. B. Davis, *Endocrinology* **109,** 218 (1981).

[4] N. Horiuchi, T. Shinki, S. Suda, N. Takahashi, S. Yamada, H. Takayema, and T. Suda, *Biochem. Biophys. Res. Commun.* **121,** 174 (1984).

[5] Y. Tanaka and H. F. DeLuca, *Proc. Natl. Acad. Sci. U.S.A.* **78,** 196 (1981).

[6] J. A. Eisman, A. J. Hamstra, B. E. Kream, and H. F. DeLuca, *Arch. Biochem. Biophys.* **176,** 265 (1976).

cytochrome *P*-450. The cytochrome *P*-450 acts as the binding site for substrate and as the catalytic site that reduces one molecule of oxygen to one molecule of H_2O and one —OH (which is then incorporated into 25-hydroxyvitamin D at the 1α position).

The 1α-hydroxylation is supported by succinate, malate, or almost any other Krebs cycle substrate. It has been reported, that antimycin A blocks the reaction completely, while cyanide, rotenone, and ditnitrophenol reduce activity to about half-maximal.[7] Further, the 1α-hydroxylase is inhibited by carbon monoxide–oxygen mixtures,[8] glutethimide and metyrapone.[9]

Assay Method

Principle. Unlabeled 25-hydroxyvitamin D is added to crude renal homogenates at a concentration sufficient to both saturate the inhibitory factor present in mammalian systems and provide optimal substrate for 1α-hydroxylation. Production of 1,25-dihydroxyvitamin D is quantified by a competitive binding protein radioassay of the purified reaction product, a method which confers great sensitivity to the measurement.

Reagents

Buffer 1
  Tris acetate, 15 m*M*, pH 7.4
  Sucrose, 0.19 *M* (in Tris acetate)
  Magnesium acetate, 2 m*M* (in Tris acetate)
Buffer 2
  Sodium succinate, 25 m*M* (in Buffer 1)
Vitamin D metabolites
  25-Hydroxyvitamin D_3, 10 m*M* (in ethanol)
  1α,25-Dihydroxy[23,24(n)-^{3}H]vitamin D_3
1,25-Dihydroxy[^{3}H]vitamin D_3 (80–120 Ci/mmol) may be purchased from the Amersham Corporation or New England Nuclear. Nonradioactive 25-hydroxyvitamin D_3 may be obtained from Upjohn Company, Kalamazoo, MI or Dr. Milan Uskokovic, Hoffmann-La Roche Drug Company, Nutley, NJ.

Preparation and Incubation of Renal Homogenate. Animals are dispatched by decapitation and the kidneys quickly removed and placed in

[7] H. F. DeLuca and H. K. Schnoes, *Annu. Rev. Biochem.* **45,** 631 (1976).
[8] D. R. Fraser and E. Kodicek, *Nature (London)* **228,** 764 (1970).
[9] J. G. Ghazarian, C. R. Jefcoate, J. C. Knutson, W. H. Orme-Johnson, and H. F. DeLuca, *J. Biol. Chem.* **249,** 3026 (1974).

ice-cold Buffer 1. A 5% (w/v) homogenate of kidney is prepared in the buffer using a Wheaton glass tissue grinder. Each 50-ml Erlenmeyer reaction vessel contains 2 ml of the homogenate (i.e., 100 mg of kidney tissue) and 1 ml of Buffer 2 containing 25 mM sodium succinate. Nonradioactive 25-hydroxyvitamin D_3 is then added to each vessel (25 μl of the 10 mM preparation). Flasks are oxygenated and incubated in a gyrorotary shaker bath at 37° for 20 min. Since the speed of the agitation may affect the enzyme activity expressed, care should be exercised to ensure that the shaker control setting is not disturbed from assay to assay. The reaction is stopped by addition of 20 ml methanol : chloroform (2 : 1, v/v). 1,25-Dihydroxy[^{3}H]vitamin D_3 (approximately 5000 cpm in 25 μl ethanol) is then added to each vessel to monitor recovery of the product through subsequent extraction and chromatographic purification steps.

Lipid Extraction. The contents of each reaction flask are transferred to a 125 ml separatory funnel and lipids are extracted by the method of Bligh and Dyer.[10] Briefly, 6 ml of Buffer 1 is added to each mixture and the funnels agitated. Subsequently, 6.5 ml of chloroform is pipetted into each funnel and the agitation repeated. When the phases are distinct (3–5 min), the lower (chloroform) layer is removed and dried under a stream of nitrogen. Dried lipid extracts can be stored under nitrogen at −20°. Recovery of 1,25-dihydroxy[^{3}H]vitamin D_3 after extraction by this method averages 88 ± 4% (range 85–96%). We have not observed significant improvement in recovery upon further extraction of the methanol–water phase.

Column Chromatography. Dried lipid extracts are solubilized in 400 μl chloroform : hexane (65 : 35, v/v) and applied to Sephadex LH-20 columns (0.7 × 14 cm). The Sephadex is prepared by swelling the beads in chloroform overnight, washing the gel three times with the chloroform : hexane (65 : 35, v/v) mixture the following morning to allow equilibration with the elution solvent, and pouring the slurry to the desired height in glass columns with glass wool plugs or Teflon stopcocks. The sample is layered onto the top of the gel (taking care not to disturb the bed) and allowed to penetrate the surface. The sample vial is then rinsed twice with another 400 μl of chloroform : hexane, and the rinse volumes are applied to the column. When these aliquots have drained into the bed, the column is eluted with 50 ml of the same solvent, and the final 36 ml of eluant, which contains the dihydroxylated vitamin D metabolites, is collected in a glass vial and dried under nitrogen. Efficiency of the column chromatography is 90 ± 4%, resulting in 79 ± 5% recovery of 1,25-dihydroxy[^{3}H]vitamin D_3.

[10] E. G. Bligh and W. J. Dyer, *Can. J. Biochem. Physiol.* **37,** 911 (1959).

Further purification of lipid extracts is achieved by using a Waters Microporasil HPLC column (0.39×30 cm). The chromatography is performed at 800 psi at a flow rate of 2 ml/min with a mobile phase composed of 10% 2-propanol in redistilled hexane. The lipid extract, purified by LH-20 chromatography, is dissolved in 400 μl of the hexane : 2-propanol (90 : 10, v/v) and injected onto the column. The column must be previously calibrated using authentic 1,25-dihydroxyvitamin D_3 to establish the elution volumes for this compound. The fraction containing 1,25-dihydroxyvitamin D_3 is collected for each sample and dried under nitrogen for subsequent assay. Efficiency of the chromatographic step averages 72 $\pm$ 5%, and sample recovery 58 $\pm$ 4%.

We believe the high-performance liquid chromatography step described is crucial for accurate assay of 1α-hydroxylase activity. Examination of a representative sample HPLC elution profile (Fig. 1) reveals a substantial amount of 25-hydroxyvitamin D_3 still present in the sample following Sephadex LH-20 chromatography. Since measurement of the 1,25-dihydroxyvitamin D_3 biosynthesized by the renal preparation is achieved by either (1) competitive binding protein radioassay of the unlabeled metabolite, as we shall describe, or (2) determination by scintillation counting of the amount of 25-hydroxy[^{3}H]vitamin D_3 converted to 1,25-dihydroxy[^{3}H]vitamin D_3, as performed by others, accurate quantification depends upon a priori removal of the 25-hydroxyvitamin D_3 substrate. This can only be achieved by using prohibitively longer LH-20 columns or by the inclusion of further purification steps following LH-20 chromatography.

Assay of 1,25-Dihydroxyvitamin D_3. The 1,25-dihydroxyvitamin D_3 in the collected fraction is measured by employing a radioligand assay.

Preparation of Binding Protein. Binding protein is prepared from the duodena of 5- to 7-week-old White Leghorn cockerels obtained from Spafas, Inc., Norwich, CT. All steps are performed at 4°. Birds are sacrificed by cervical dislocation, the duodenal loops rapidly removed, and the contents expelled by compressing the tissue with a closed scissors. Ten milliliters of ice-cold phosphate buffer (0.05 M potassium phosphate, pH 7.4, with 0.05 M potassium chloride) is forced through the lumen with a syringe, and the tissue contents are again expelled by compression. The intestine is cut along its length and washed three times in cold phosphate buffer. The mucosa is scraped from the intestinal wall and collected in a preweighed beaker containing 100 ml buffer. After determining the weight of collected material, buffer is added to raise the dilution to 5 volumes (ml/g). The mucosal material is centrifuged at 2000 g for 10 min in a Sorvall RC-5B refrigerated centrifuge, and the supernatant removed by aspiration. The pellet is resuspended and homogenized in 2 volumes of

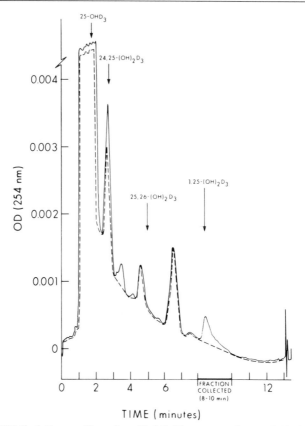

Fig. 1. HPLC elution profiles of purified lipid extracts of normal (dashed line) and vitamin D-deficient (solid line) kidney homogenates. Samples were loaded on a Waters Microporasil column and eluted with hexane : 2-propanol (90 : 10) at 800 psi, 2 ml/min. Arrows indicate the previously calibrated locations of authentic vitamin D metabolite standards. The peak appearing immediately prior to the position of 1,25-dihydroxyvitamin D_3 elution has not been identified. Although a peak is noted at the position of 1,25-dihydroxyvitamin D_3 in the D-deficient kidney homogenate, no peak is observed in the profile of normal kidney. This is consistent with increased 1,25-dihydroxyvitamin D_3 synthesized ($\sim$14 fmol/mg kidney/min; detection of a peak is limited to 7 fmol or greater) under conditions of vitamin D deficiency. The absence of a peak in the normal profile is expected since our measured range for normal kidney has been 3.5–6.0 fmol 1,25-dihydroxyvitamin D_3/mg kidney/min. These measurements were obtained from mouse models.

buffer using a Sorvall Omni-Mixer. It is further homogenized with 1 pass of 10 sec using a Teflon glass homogenizer (clearance 0.015–0.023 cm) and the homogenate centrifuged at 105,000 g for 75 min in a Sorvall OTD Ultracentrifuge. The clear cytosol supernatant under the thin lipid layer is

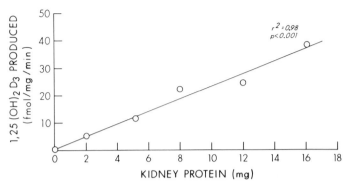

FIG. 2. Dependency of *in vitro* production of 1,25-dihydroxyvitamin D_3 on the tissue protein concentration of the renal homogenate of mice. All flasks were incubated for 20 min at 37° with 80 μM 25-hydroxyvitamin D_3 provided as substrate. Reproduced with permission from Lobaugh and Drezner.[12]

collected with a Pasteur pipette. An aliquot of the pooled supernatant is taken to determine protein concentration by the method of Lowry *et al.*[11] The material is subsequently lyophilized and stored in liquid nitrogen. Binding activity is determined according to traditional techniques, and the protein remains stable for at least 2 months with proper storage.

Competitive Binding Radioassay. Dried samples are reconstituted within 1 week after purification (having been stored at −20°) in 1 ml of HPLC grade 2-propanol. A 100-μl aliquot is transferred to a scintillation vial for determination of sample recovery. Assay tubes contain 100 μl of standard (0-440 pg 1,25-dihydroxyvitamin D_3 in 2-propanol) or sample, 15 μl of 1,25-dihydroxy[^{3}H]vitamin D_3 (~12,000 cpm; specific activity ~92 Ci/mmol) in 2-propanol and 1 ml of binding protein (0.5–1.0 mg/ml in 0.05 M potassium phosphate buffer, pH 7.4, containing 0.05 M potassium chloride). The reaction mixtures are incubated for 60 min at 25°, and after incubation, free and bound 1,25-dihydroxy[^{3}H]vitamin D_3 are separated by addition of 0.3 ml Dextran T-70-coated charcoal. The supernatants, containing bound radiolabel, are then poured into glass scintillation vials and counted in Aquasol. Samples are routinely run in duplicate and at multiple dilutions (the range depending upon the magnitude of expected 1α-hydroxylase activity). Assay sensitivity is reproducibly 10 pg (which represents a displacement of 150–300 cpm). This sensitivity together with the average sample recovery of 58% allow for measurement of about 0.5 fmol 1,25-dihydroxyvitamin D_3/mg kidney/min.

[11] O. H. Lowry, N. J. Rosebrough, A. L. Farr, and R. J. Randall, *J. Biol. Chem.* **193**, 265 (1951).

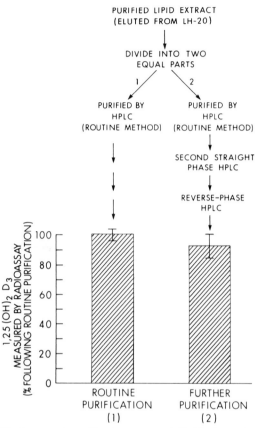

FIG. 3. Effect of further sample purification steps on 1,25-dihydroxyvitamin D_3 measured by the competitive binding protein radioassay. Routine purification (1), as described in this method, included lipid extraction, Sephadex LH-20 chromatography, and HPLC on a Waters Microporasil column using hexane : 2-propanol (90 : 10) as mobile phase. Purification protocol 2 added recycling through the Microporasil column and a reverse-phase HPLC step with methanol : water (75 : 25) as mobile phase on a Waters C_{18} micro-Bondapak column. No significant difference in the measurable 1,25-dihydroxyvitamin D_3 was observed between purification methods ($n = 4$).

1α-Hydroxylase activity, as measured by this method, varies linearly with the concentration of protein in the renal homogenate (Fig. 2). The effect of the inhibitor protein is overcome by use of high unlabeled substrate concentrations. While previous studies[12] have established that use of 80 μM 25-hydroxyvitamin D_3 (~25 μl of 10 mM substrate) provides a

[12] B. Lobaugh and M. K. Drezner, *Anal. Biochem.* **129**, 416 (1983).

saturating concentration for the enzyme and that a 20 min incubation permits measurement of the initial rate of enzyme activity, it is important to recognize the innate limitations that these conditions impose upon the assay. While measurement of V_{max} is assured regardless of the amount of inhibitory protein present in the homogenate, analysis of K_m cannot be undertaken. Thus, changes in enzyme activity due to alterations of K_m would not be detected. Nonetheless, the expected changes in 1α-hydroxylase activity in response to alterations of the known modulators of enzyme function (PTH, hypophosphatemia, vitamin D deficiency, 1,25-dihydroxyvitamin D excess, etc.) have routinely been reflected by this method.

Further purification of the reaction product, by rechromatography on the column or by reverse-phase chromatography over a Waters C_{18}-micro-Bondapak column does not significantly affect the amount of 1,25-dihydroxyvitamin D_3 measured by the radioassay. Thus, samples subjected to extensive repurification have measurable 1,25-dihydroxyvitamin D_3 that is $92.4 \pm 7.8\%$ of that in samples purified by the routine method (Fig. 3). In addition, the coeluting 10-nor-10-oxo-25-hydroxyvitamin D_3 exhibits low cross-reactivity with the binding protein used in the radioligand assay described here (and presents minimal interference to accurate measurement of enzyme activity). Therefore, sensitive measurement of 1α-hydroxylase can be achieved after a single HPLC purification step.

[18] Quantitation of Vitamin D_2, Vitamin D_3, 25-Hydroxyvitamin D_2, and 25-Hydroxyvitamin D_3 in Human Milk

By BRUCE W. HOLLIS and NANCY E. FRANK

Valid assessment of the antirachitic activity present in native milk has been the goal of many investigators spanning a substantial number of years. The earliest attempts to quantitate the vitamin D content of milk were undertaken utilizing bioassay techniques.[1,2] These studies provided crude estimations with respect to the antirachitic potential of native milk but were subject to tremendous variation. A subsequent era in the estimation of vitamin D levels in milk involved chemical methods, specifically

[1] R. S. Harris and J. W. M. Bunker, *Am. J. Public Health* **29,** 744 (1939).
[2] L. J. Polskin, K. Benjamin, and A. E. Sobel, *J. Nutr.* **30,** 451 (1945).

colorimetric reactions.[3] These chemical determinations, while more sensitive than bioassays, still remained very nonspecific and also were subject to considerable variation. In fact, the nonspecificity inherent to this type of analysis led to the quantitation of "so-called" vitamin D sulfate in milk.[4-6] These data dominated the literature for more than a decade until it was recently dispelled as inaccurate.[7]

The current era of analysis concerning the vitamin D content of milk is based on high-performance liquid chromatography (HPLC) coupled with competitive protein binding assay (CPBA). Osborn and Norman[8] published an initial report which stated that human milk contained substantial amounts of a 25-hydroxyvitamin D-like substance. Their results were based on competitive protein binding analysis. However, these results were in error because the authors failed to adequately purify their milk extracts prior to CPBA and as a result artifacts were measured. Hollis et al.[9] provided the first accurate quantitation of vitamin D and its major metabolites in human and bovine milk utilizing HPLC coupled with CPBA. Subsequent work by others supported these original results.[10,11] All of these later studies, which were based on HPLC and CPBA, supported the concept that vitamin D and its 25-hydroxylated metabolite, 25-hydroxyvitamin D, represented the only quantiatively important vitamin D compounds in native milk.[9,11] As a result of these findings, assessment of the dihydroxylated forms of vitamin D in milk is not considered of routine importance and thus will not be included as a topic in this review. We will now describe an update of previously published procedures which individually measures vitamins D_2, vitamin D_3, 25-hydroxyvitamin D_2 (25-OHD$_2$), and 25-hydroxyvitamin D_3 (25-OHD$_3$) in a small milk sample and is much simplified and improved compared to earlier versions.[9,12]

Materials and Methods

Sterols. [1α,2α-³H]Vitamin D_3 (12.3 Ci/mmol) was obtained from The Radiochemical Center (Amersham, Arlington Heights, IL). 25-OH[26,27-

[3] C. H. Nield, W. C. Russel, and A. Zimmerli, *J. Biol. Chem.* **136**, 73 (1940).
[4] Y. Sahashi, T. Suzuki, M. Higaski, and T. Asano, *J. Vitaminol.* **13**, 33 (1967).
[5] N. LeBoulch, C. Gulat-Marnay, and Y. Raoul, *Int. J. Vitam. Nutr. Res.* **44**, 167 (1974).
[6] D. R. Lakdawala and E. W. Widdowson, *Lancet* **1**, 167 (1977).
[7] B. W. Hollis, B. A. Roos, H. H. Draper, and P. W. Lambert, *J. Nutr.* **111**, 384 (1981).
[8] T. W. Osborn and A. W. Norman, *in* "Vitamin D: Biochemical, Chemical and Clinical Aspects Related to Calcium Metabolism" (A. W. Norman, K. Schaefer, H. G. Grigoleit, and D. von Iterrath, eds.), p. 523. de Gruyter, Berlin, 1977.
[9] B. W. Hollis, B. A. Roos, H. H. Draper, and P. W. Lambert, *J. Nutr.* **111**, 1240 (1981).
[10] L. E. Reeve, R. W. Chesney, and H. F. DeLuca, *Am. J. Clin. Nutr.* **36**, 122 (1982).
[11] Y. Weisman, J. C. Bawnik, Z. Eisenberg, and Z. Spirer, *J. Pediatr.* **100**, 745 (1982).
[12] B. W. Hollis, *Anal. Biochem.* **131**, 211 (1983).

^{3}H]D$_3$ (90 Ci/mmol) and 25-OH[26,27-^{3}H]D$_2$ (90 Ci/mmol) were gifts from Drs. R. L. Horst, T. A. Reinhardt (Ames, IO), and J. L. Napoli (Dallas, TX). Vitamins D$_2$, D$_3$ (Sigma, St. Louis, MO), 25-OHD$_3$ (Upjohn Co., Kalamazoo, MI), and 25-OHD$_2$ (Drs. Horst and Reinhardt) were obtained to serve as standards in the forthcoming procedures.

Solvents. HPLC grade methanol, dichloromethane, chloroform, ethyl acetate, hexane, isopropanol, and acetonitrile were purchased from Fisher Scientific (Pittsburgh, PA). All other reagents used in the procedure were of reagent grade unless otherwise noted.

Milk Extraction. Two milliliters of whole milk was placed in a 50-ml screw-top glass centrifuge tube. To each milk sample was added 1000 cpm of the following radioactive standards for monitoring the analytical recoveries of the assay: [^{3}H]vitamin D$_3$, 25-OH[^{3}H]D$_2$, and 25-OH[^{3}H]D$_3$. After vortex mixing of the milk samples, the lipids were extracted using potassium carbonate : ethyl acetate. To accomplish this extraction 0.17 milk volume of saturated potassium carbonate was added to each sample and mixed well. Next, 0.8 milk volume of distilled water was added, mixed well, and followed by the addition of 3 milk volumes ethyl acetate with further mixing. The samples were then centrifuged at 1000 g for 5 min at 20° followed by removal of the upper organic phase by aspiration. Three additional milk volumes of ethyl acetate were reintroduced to each sample and the procedure repeated. The organic phases were combined and dried under N$_2$ in preparation for the alkaline backwash.

Alkaline Backwash. The organic residue from the extraction procedure was resuspended in 2.5 ml of dichloromethane. The methylene chloride solution was then mixed with 5 ml of 0.2 M Na$_2$HPO$_4$ (pH 10.5). Following mixing, 5 ml of a mixture of chloroform : hexane (5 : 1, v/v) was added with further mixing. The samples were then centrifuged at 1000 g for 5 min at 20° followed by removal of the lower organic phase by aspiration. The organic layer was then dried under N$_2$ and stored in isopropanol at −20°.

Preparative Chromatography. Silica Sep-Pak preparative chromatography cartridges (Waters Associates, Milford, MA) are used to eliminate some contaminating lipids and also to separate vitamin D from 25-OHD (Fig. 1). If the organic extracts had been stored in isopropanol they should again be brought to complete dryness under N$_2$. The Sep-Pak cartridges are prepared and equilibrated by eluting 3 ml methanol, 4 ml isopropanol, and finally 7 ml hexane. The samples are applied to the Sep-Pak in 0.25 ml × 2 washes of 25% dichloromethane in hexane. The cartridges are then eluted under vacuum using a Sep-Pak cartridge rack (Waters Assoc.) using the following sequence : 9 ml hexane (discard), 3 ml 7% ethyl acetate in hexane (discard), 15 ml 7% ethyl acetate in hexane (vitamins D$_2$ and D$_3$), 2.5 ml 15% ethyl acetate in hexane (discard), and 9.5 ml 3%

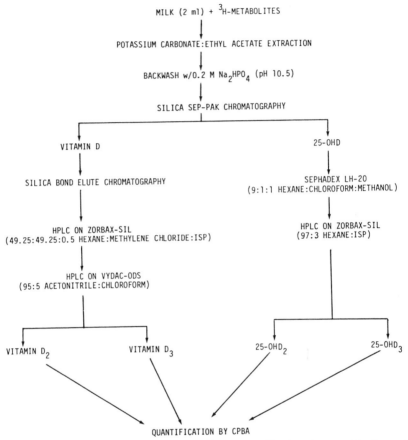

FIG. 1. Schematic of the purification and ultimate quantification steps used for the assay of vitamins D_2, D_3, 25-OHD$_2$, and 25-OHD$_3$ in human milk.

isopropanol in hexane (25-OHD$_2$ and 25-OHD$_3$). The fractions collected are dried under N_2 in preparation for further chromatography.

The vitamin D fraction is further purified using a silica Bond Elute cartridge (Analytichem International, Harbor City, CA). The Bond Elute columns are prepared and equilibrated by eluting with 4 ml 20% isopropanol in dichloromethane and finally 5 ml 0.2% isopropanol in dichloromethane. The vitamin D fraction is applied to the Bond Elute cartridge in 0.125 ml × 2 washes of 0.2% isopropanol in dichloromethane. The cartridge are then eluted under vacuum using the Sep-Pak cartridge rack with the following sequence: 2 ml 0.2% isopropanol in dichloromethane (discard), 8 ml 0.2% isopropanol in dichloromethane (vitamin D). The collection is dried under N_2 and stored in isopropanol at $-20°$ until further analysis.

The 25-OHD fraction eluted from the Sep-Pak cartridges was further purified on a column (0.7 × 12 cm) of Sephadex LH-20 (Pharmacia, Piscataway, NJ) developed in and eluted with hexane : chloroform : methanol (9 : 1 : 1, v/v) as previously described.[12]

The 25-OHD fraction, which contains both 25-OHD$_2$ and 25-OHD$_3$, from this chromatographic system was collected, dried under N_2, and stored at $-20°$ in isopropanol until further analysis.

High-Performance Liquid Chromatography. Final purification of vitamins D$_2$, D$_3$, 25-OHD$_2$, and 25-OHD$_3$, prior to quantitation by competitive CPBA was performed using various HPLC systems. Normal-phase HPLC was performed using a 0.4 × 25 cm Zorbax-Sil column (Dupont, Wilmington, DE) while reverse-phase HPLC was carried out using a 0.4 × 25 cm Vydac-ODS column (Separations Group, Hisperia, CA) on a Beckman Model 344 high-performance liquid chromatograph.

The vitamin D fraction from the Bond Elute cartridge was purified further using two separate HPLC systems (Fig. 1). The first system was normal-phase developed in hexane : dichloromethane : isopropanol (49.25 : 49.25 : 0.5, v/v) with a flow rate of 2 ml/min.[13] Vitamins D$_2$ and D$_3$ coeluted on this system with their retention time being approximately 8 min. The vitamin D peak from this system was collected and applied to the final HPLC system which was reverse-phase and separated vitamin D$_2$ from vitamin D$_3$. This system was developed in acetonitrile : chloroform (95 : 5, v/v) with a flow rate of 1.75 ml/min. Under these conditions vitamin D$_2$ eluted at approximately 10 min while vitamin D$_3$ eluted at approximately 12 min as depicted on Fig. 2A. The fractions containing vitamins D$_2$ and D$_3$ were collected independently, dried under N_2, and suspended in absolute ethanol for assay and recovery estimates. The 25-OHD fraction from the Sephadex LH-20 was purified by HPLC using a single normal-phase chromatographic system developed in hexane : isopropanol (97 : 3, v/v) with a flow rate of 2 ml/min. This system separates 25-OHD$_2$ from 25-OHD$_3$ which have retention times of approximately 8.5 and 10.5 min, respectively, under these conditions (Fig. 3A). These two fractions were collected independently, dried under N_2, and suspended in absolute ethanol for assay and recovery estimates.

Competitive Protein Binding Analysis. Vitamins D$_2$, D$_3$, 25-OHD$_2$, and 25-OHD$_3$ were quantitated by CPBA using previously described procedures.[12–15] Briefly, these assays utilize the rat plasma vitamin D-binding

[13] B. W. Hollis and W. B. Pittard, III, *J. Clin. Endocrinol. Metab.* **59,** 652 (1984).
[14] B. W. Hollis, B. A. Roos, and P. W. Lambert, *Steroids* **37,** 609 (1981).
[15] B. W. Hollis, *J. Steroid Biochem.* **21,** 81 (1984).

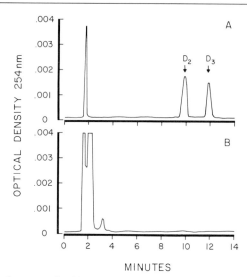

MINUTES

FIG. 2. High-performance liquid chromatography of standard vitamins D_2 and D_3 and milk extract profile. Column calibration was on a single Vydac ODS column (0.4 × 25 cm) by injecting 10 ng of both vitamins D_2 and D_3 in 100 μl of acetonitrile and monitoring optical density at 254 nm (A). The column was developed in and eluted with acetonitrile : chloroform (95 : 5, v/v) at a flow rate of 1.75 ml/min. A milk sample (2 ml) was extracted and processed thru the purification procedure and then applied to HPLC (B). Optical density was monitored at 254 nm.

protein diluted 1 : 35,000 in 0.05 M borate buffer (pH 8.4) containing 0.02% gelatin. Each assay mixture was placed into 12 × 75-mm borosilicate glass tubes and consisted of (1) 0.5 ml of 1 : 35,000 diluted rat plasma in above buffer; (2) standard vitamin D_2, D_3, 25-OHD_2, 25-OHD_3, or unknowns in 25 μl of ethanol; and (3) 5000 cpm of 25-OH[^{3}H]D_3 in 25 μl ethanol, however, addition of this final component to the assay tubes is delayed due to the use of nonequilibrium conditions.[14,15] For the vitamin D_2 and D_3 assays a 24-hr preincubation is used prior to the addition of 25-OH[^{3}H]D_3 which is permitted to incubate for an additional 1 hr prior to separation of bound and free sterol. The 25-OHD_2 and 25-OHD_3 assays are performed in a similar fashion except for only a 2-hr preincubation prior to the addition of tracer. Bound sterol was separated from free sterol using dextran-coated charcoal as previously described.[12] The standard curve range for vitamins D_2, D_3, 25-OHD_2, and 25-OHD_3 were 0.02–0.4, 0.01–0.25, 0.01–0.05, and 0.01–0.05 ng, respectively. The entire assay procedure is depicted in Fig. 1.

Calculations. Final concentrations of each sterol can be calculated by hand from a standard curve constructed from plotting cpm 25-OH[^{3}H]D_3

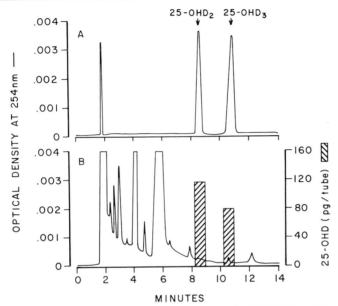

FIG. 3. High-performance liquid chromatography of standard 25-OHD₂ and 25-OHD₃ and measured 25-OHD₂ and 25-OHD₃ values of the HPLC fractions from human milk. Column calibration was achieved on a Zorbax-Sil column (0.4 × 25 cm) by injecting 15 ng of both 25-OHD₂ and 25-OHD₃ in 20 μl of hexane : isopropanol (97 : 3, v/v) and monitoring optical density at 254 nm (A). The column was developed in and eluted with the same solvent system at a flow rate of 2.0 ml/min. A milk sample (2 ml), which contained only endogenous 25-OHD₂ and 25-OHD₃, was first extracted and processed through the purification procedure and then applied to HPLC (B). One-minute fractions of effluent were collected into incubation tubes, and 25-OHD values measured by CPBA as described in the text.

bound vs standard concentration. Alternatively, the standard curve data may be transformed and calculated using a logit/log transformation.[16] The amount of unknown per tube is then corrected for dilution and final recovery with the results being expressed in picograms per milliliter of milk.

Discussion

The present procedure described in this text provides many improvements as compared to our previous publications.[9,12] First, the new organic extraction procedure using potassium carbonate : ethyl acetate provides quantitative recoveries of both vitamin D and 25-OHD from milk. This

[16] T. Chard, "An Introduction to Radioimmunoassay and Related Techniques," p. 440. North-Holland Publ., Amsterdam, 1978.

extraction procedure also provides the advantage of extracting fewer polar lipids than does previously utilized methods.[9,12] Fewer polar lipids in the organic extract allowed direct alkaline backwash of the extract without formation of emulsions. The lack of emulsions also allowed us to eliminate the lipid precipitation steps necessary in prior methods.[9,12]

The alkaline backwash procedure is a very important step in the overall assay and reliable results cannot be obtained without it. The primary purpose of the backwash is to remove free fatty acids from the initial lipid extract as they are known to interfere with the protein binding analysis of vitamin D and 25-OHD.[17] We have also observed that methylene chloride is the only suitable organic medium that allows adequate removal of free fatty acids from a milk extract. Therefore, when other organic mediums such as ether are used to accomplish this purpose caution should be exercised.[18]

Preparative chromatography of the milk extract on silica Sep-Pak cartridges has been slightly modified as compared to our earlier procedure.[12] This chromatographic step is effective in separating vitamin D from 25-OHD and also in removing substantial amounts of extraneous lipid material. The use of the Sep-Pak cartridge rack has also contributed to a significant increase in sample processing capacity. Incorporation of an additional preparative chromatographic step utilizing silica Bond Elute cartridges has also decreased the lipid load prior to introducing the sample extract onto the HPLC column. This decrease in lipid concentration allows a more rapid sample processing time on the HPLC and tends to extend column life.

One of the more significant enchancements in the present assay system involves the baseline separation of vitamins D_2 and D_3 on a single reverse-phase HPLC column. In order to accomplish this separation in the past two reverse-phase columns in tandem had to be used.[12] This previous separation procedure was tedious as well as being very hard on the HPLC equipment, often generating pressures of 3500 psi at flow rates of only 1.25 ml/min. The new separation procedure is accomplished in 14 min and at low column backpressures, <1000 psi. Also, the new separation system eliminates the use of water in the eluting mobile phase. Comparisons of the new and old separation procedures are depicted in Figs. 2 and 4, respectively. Through the entire extraction and purification procedure one can expect to recover approximately 40 and 50% of vitamin D and 25-OHD, respectively.

[17] B. W. Hollis, P. W. Lambert, and R. L. Horst, in "Perinatal Calcium and Phosphorus Metabolism" (M. F. Holick et al., eds.), p. 157. Elsevier, Amsterdam, 1983.
[18] J. E. Bishop, A. W. Norman, J. W. Coburn, P. A. Roberts, and H. L. Henry, Miner. Electrolyte Metab. 3, 181 (1980).

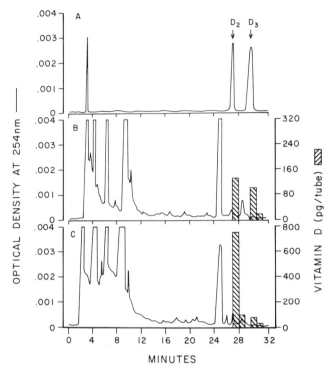

FIG. 4. High-performance liquid chromatography of standard vitamins D_2 and D_3 and measured vitamins D_2 and D_3 values of the HPLC fractions from human milk. Column calibration was achieved on two ultrasphere ODS columns (0.4 × 25 cm) in series by injecting 20 ng of both vitamins D_2 and D_3 in 20 μl of methanol and monitoring optical density at 254 nm (A). The columns were developed in and eluted with methanol: water (98 : 2, v/v) at a flow rate of 1.25 ml/min. Milk samples (2 ml), which contained only endogenous vitamins D_2 and D_3, were first extracted and processed thru the purification procedure and then applied to HPLC (B and C). One-minute fractions of effluent were collected into incubation tubes, and vitamin D values measured by CPBA as described in the text.

Final quantitation of all vitamin D compounds is performed using CPBA as previously described.[12,15] These assays are all performed in a nonequilibrium fashion as to greatly increase their sensitivity. Another advantage of nonequilibrium analysis is that since one can introduce less sample extract into the assay tube for the determination of the desired compound lipid interference is less likely to result and thus contribute to spurious results.[8]

The present procedure has been used in a wide range of studies to determine not only the normal vitamin D content of milk but also to determine how the antirachitic potential of human milk responds to vari-

ous stimulus including dietary supplementation and ultraviolet photo-
therapy.[9,17,19–21]

Acknowledgments

This work was supported in part by Grant HD-16510 and Research Career Development
Award HD-000479 from the NIH. I gratefully acknowledge Pamela Woodruff for the typing
of this manuscript.

[19] B. W. Hollis, F. R. Greer, and R. C. Tsang, *Calcif. Tissue Int.* **34**, 552 (1982).
[20] F. R. Greer, B. W. Hollis, and J. L. Napoli, *J. Pediatr.* **105**, 61 (1984).
[21] F. R. Greer, B. W. Hollis, D. J. Cripps, and R. C. Tsang, *J. Pediatr.* **105**, 431 (1984).

[19] 1,25-Dihydroxyvitamin D Microassay Employing Radioreceptor Techniques

By TIMOTHY A. REINHARDT and BRUCE W. HOLLIS

It is well established that 1,25-dihydroxyvitamin D [1,25-$(OH)_2$D] is a
steroid hormone produced by the kidney, whose production is tightly
regulated and affected by a large number of disease states involving altered vitamin D metabolism.[1–8] Because changes in circulating 1,25-
$(OH)_2$D concentrations are pivotal in the physiopathological importance
of a number of diseases of calcium and vitamin D metabolism, considerable effort has been directed toward the development of specific assays

[1] J. L. Napoli and H. F. DeLuca, *in* "Berger's Medicinal Chemistry" (M. E. Wolff, ed.),
Part II, p. 705. Wiley, New York, 1979.
[2] M. R. Haussler and T. A. McCain, *N. Engl. J. Med.* **297**, 974 (1977).
[3] A. W. Norman, "Vitamin D: The Calcium Homeostatic Steroid Hormone." Academic
Press, New York, 1979.
[4] R. W. Gray, *Calcif. Tissue Int.* **33**, 477 (1981).
[5] R. L. Horst, H. F. DeLuca, and N. A. Jorgensen, *Metab. Bone Dis. Relat. Res.* **1**, 29
(1978).
[6] A. E. Broadus, R. L. Horst, R. Lang, E. T. Littledike, and H. Rasmussen, *N. Engl. J.
Med.* **302**, 421 (1980).
[7] G. L. Barbour, J. W. Coburn, E. Slatopolsky, A. W. Norman, and R. L. Horst, *N. Engl.
J. Med.* **305**, 440 (1981).
[8] T. A. Reinhardt and H. R. Conrad, *J. Nutr.* **110**, 1589 (1980).

for 1,25-(OH)₂D.[9-16] The development by Brumbaugh *et al.*[10] of the first radioreceptor assay to measure plasma 1,25-(OH)₂D levels represented a major advance in vitamin D technology. The application by Eisman *et al.*[11] of high-performance liquid chromatography to the purification of 1,25-(OH)₂D prior to assay represented the next simplification in the assay of 1,25-(OH)₂D in plasma and has provided the basis for most of the assays developed in succeeding years.[11-15]

However, the measurement of 1,25-(OH)₂D has continued to be difficult, requiring large sample volumes, tedious extraction and purification procedures, and a large dependence on expensive instrumentation such as high-performance liquid chromatography or tissue culture equipment.

Our purpose in developing the following assay was severalfold: (1) to employ a widely available (thymus glands from slaughter houses), inexpensive, and stable binding protein which exhibits high sensitivity and high specificity for both 1,25-(OH)₂D₃ and 1,25-(OH)₂D₂, (2) to reduce the sample size required for assay, (3) to reduce solvent requirements, (4) to reduce assay time per sample, and (5) most importantly, to eliminate the high technical and instrumentational requirements associated with most present 1,25-(OH)₂D assays.

Reagents

Synthetic 1,25-(OH)₂D₃ and 1,25-(OH)₂D₂ were generous gifts from Dr. Milan Uskokovic and Dr. John Partridge of Hoffmann-La Roche, Inc. (Nutley, NJ). 1,25-Dihydroxy[23,26-³H]vitamin D₃ (80 Ci/mmol) and 1,25-(OH)₂[26,27-³H]D₂ (80 Ci/mmol) were prepared in our laboratory by C³H₃MgI reduction of 27-nor-25-keto-1-OHD₃ and 27-nor-25-keto-1-

[9] T. A. Reinhardt, R. L. Horst, J. W. Orf, and B. W. Hollis, *J. Clin. Endocrinol. Metab.* **58,** 91 (1984).

[10] P. F. Brumbaugh, D. H. Haussler, K. M. Bursac, and M. R. Haussler, *Biochem. J.* **13,** 4091 (1974).

[11] J. A. Eisman, A. J. Hamstra, B. E. Kream, and H. F. DeLuca, *Arch. Biochem. Biophys.* **176,** 235 (1976).

[12] R. L. Horst, E. T. Littledike, J. L. Riley, and J. L. Napoli, *Anal. Biochem.* **116,** 189 (1981).

[13] S. Dokoh, J. W. Pike, J. S. Chandler, J. M. Mancini, and M. R. Haussler, *Anal. Biochem.* **116,** 211 (1981).

[14] J. E. Bishop, A. W. Norman, J. W. Coburn, P. A. Roberts, and H. L. Henry, *Miner. Electrolyte Metab.* **3,** 81 (1980).

[15] M. R. Hughes, D. J. Baylink, P. G. Jones, and M. R. Haussler, *J. Clin. Invest.* **58,** 61 (1976).

[16] S. C. Manolagas, F. L. Fuller, J. E. Howard, A. S. Brinkman, and L. J. Deftos, *J. Clin. Endocrinol. Metab.* **56,** 251 (1983).

OHD$_2$, respectively. The purity and concentrations of 1,25-(OH)$_2$D$_3$ and 1,25-(OH)$_2$D$_2$ were confirmed by UV spectroscopy using an extinction coefficient (E_{264}) of 18,300 M^{-1} cm^{-1} for 1,25-(OH)$_2$D$_3$ and E_{264} of 19,400 M^{-1} cm^{-1} for 1,25-(OH)$_2$D$_2$. All solvents used were HPLC grade and, unless noted, all other reagents were reagent grade.

Preparation of Thymus Cytosol

Thymus gland cytosol is prepared as previously described with modifications.[9,17] Thymus glands are removed from 1- to 6-month-old calves at the time of euthanasia or slaughter and immediately placed in ice-cold phosphate-buffered saline for transport to the laboratory. Thymus glands are then trimmed of all blood and blood clots and cut into small pieces (~1 cm^3). Next, cubed thymus tissue is placed into a Buchner funnel and washed with 1 liter of ice-cold phosphate-buffered saline for every 100 g of thymic tissue. At this point, the tissue may either be processed immediately for cytosol or quick frozen in liquid nitrogen and stored at $-20°$ until used for the preparation cytosol. Frozen thymic tissue maintains 1,25-(OH)$_2$D receptor activity for up to 1 year.

Cytosol is then prepared from fresh or frozen glands as follows (all steps were carried out at 4°): thymus glands are homogenized (25% w/v) in a buffer containing 50 mM K$_2$HPO$_4$, 400 mM KCl, 5 mM dithiothreitol, and 1.5 mM EDTA, pH 7.5 by five 30-sec bursts of a Polytron PT-20 tissue disrupter (Brinkmann Instruments, Westbury, NY) at setting 7–8. The tissue is cooled on ice between homogenizations. Homogenates are centrifuged for 15 min at 20,000 g to remove large particulate matter. The supernatant is then centrifuged for 1 hr at 300,000 g and the cytosol (minus pellet and floating lipid layer) is removed and fractionated by the slow addition of solid (NH$_4$)$_2$SO$_4$ (enzyme grade, Schwarz-Mann, Orangeburg, NY) to 35% saturation. [Note: nonenzyme grade (NH$_4$)$_2$SO$_4$ may be used at this point if cytosol pH is rigorously monitored and kept at pH 7.5.] Cytosol is then stirred slowly for 30 min while maintaining the temperature at 4°. Cytosol is then aliquoted into 15-ml centrifuge tubes (7 ml) and centrifuged at 20,000 g for 20 min. The supernatant is discarded and the tubes are allowed to drain on tissue paper for 5 min. At this point, the pellets may be stored one of two ways: (1) the pellets are stored under nitrogen at $-20°$, or (2) for additional stability, the pellets are lyophilized overnight and then stored at $-20°$ under argon. Either method has proven satisfactory in our hands, with the second method proving greater stability at room temperature (greater than 60 hr without loss of activity).

[17] T. A. Reinhardt, R. L. Horst, E. T. Littledike, and D. C. Beitz, *Biochem. Biophys. Res. Commun.* **106,** 1012 (1982).

Prior to assay, one pellet is redissolved in 7 ml of 50 mM K$_2$HPO$_4$, 150 mM KCl, 5 mM DTT, and 1.5 mM EDTA, pH 7.5 (assay buffer) by mixing on ice using a small magnetic stir bar and magnetic stirring plate. Excessive vortexing should be avoided as the foaming this generates is destructive to the receptor. Typically, a small portion of the protein pellet resists solubilization. This insoluble matter is removed by centrifugation and in no way affects the assay. This receptor solution is then diluted 1 : 7 with the assay buffer and used in the assay at 470 μl/tube. The correct dilution of the stock receptor solution for assay should be determined empirically for each new batch of pellets prepared; however, we find very little batch-to-batch variability using thymuses from animals 1–6 months old.

Dextran-Coated Charcoal Preparation

Six grams of charcoal (Norit A, Sigma Chemical Co., St. Louis, MO) and 0.6 g Dextran T-70 (Pharmacia Fine Chemicals, Piscataway, NJ) are suspended in 500 ml of 0.1 M boric acid plus 0.05% BSA, pH 8.6, and stirred slowly overnight at 4°. The next morning the suspension is centrifuged at 1500 g for 20 min, and the supernatant carefully decanted and discarded. Pelleted dextran-coated charcoal is then resuspended in 500 ml of the buffer described. This dextran-coated charcoal is stable for at least 2 weeks at 4°.

Sep-Pak Preparation

New and used Sep-Paks (Waters Associates, Inc., Milford, MA) are prepared as follows prior to use. C$_{18}$ Sep-Paks are washed sequentially with 5 ml hexane, 5 ml chloroform, 5 ml methanol, and 5 ml distilled water. Silica Sep-Paks are washed sequentially with 5 ml of methanol, 5 ml chloroform, 5 ml hexane, and 5 ml hexane : isopropanol (96 : 4). The conditioning procedure allows the reuse of Sep-Pak cartridges 2–3 times without loss of capacity or changes in elution patterns. Labs with humidity problems may find that the silica Sep-Paks are good for only one run. In addition, reuse of either Sep-Pak type beyond 2–3 times may lead to failure due to the chloroform washes breaking down the plastic the Sep-Paks are packed in, thus changing the bed structure.

Sample Extraction and Purification

Serum or plasma samples (0.2–1.0 ml) are placed in 12 × 75 mm borosilicate glass tubes and adjusted to 1 ml final volume with saline. Eight hundred cpm of 1,25-(OH)$_2$[^{3}H]D$_3$ (80 Ci/mmol) in 20 μl of ethanol is added to each sample and to a scintillation vial for monitoring of recoveries. The samples are then vortexed and allowed to stand for 10 min.

Extraction of vitamin D metabolites is done by the addition of one volume of acetonitrile to each sample. The samples are vortexed vigorously for 20 sec, followed by centrifugation for 15 min at 1500 g. Following centrifugation, the supernatant is decanted into a tube containing 0.5 vol of 0.4 M K_2HPO_4, pH 10.6, and vortexed. This extract is then applied directly to a prewashed C_{18} Sep-Pak. Excess salt is removed by washing the Sep-Pak with 5 ml of distilled water and polar lipids are removed with a 2.5 ml methanol : water (70 : 30) wash. The vitamin D metabolites are then eluted with 4 ml of acetonitrile and this fraction is dried under a stream of nitrogen. The dried extract is then redissolved in 0.5 ml of hexane : isopropanol (96 : 4) and applied to a silica Sep-Pak. An additional 0.5 ml of solvent is used to wash the extract tube and applied to the Sep-Pak. The cartridge is washed with 9.0 ml of starting solvent which elutes 25-OHD. Next, the cartridge is washed with 8 ml hexane : isopropanol (94 : 6), which removes 60–70% of the 24,25-$(OH)_2D$ and additional nonpolar lipids. The 1,25-$(OH)_2D$ fraction is then eluted with 9 ml of hexane : isopropanol (84 : 16) and dried under a stream of nitrogen. The 1,25-$(OH)_2D$ fraction is redissolved in 100 μl of absolute ethanol and 25 μl is counted to estimate recovery and two 25-μl aliquots are used to perform the assay in duplicate.

Radioreceptor Assay: Nonequilibrium

Standards and samples in duplicate (0, 1, 2, 4, 8, 16, 32, 64, and 800 pg for nonspecific binding determinations) are added in 25 μl of ethanol to 12 × 75 mm borosilicate glass tubes on ice. This is followed by the addition of 470 μl (~0.7 mg protein/tube) receptor (reconstituted and diluted as described above) to the standards and samples. The samples and standards are then vortexed and placed in a 25° water bath for 45 min with gentle shaking. After 45 min, the tubes are transferred to an ice bath and allowed to cool for 5 min followed by the addition of 5000 cpm of 1,25-$(OH)_2[^3H]D_3$ (80 Ci/mmol) in 25 μl of ethanol. The tubes are vortexed and the incubation is continued for an additional 15 min at 25° with gentle shaking. At the end of this second incubation, the tubes are allowed to cool for 5 min in an ice bath and 200 μl of the dextran-coated charcoal suspension described above is added to each tube, followed by vortexing. The tubes are vortexed again after 10 min and after 20 min of charcoal treatment, bound and free 1,25-$(OH)_2D$ is separated by centrifugation at 2000 g for 15 min. The supernatant containing the bound hormone is decanted into scintillation vials and counted.

We have found that the optimal assay size using this procedure is 68 tubes. This represents 24 samples and 10 standards, all done in duplicate.

It has been our experience that large assays lead to systematic errors due to large differences in time between the first tube being treated with charcoal and the last tube being treated. Therefore, in order to maintain a high degree of quality control, one should not exceed these assay size constraints without first testing them in one's own lab. Additionally, one or two internal lab standards should be run in every assay.

Calculations

The 1,25-(OH)$_2$D values (pg/tube) are calculated using a logit/log plot of the binding assay data.

$$\text{logit} = \ln\left(\frac{B/B_0}{1 - B/B_0}\right) \tag{1}$$

For our purposes, this is done using a Hewlett-Packard HP41CV calculator and the RIA program contained in their Clinical Lab and Nuclear Medicine Program Pak (Hewlett-Packard, Palo Alto, CA). This procedure provides optimal linearization of our binding assay data. The 1,25-(OH)$_2$D concentrations (pg/ml) of plasma or serum are then obtained by correcting pg/tube data for recovery starting sample volume and plasma equivalent per assay tube.

Assay Characteristics and Validation

Extraction and purification of 1,25-(OH)$_2$D from plasma or serum samples is accomplished by protein precipitation with acetonitrile followed by solid-phase extraction of the 1,25-(OH)$_2$D from the acetonitrile-K$_2$HPO$_4$-treated supernatant on a C$_{18}$ Sep-Pak. Polar plasma lipids, pigments, and salts are removed by the water and methanol : water (70 : 30) washes. Approximately 85% of the sample 1,25-(OH)$_2$D is recovered in the 4 ml of acetonitrile used for the final elution of the C$_{18}$ Sep-Pak. The addition of 0.4 M K$_2$HPO$_4$ (pH 10.6) to the original acetonitrile extract, prior to application of the sample to the C$_{18}$ Sep-Pak, is particularly important when assaying samples from patients with varying degrees of hyperlipidemia. This addition of basic K$_2$HPO$_4$ greatly enhances the removal of lipids from the 1,25-(OH)$_2$D fraction which might otherwise interfere in the assay.

Following solid-phase extraction, nonpolar lipids, 25-OHD and 24,25-(OH)$_2$D, are removed from the 1,25-(OH)$_2$D fraction using silica Sep-Paks. The data in Fig. 1 show that 25-OHD is completely removed in the first 9–10 ml of elution with hexane : isopropanol (96 : 4). The majority of 24,25-(OH)$_2$D remaining in the sample is then eluted with 8 ml of hexane : isopropanol (94 : 6) followed by the elution of both 1,25-(OH)$_2$D$_3$ and

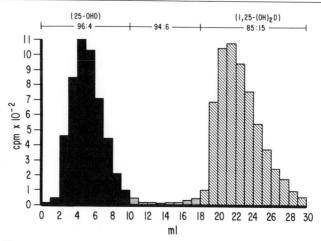

FIG. 1. Chromatogram of silica Sep-Pak elution pattern for 25-OHD₃ and 1,25-(OH)₂D₃. The sample 25-OHD is eluted in the first 10 ml of hexane : isopropanol (96 : 4). The next 8 ml of hexane : isopropanol (94 : 6) elutes the majority of 24,25-(OH)₂D. Finally, the sample 1,25-(OH)₂D is eluted with 9 ml of hexane : isopropanol (84 : 16).

$1,25\text{-}(OH)_2D_2$ with 9 ml of hexane : isopropanol (84 : 16). An average of 70% of the sample $1,25\text{-}(OH)_2D$ is recovered at this point for assay.

The sensitivity of the thymus-based radioreceptor assay was optimized using a nonequilibrium incubation technique. Using this technique, the sensitivity of the assay, defined at 2 SD of the zero tube, was 1.5 pg/tube. Fifty percent displacement of bound trace occurs approximately at 8–9 pg/tube (Fig. 2). The unique character of the thymus $1,25\text{-}(OH)_2D$ receptor used in this assay is that it recognizes both $1,25\text{-}(OH)_2D_3$ and $1,25\text{-}(OH)_2D_2$ equally, while it detects other vitamin D metabolites poorly. Cross-reactivities compared to $1,25\text{-}(OH)_2D$ are 0.1% for 25-OHD, 0.02% for $24,25\text{-}(OH)_2D$, and 0.008% for $25,26\text{-}(OH)_2D_3$. Because of the low recoveries of these metabolites in the $1,25\text{-}(OH)_2D$ fraction, cross-reactivities, and a maximum of 0.33 ml plasma equivalent per assay tube, we have determined that normal and above normal circulating concentrations of 25-OHD, $24,25\text{-}(OH)_2D$, and $25,26\text{-}(OH)_2D$ do not interfere with $1,25\text{-}(OH)_2D$ measurement. Experiments in which normal human plasma samples were spiked with $24,25\text{-}(OH)_2D$ or $25,26\text{-}(OH)_2D$ at 5 times normal levels and assayed for $1,25\text{-}(OH)_2D$ showed no difference between the values obtained for $1,25\text{-}(OH)_2D$ between spikes and control samples.

In validating our assay method we have determined that the analytical recovery of $1,25\text{-}(OH)_2D_2$ and $1,25\text{-}(OH)_2D_3$ added to plasma samples (8– 64 pg of $1,25\text{-}(OH)_2D$ added per ml plasma) averaged 99.0 ± 6.2% for

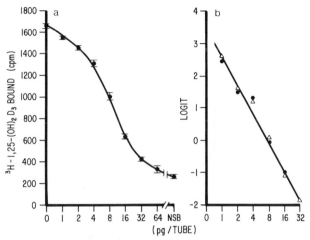

FIG. 2. Standard curve obtained for 1,25-(OH)$_2$D using the thymus 1,25-(OH)$_2$D receptor and nonequilibrium incubation conditions (a). (b) Linearization of standard curves for 1,25-(OH)$_2$D afforded by a logit–log transformation. (●) 1,25-(OH)$_2$D$_3$; (△) 1,25-(OH)$_2$D$_2$.

added 1,25-(OH)$_2$D$_2$ and 93.3 ± 7.9% (mean ± SEM) for added 1,25-(OH)$_2$D$_3$, thus, indicating that our assay accurately measures total 1,25-(OH)$_2$D and that 1,25-(OH)$_2$[^{3}H]D$_3$ accurately estimates extraction and purification recoveries for both 1,25-(OH)$_2$D$_2$ and 1,25-(OH)$_2$D$_3$. This is in contrast to other reported assays which underestimate 1,25-(OH)$_2$D$_2$ concentrations by 18–61%.[13,15,16,18] In light of our recent finding that 25-OHD$_2$ constitutes as much as 39% of the total 25-OHD circulating in patients in which 1,25-(OH)$_2$D assays are most often applied, it is clear that it is important to obtain accurate total 1,25-(OH)$_2$D measurements as are obtained by this assay.[9]

The ultimate validation of this simplified method for 1,25-(OH)$_2$D assays was obtained by direct comparison of the results obtained with this assay with well-established assay employing rigorous purification of sample 1,25-(OH)$_2$D using Sephadex LH-20 and HPLC chromatography prior to assay.[12] Human plasma samples from patients with a wide variety of clinical problems were assayed by both methods and, as shown in Fig. 3, there was an excellent correlation between the results obtained by the established method and our simplified method. The correlation coefficient for this comparison was 0.96, with a slope of 1.05 and an intercept of −0.19.

[18] G. Jones, B. Byrnes, F. Palna, D. Segev, and Y. Mazur, *J. Clin. Endocrinol. Metab.* **50**, 773 (1980).

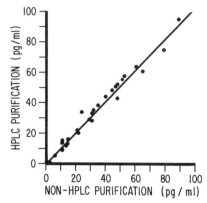

FIG. 3. Comparison of 1,25-(OH)₂D values obtained using the assay described with values obtained using a well-established assay which employs both Sephadex LH-20 chromatography and HPLC purification of the sample 1,25-(OH)₂D prior to assay. (From Reinhardt et al.[9])

The present 1,25-(OH)$_2$D assay has been validated in a number of ways. Dilution tests using human plasma yield linear results over a wide range of dilutions. The reproducibility of the assay was confirmed by its low intra (6.8%) and interassay (11.8%) coefficients of variations. The table summarizes the results obtained by the application of this assay to clinical studies. The mean values for 1,25-(OH)$_2$D obtained by using this simplified methodology are in excellent agreement with 1,25-(OH)$_2$D values obtained by labs using more technically demanding methodology.[4,6,7,10–16]

CONCENTRATIONS OF 1,25-(OH)₂D IN NORMAL SUBJECTS AND PATIENTS
DETERMINED BY THE NON-HPLC ASSAY[a]

	Concentration of 1,25-(OH)₂D (pg/ml)						
Sample	Normals	Children	Infant cord blood	Chronic renal failure	Anephrics	Primary hyperpara-thyroidism	Rats
Mean	37.4	50.5	22.9	10.6	Undetectable[b]	68.9	86.6
SEM	± 2.2	± 5.5	± 4.4	± 1.5	—	± 5.5	± 4.0
Range	21–65	29–92	10–45	<5–16	—	39–115	72–98
n	22	14	7	7	10	13	6

[a] Adapted from Reinhardt et al.[9]
[b] Detection limit of the assay is 5 pg of 1,25-(OH)₂D per ml plasma.

In summary, the radioreceptor assay for 1,25-(OH)$_2$D described here simplifies the quantitation of 1,25-(OH)$_2$D. The unique features of the assay are a widely available, inexpensive, and stable receptor protein with both high sensitivity and specificity; small sample requirements, simultaneous measurement of both 1,25-(OH)$_2$D$_2$ and 1,25-(OH)$_2$D$_3$, and the elimination of the need for HPLC or tissue culture equipment and technical expertise. The assay presented is rapid and useful for quantitation of 1,25-(OH)$_2$D in human and experimental animal samples. Finally, the small sample requirements of the assay may be useful in pediatric studies where sample size is limited.

[20] Assay for 1,25-Dihydroxyvitamin D Using Rabbit Intestinal Cytosol-Binding Protein

By W. E. DUNCAN, T. C. AW, and J. G. HADDAD

Most competitive binding assays for 1,25-hydroxyvitamin D$_3$ [1,25-(OH)$_2$D$_3$] have used intestinal cytosol from rachitic chickens as the source of binding protein.[1–6] These chicks are usually raised for 4–12 weeks on a rachitogenic diet prior to preparation of the binding protein. Such chicks are expensive to maintain and, in the experience of some investigators, a significant number die before the vitamin D-deficient intestinal mucosa can be obtained. We describe here a competitive binding assay for 1,25-(OH)$_2$D$_3$ using easily obtained binding protein from the small intestines of vitamin D-sufficient rabbits.

[1] P. F. Brumbaugh, D. H. Haussler, R. Bressler, and M. R. Haussler, *Science* **183**, 1089 (1974).

[2] J. A. Eisman, A. J. Hamstra, B. E. Kream, and H. F. DeLuca, *Arch. Biochem. Biophys.* **176**, 235 (1976).

[3] S. Dokoh, R. Morita, M. Fukunaga, I. Yamamoto, and K. Torizuka, *Endocrinol. Jpn.* **25**, 431 (1978).

[4] R. M. Shepard, R. L. Horst, A. J. Hamstra, and H. F. DeLuca, *Biochem. J.* **182**, 55 (1979).

[5] J. P. Mallon, J. G. Hamilton, C. Nauss-Karol, K. J. Karol, C. J. Ashley, D. S. Matuszewski, C. A. Tratnyek, G. F. Bryce, and O. N. Miller, *Arch. Biochem. Biophys.* **201**, 277 (1980).

[6] M. J. M. Jongen, W. J. F. van der Vijgh, H. J. J. Williams, and J. C. Netelenbos, *Clin. Chem. (Winston-Salem, N.C.)* **27**, 444 (1981).

Cytosol Preparation

Male, New Zealand, white rabbits were anesthesized by an iv injection of sodium pentobarbital (50 mg/kg), and five 15-cm segments of small intestine immediately distal to the gastroduodenal junction were excised. All subsequent steps were performed in a 5° cold room or on ice. After flushing the intestine with iced 10 mM sodium phosphate in 0.15 M NaCl, pH 7.4 (PBS), the mucosa was scraped free from the underlying serosa and washed twice with cold PBS. The washed mucosa was homogenized with a motor-driven Teflon pestle in two volumes of buffer containing 26 mM Tris, 5 mM dithiothreitol, 1 mM EDTA, 10 mM sodium molybdate, and 300 mM KCl, pH 7.4 (hypertonic buffer). The homogenate was centrifuged at 25,000 g for 10 min and then the supernatant was centrifuged at 100,000 g for 60 min. The resulting supernatant fraction was carefully removed, avoiding the lipid layer, and stored under argon at $-80°$. The average yield of cytosol was 28 ± 3 ml containing 12 ± 1 mg protein/ml (mean $\pm$ SEM for 5 rabbits). Protein concentrations were determined by a protein dye binding method[7] using bovine serum albumin as the standard.

1,25-(OH)$_2$D Extraction and Chromatography

Plasma (2–3 ml) was incubated with 700 dpm of tritiated 1,25-(OH)$_2$D$_3$ (1,25-dihydroxy[23,24(n)-^{3}H]cholecalciferol, 91 Ci/mmol, Amersham Corporation, Arlington Heights, IL) at room temperature for 30 min in a Teflon-lined, screw-capped tube. Five volumes of diethyl ether were added, and the tubes were shaken horizontally for 60 min in an Eberbach shaker. The aqueous bottom layer was frozen at $-80°$ in a dry ice/acetone bath, and the ether was decanted and dried under a stream of nitrogen. The ethereal residue was solubilized in 1 ml of anhydrous diethyl ether and applied to 3 cm columns of silicic acid prepared in the same solvent in disposable glass Pasteur pipets. Alternatively, samples were applied to Sep-Pak silicic acid cartridges (Waters Associates, Milford, MA) and processed similarly. Six milliliters of ether was washed through the columns and discarded prior to elution with 8 ml of methanol : ether (5 : 95, v/v). The latter eluates were dried under a stream of nitrogen or in a heater-vacuum-vortex apparatus (Vortex Evaporator, Buchler, Fort Lee, NJ) at 37°.

The methanol–ether eluates were subjected to high-performance liquid chromatography on a Zorbax-Sil column with n-hexane : isopropanol (88 : 12, v/v) at 1,056 psi and a 2 ml/min flow rate. Under these conditions,

---

[7] M. Bradford, *Anal. Biochem.* **72**, 248 (1976).

1,25-$(OH)_2D_3$ standards elute at 8.0 min. The 1,25-$(OH)_2D_3$ region was collected, dried under nitrogen, and solubilized in ethanol for assay and estimation of recovery.

Assay Conditions

Dilutions of intestinal cytosol (1 ml, 0.2 mg of protein) were added to 25 μl of ethanol containing 56 pM tritiated 1,25-$(OH)_2D_3$ with either an aliquot of the unknown sample or increasing concentrations of nonradioactive sterol solubilized in 25 μl of absolute ethanol. This mixture was incubated for 3 hr at 4°. The 1,25-$(OH)D_3$ solutions were standardized according to their extinctions at 264 nm in absolute ethanol. Free and bound sterol were separated by adding 0.25 ml of a cold dextran-coated charcoal solution [1.25 mg charcoal (Norit-A) and 0.125 mg of Dextran 20 suspended in hypertonic buffer per tube] to the cytosol–sterol solution. After incubation for 10 min at 4° and centrifugation at 1000 g for 10 min at 4°, 1 ml of the supernatant was transferred to a counting vial containing 7 ml of Budget-Solve scintillation liquid (Research Products International Corp., Mount Prospect, IL) and counted in a Packard Tri-Carb 460 CD liquid scintillation counter.

Figure 1 illustrates a competitive binding standard curve for 1,25-$(OH)_2D_3$. Addition of 1–2.5 pg of 1,25-$(OH)_2D_3$ reproducibly displaced the tracer sterol from this binding protein. The binding activity of the

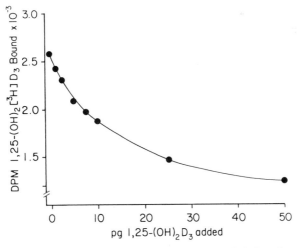

FIG. 1. Competitive binding assay standard curve for 1,25-$(OH)_2D_3$ using the vitamin D-sufficient rabbit intestinal cytosol-binding protein. The results are shown as the mean of duplicate determinations. Data from Duncan *et al.*[8]

TABLE I

HUMAN PLASMA 1,25-(OH)$_2$D CONCENTRATIONS
USING THE RABBIT INTESTINAL CYTOSOL RECEPTOR[a]

Group	N[b]	1,25-(OH)$_2$D (pg/ml)
Normal subjects (18–45 years)	20	34.7 ± 10.7
Chronic renal disease	5	13.6 ± 5.5[c]
Primary hyperparathyroidism	6	69.5 ± 40.8[d]

[a] From Duncan et al.[8]
[b] Number of determinations.
[c] $p < 0.001$ compared to normal subjects.
[d] $p < 0.01$ compared to normal subjects.

intestinal cytosol was stable after 8 months of storage at $-80°$. Human plasma samples were extracted and chromatographed twice, with 60–80% recovery of the tracer sterol. Compared to normal subjects, patients with chronic renal disease (glomerular filtration rates less than 15 ml/min) had lower concentrations and patients with primary hyperparathyroidism had higher concentrations of 1,25-(OH)$_2$D (Table I).[8] To optimize the yield of binding protein, we examined the binding capacity for 1,25-(OH)$_2$D of intestinal cytosol from rabbits of various ages (Table II).[9] While significant increases in receptor content (N_{max}) were observed in very young animals (4 and 6 weeks old), the total yield of binding protein was less than that obtained from older animals. Using 200 μg of cytosol protein per tube, we found that a 2.0–3.5 kg rabbit (6–8 months old) yields enough binding for approximately 1500 assay tubes.

Discussion

We have previously demonstrated that the rabbit intestinal cytosol receptor for 1,25-dihydroxyvitamin D$_3$ is identical to the intestinal receptor from chicks, mice, rats, and humans with respect to sedimentation coefficient, dissociation constant, relative affinity for vitamin D metabolites, and affinity for diethylaminoethyl (DEAE) exchange resin.[8] Because of the plentiful supply of cytosol receptor in rabbit intestine, these animals provide a convenient source of binding protein for the 1,25-(OH)$_2$D

[8] W. E. Duncan, T. C. Aw, P. G. Walsh, and J. G. Haddad, Anal. Biochem. 132, 209 (1983).
[9] W. E. Duncan, P. G. Walsh, M. A. Kowalski, and J. G. Haddad, Comp. Biochem. Physiol. A 78A, 333 (1984).

TABLE II

BINDING CAPACITY FOR 1,25-$(OH)_2D$ OF RABBIT
INTESTINAL CYTOSOL[a,b]

Rabbit (N)	N_{max} (fmol/mg protein)
Term fetus[c]	Not detected
2 weeks (3)	137 ± 13
4 weeks (3)	289 ± 19***
6 weeks (3)	239 ± 36*
10 weeks (2)	204 ± 13
24 weeks (3)	124 ± 22
32 weeks (3)	86 ± 15**
Adult (2)	153 ± 29

[a] Data from Duncan et al.[9]

[b] Each intestinal cytosol was analyzed by saturation analysis in three separate experiments. The experimental results are expressed as the mean ± SD of cytosol values from separate animals. The binding capacity (N_{max}) was determined by triplicate Scatchard analyses of cytosol prepared from each animal. The number of animals are in parentheses. Significant difference from each adult value: * $p < 0.05$; ** $p < 0.025$; *** $p < 0.005$. In rabbits less than 10 weeks of age, the proximal one-half of the small intestine was excised.

[c] Thirteen, pooled fetal intestines.

assay. Addition of 1–2.5 pg of 1,25-$(OH)_2D_3$ displaced the tracer sterol from the intestinal cytosol binding protein in the competitive protein binding assay, thus producing as sensitive a standard curve as those assays using vitamin D-deficient chick intestinal cytosol binding protein. Recently, we obtained reference 1,25-$(OH)_2D_2$ for comparison, and have observed equal potency for it and 1,25-$(OH)_2D_3$ in the displacement of 1,25-$(OH)_2[^3H]D_3$ from the rabbit intestinal receptor. This standard curve was reproducible for at least 8 months after preparation of the cytosol. Our results, employing rabbit intestinal cytosol as the binding protein in the assay of human plasma 1,25-$(OH)_2D$ concentrations, agree with those reported by other investigators who employ the intestinal cytosol from rachitic chicks. Vitamin D-sufficient rabbits provide a convenient, plentiful, and inexpensive source of binding protein for use in the competitive protein binding assay for 1,25-$(OH)_2D$.

Acknowledgments

The opinions or assertions contained herein are the private views of the authors and are not to be construed as official or as reflecting the views of the Department of Army or the Department of Defense.

The authors wish to thank Mrs. Estelle Coleman and Mrs. Judy Dubbs for help in preparation of this manuscript. This work was supported by National Research Service Award 5T32 AM 07314, Grant AM 28292, and CRC Grant RR 40 from the National Institutes of Health.

[21] Cytoreceptor Assay for 1,25-Dihydroxyvitamin D

By STAVROS C. MANOLAGAS

Vitamin D_3 (cholecalciferol) made in the skin and vitamin D_2 (ergocalciferol) ingested with food undergo two successive hydroxylations, first in the liver and then in the kidney, to form the hormone 1,25-dihydroxyvitamin D [1,25-$(OH)_2D$]. This hormone plays an important role in mineral and skeletal homeostasis and perhaps in hematopoietic cell differentiation and immune phenomena. Like other steroid hormones, 1,25-$(OH)_2D$ acts on its target tissues by binding initially to specific intracellular receptors. Changes in the circulating levels of this hormone occur either during physiological states of increased calcium requirements or in several disease states including renal osteodystrophy, parathyroid abnormalities, rickets, osteomalacia, sarcoidosis, and perhaps old age osteoporosis. In addition, 1,25-$(OH)_2D_3$ is employed as a therapeutic modality. Hence, the measurement of the concentrations of this hormone in blood has become important for research as well as for clinical medicine. Until recently, this measurement posed a major technical problem because it required a variety of steps including column chromatography and high-performance liquid chromatography in order to isolate 1,25-$(OH)_2D$ prior to assays.

The cytoreceptor assay[1,2] is a new methodological approach to the 1,25-$(OH)_2D$ measurement in blood and obviates several of the technical problems of the previous methods. This assay is based on the biological principle that 1,25-$(OH)_2D$ diffuses freely across cell membranes and is retained inside cells bound to its specific receptor while other metabolites of D are largely prevented from entering the cells because they remain bound to the serum D binding protein.[1] In the cytoreceptor assay, the

---

[1] S. C. Manolagas and L. J. Deftos, *Biochem. Biophys. Res. Commun.* **95**, 596 (1980).
[2] S. C. Manolagas and L. J. Deftos, *Lancet* **2**, 401 (1980).

quantitation of 1,25-(OH)₂D is accomplished by comparing the displacement of radioactive 1,25-(OH)₂D₃ retained inside intact cells, by the 1,25-(OH)₂D of unknown samples to the displacement of standard amounts of 1,25-(OH)₂D₃. The procedure described here includes minor modifications of the originally published method.[3,4] These modifications have simplified and optimized the performance of this assay.

Materials and Reagents

1. 1,25-(OH)₂-[26,27-³H]D₃ (specific activity 160 Ci/mmol) (Amersham, Arlington Heights, IL). Kept in 1 : 1 toluene–ethanol at −20°.

2. Unlabeled 1,25-(OH)₂D₃ (Hoffman-LaRoche, Nutley, NJ). Standard solution is prepared at a concentration of 1280 pg/ml in ethanol and kept stored at −20° in sealed amber ampules.

3. Human α-globulin (Fraction IV, Miles Laboratories, Inc.).

4. Bovine serum albumin, Fraction V (Sigma, St. Louis, MO).

5. Benzene (HPLC grade, Baker); methanol (HPLC grade, VWR); absolute ethanol (USP grade).

6. Solubilizing solution, Omnisol (WestChem, San Diego, CA).

7. Liquid scintillation cocktail, BetaMax (WestChem, San Diego, CA).

8. Disposable extraction columns, Extrelut 3 (E.M. Science, Gibbstown, NJ).

9. Incubation medium is MEM culture medium with Hanks' salts (Irvine Scientific, Santa Ana, CA) adjusted to 25 mM HEPES and pH of 7.4. The medium is filter sterilized and subaliquoted (sterilely) into 50 ml portions and stored at 4°.

10. Washing solution is normal saline (0.9% NaCl) enriched with 0.2% BSA (see #4) kept at 4°.

11. Evaporating apparatus.

12. Nitrogen gas tank.

13. Solution of 1 N NaOH kept at room temperature.

14. Cells: The cells used are rat osteogenic sarcoma lines (ROS).[5] These cells are cultured on the surface of plastic flasks T175 (Falcon) at 37°C in 5% CO₂ in air with Coon's F12 medium (Irvine Scientific), 10% fetal calf serum (GIBCO), and penicillin–streptomycin 10 μg/ml. Every 3 days confluent cells are split 1 : 2 to 1 : 3 and are subcultured in new flasks.

---

[3] S. C. Manolagas, F. L. Culler, J. E. Howard, A. S. Brickman, and L. J. Deftos, *J. Clin. Endocrinol. Metab.* **56,** 751 (1983).

[4] S. C. Manolagas, *in* "Assay of Calcium-Regulating Hormones" (D. D. Bikle, ed.), p. 139, Springer-Verlag, Berlin and New York, 1978.

[5] R. J. Majeska, S. B. Rodan, and G. A. Rodan, *Endocrinology (Baltimore)* **107,** 1434 (1980).

For the final subculture before harvesting, the culture medium is enriched with 10^{-7} M triamcinolone acetonide (Sigma). After 3 days of growth in the presence of triamcinolone, the medium is aspirated and the flasks are rinsed with 10 ml phosphate-buffered saline (PBS). The cells are then freed by 5-min treatment with 4 ml trypsin–EDTA solution in PBS buffer; this solution is prepared by 1 : 8 dilution of a concentrated preparation of trypsin–EDTA ($10\times$, Irvine Scientific). After the 5-min treatment, fresh medium is added and the cells are transferred into plastic conical tubes where they are spun at 200 g for 5 min. The supernatant is then aspirated, and the pelleted cells are resuspended in regular medium containing 10% DMSO. While cells are being harvested from consecutive flasks the freed cells are kept at 4°; the total yield is finally pooled and mixed gently to obtain a homogeneous suspension. An aliquot of this suspension is counted with a hemocytometer to determine the total cell number and the suspension is adjusted to 5×10^6 cells/ml; aliquots of this are then frozen at $-70°$ in plastic Nunc cryotubes (Irvine Scientific) in a Revco ultralow freezer. Alternatively, cells can be frozen in liquid nitrogen.

1,25-(OH)$_2$D Extraction Procedure

1. Aliquot 2 ml of each sample (serum or plasma) into labeled 13×100 mm glass test tubes. Although 2 ml represents optimal sample volume, the assay can be performed with samples as small as 1 ml. Include duplicate distilled H$_2$O blanks and control serum samples.

2. Add 20 μl of 1,25-(OH)$_2$[^{3}H]D$_3$ in ethanol (containing approximately 1500 cpm) to each sample and to three glass scintillation vials for recovery total activity (RTA). Vortex and let stand at least 30 min.

3. Add 230 μl absolute EtOH to RTA vials and keep for counting in 8 ml scintillation cocktail.

4. Add 50 μl of 1 N NaOH/ml sample and vortex. Bring sample volume up to 3 ml using distilled H$_2$O; vortex.

5. Apply samples to Extrelut 3 columns using glass Pasteur pipettes. Allow 5–10 min for the sample to settle into column.

6. Add 15 ml benzene to each column. Collect eluents in glass scintillation vials.

7. When columns stop dripping, remove vials and dry eluents under N$_2$ gas in a 37° waterbath until *completely* dry (approximately 1 hr). Add 1 ml methanol to each vial, vortex, then dry again.

8. Reconstitute dried samples with 1 ml absolute EtOH and vortex well. Cap tightly and store at $-20°$, if not to be used in the assay immediately.

The expected extraction efficiency of the Extrelut 3 column-benzene system for 1,25-(OH)$_2$D$_3$ is illustrated in Fig. 1. As seen in the figure, other

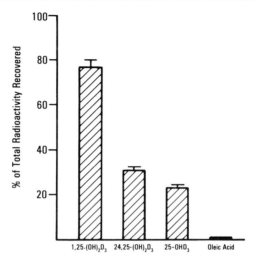

FIG. 1. Extraction of vitamin D_3 metabolites using Extrelut 3 minicolumns and benzene. Percentages of the total 1,25-$(OH)_2$[³H]D_3, 24,25-$(OH)_2$[³H]D_3, and 25-OH[³H]D_3 added in the samples which were recovered in the eluent of the column are indicated. Percentage of total [³H]oleic acid recovered during this procedure is also shown. Each bar represents the mean from three determinations; vertical lines are standard deviations of the mean.

major metabolites of vitamin D_3 are extracted less well in this system while lipids (oleic acid) are largely retained in the column. (See note added in proof, p. 198.)

Assay Procedure

1. Aliquot 230 μl of each of the ethanol-reconstituted sample extracts into one glass scintillation vial and into three glass test tubes (12 × 75 mm). The scintillation vial aliquot is used to determine recoveries from the extraction procedure.

2. Make serial dilutions of the 1280 pg/ml standard with absolute ethanol in one set of glass test tubes (e.g., 0.5 ml of standard + 0.5 ml ethanol) to obtain solutions containing 64, 32, 16, 8, 4, and 2 pg of 1,25-$(OH)_2D_3$/ 100 μl of ethanol.

3. Aliquot 100 μl × 3 of either absolute ethanol (O standard) or the standards (2, 4, 8, 16, 32, 64, 128 pg/100 μl) into 12 × 75 mm glass tubes.

4. Dry standards and samples under N_2 gas at room temperature. Check each tube for complete dryness.

5. While samples are drying, prepare the incubation medium by adding 0.5 mg of human α-globulin/ml of MEM Hanks' medium (room temperature). Add absolute ethanol to obtain a final concentration of 3% and mix well. You need 0.2 ml × number of tubes.

6. Add 50 μl of the incubation medium to each tube and let sit for 15 min.

7. Place appropriate number of vials of freeze-stored cells (you need approximately 0.5×10^6 cells $\times$ number of assay tubes) in a 37° waterbath until thawed enough to mix contents (2–3 min). Pour immediately into a 50-ml-plastic centrifuge tube containing 6 ml incubation medium for each vial of cells to be used. Spin at 200 g for 5 min at room temperature to pellet the cells. Aspirate medium and resuspend the cell pellet in fresh incubating medium; this suspension should contain approximately 0.5×10^6 cells/100 μl. Mix gently using a Pasteur pipette to ensure an even suspension (until visible clumps of cells have been dissolved). The cell suspension should be used within a few minutes after it is prepared and not be refrozen, otherwise the integrity of the cells cannot be maintained.

8. Add 100 μl of cell suspension to the assay tubes using a Hamilton repeating syringe and start incubation at room temperature (approximately 23°). During the incubation, maintain gentle agitation to prevent cells from settling in the bottom of the tube; this can be accomplished by placing the rack with the assay tubes on a shaking platform (approximately 150 rpm). Continue incubation for 45 min.

9. Prepare tracer by pipetting appropriate volume of the stock 1,25-$(OH)_2[^3H]D_3$ solution into a glass tube. You need $\simeq 4500$ cpm per assay tube. Dry the tracer solution under N_2 gas at room temperature and then resuspend into incubation medium to a final concentration of 4500 cpm/20 μl.

10. At the end of the first 45 min incubation add 20 μl of the tracer solution into each assay tube. Continue incubation on the shaking platform for another 30 min.

11. At the end of the second incubation, transfer tubes to an ice water bath (2°) and let cool for 15 min.

12. Add 2.5 ml of cold (4°) washing solution (normal saline + BSA) to each tube and leave in ice water bath for another 15 min.

13. Spin tubes in refrigerated centrifuge for 10 min at 1000 g to pellet the cells. Decant supernatant and blot tubes onto absorbent paper.

14. Add another 2.5 ml of washing solution to each tube and vortex gently. Spin again for 10 min at 1000 g. Decant supernatant as before and drain tubes well to minimize amount of liquid remaining.

15. Add 0.3 ml solubilizing solution (Omnisol) to each tube and vortex thoroughly. Leave at room temperature for 15 min.

16. Add 2 ml of scintillation cocktail (BetaMax) to each tube and vortex well to mix phases. Decant into scintillation vials containing 6 ml scintillation cocktail.

17. Cap vials tightly and shake content to mix. Chill vials to 4° before counting in a refrigerated beta counter.

Figure 2 depicts a representative standard curve. For comparison a

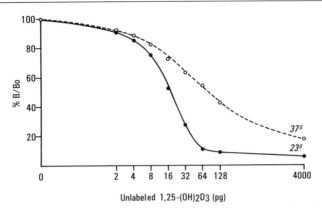

Unlabeled 1,25-(OH)2D3 (pg)

FIG. 2. Representative standard curves of the cytoreceptor assay. Incubations were performed at either 23° (●) or 37° (○). Points indicate the displacement of 1,25-(OH)$_2$[^{3}H]D$_3$ by standard amounts of unlabeled 1,25-(OH)$_2$D$_3$. Each point represents the mean of triplicate determinations.

standard curve obtained at 37° is also shown; in previous descriptions of the method, we had recommended 37° for this incubation. The higher sensitivity of the curve obtained at room temperature (23°) compared to the sensitivity of the curve obtained at 37° probably reflects improved viability and integrity of the frozen-thawed cells under the former conditions; the radioactivity taken up by the cells during incubation is also increased at room temperature.

Calculations

Recovery Factor

Recovery factor (RF) = total radioactivity added for recovery counts/ recovered radioactivity.

Assay Calculation

1. Average all replicates to obtain mean counts.
2. Divide standard or unknown (cpm) by zero standard (cpm) and multiply by 100 to obtain percentage bound (B)/maximum binding (B_0).

$$\% \ B/B_0 = \frac{\text{standard (cpm) or sample (cpm)}}{\text{zero standard (cpm)}} \times 100$$

3. Plot concentration of standards against percentage B/B_0 using semi-log graph paper.
4. Tube dose for sample is read from standard curve. To obtain final value (pg/ml) multiply tube dose (pg) by the recovery factor and divide by the initial sample volume.

TABLE I
EXPECTED RANGES FOR THE ASSAY

	Normal adults	Normal children	Primary hyperpara-thyroidism	Renal failure (hemodialysis)
Mean	35.9	43.3	63.0	7.4
SD	11.4	13.9	29.0	8.2
Number of determinations	80	40	21	130
Observed range	19–57	22–79	25–146	<4–16

$$\text{pg/ml} = \frac{\text{Tube dose (pg)} \times \text{RF}}{\text{initial sample volume (ml)}}$$

Expected ranges for the assay are summarized in Table I.

The intraassay variation has been determined with one clinical sample using one standard curve in one assay. The mean, standard deviation (SD), and coefficient of variation of nine replicates were mean = 54.1 pg/ml, SD = 3.4, and CV = 6.28%.

The interassay variation determined in four serum pools is summarized in Table II. The sensitivity of the method is 2 pg and the specificity for various metabolites of vitamin D has been determined as follows: 1,25-$(OH)_2D_3$ 100%; 1,25-$(OH)_2D_2$ 75%; 25-OHD_3 0.06%, 24,25-$(OH)_2D_3$ 0.003%. Hence, the serum 1,25-$(OH)_2D$ level determined by this method represents both 1,25-$(OH)_2D_3$ and 1,25-$(OH)_2D_2$. However, greater than 95% of all vitamin D metabolites in humans occur in the D_3 form, except for extremely high vitamin D_2 intake. From the major metabolites of vitamin D the 25-OHD is the compound more likely to interfere with the 1,25-$(OH)_2D$ measurement in the cytoreceptor assay because it circulates at higher concentration than any other metabolite (approximately 25 ng/ml) and has also relatively high affinity for the receptor. We have established that it requires a concentration greater than 125 ng/ml of 25-OHD to produce significant interference with the 1,25-$(OH)_2D$ measurement[3]; such high levels of 25-OHD, however, occur only in the rare instances of

TABLE II
INTERASSAY VARIATION IN SERUM POOLS

Sample	Number of determinations	Mean value (pg/ml)	SD	CV (%)
1	21	76.4	11.4	14.9
2	40	38.9	6.7	17.1
3	22	46.3	7.0	15.1
4	40	39.0	6.6	16.9

TABLE III
INTERLABORATORY REPRODUCIBILITY

	Cytoreceptor Lab 1	Cytoreceptor Lab 2	HPLC Lab 1	HPLC Lab 2	CV (%)
		Normals			
	27	28	9	31	42
	63	79	50	96	28
	41	40	41	50	10.4
	17	16	12	25	31.1
	22	19	19	17	9.7
	38	30	23	34	20.5
	47	41	28	37	20.7
	27	20	22	22	13
	53	49	72	48	20.1
	35	35	20	27	24.8
$x \pm$ SEM	37	35.7	29.6	38.7	22.03
	±4.6	±5.8	±6.1	±7.2	±3.2
		Pregnancy			
	71	64	67	58	8.2
	51	45	27	40	24.9
	57	45	22	31	39.9
	89	78	69	86	11.2
	73	63	67	48	16.8
	50	38	28	36	23.9
	85	69	54	54	18.9
$x \pm$ SEM	68	57.4	47.7	51.8	20.54
	±6.0	±5.6	±8.0	±7.2	±4.0
		Renal failure			
	14	<4[a]	<5[a]	11	56.6
	15	<4[a]	17	<6[a]	61.4
	17	<4[a]	15	9	53.2
	19	<4[a]	<5[a]	12	69
	14	9	19	23	38
	12	6	<5[a]	13	46.3
$x \pm$ SEM	15.16	5.16	11	12.33	54
	±1.0	±0.8	±2.7	±2.3	±4.0

[a] For values less than the detection limit we have used the detection limit to obtain the mean ± SEM.

vitamin D intoxication. We have also established in our laboratory that other steroid hormones do not interfere with the 1,25-(OH)$_2$D measurements even at pharmacologic levels.

The interlaboratory reproducibility of the cytoreceptor assay and its comparison with HPLC assays for 1,25-(OH)$_2$D has been tested by four

TABLE IV
COEFFICIENT OF CORRELATION[a]

	Cytoreceptor Lab 1	Cytoreceptor Lab 2	HPLC Lab 1	HPLC Lab 2
Cytoreceptor Lab 1	—	$r = 0.95$	$r = 0.87$	$r = 0.86$
Cytoreceptor Lab 2		—	$r = 0.86$	$r = 0.94$
HPLC Lab 1			—	$r = 0.81$
HPLC Lab 2				—

[a] Reproduced from Manolagas et al.[6] with the permission of the editor of *Lancet*.

laboratories[6] (two used the cytoreceptor method and two used standard chromatographic assays). Replicates from 23 serum samples from normal individuals ($n = 10$), pregnant women who are known to have elevated levels ($n = 7$), and patients with end-stage renal failure ($n = 6$) were assayed in each laboratory with the respective method. The values obtained by this interlaboratory evaluation are shown in Table III.

The coefficient of correlation of the 1,25-$(OH)_2D_3$ levels ($n = 23$) determined in each laboratory with the levels determined by each one of the other three is shown in Table IV.

Acknowledgments

The author wishes to thank Ms. D. Curran and C. Bochra for technical assistance and Dr. R. Reitz of the Endocrine Metabolic Center, Oakland, CA, Dr. R. Horst of the National Animal Disease Center, Ames, IA, and Dr. J. Haddad, University of Pennsylvania, for the interlaboratory evaluation of the method.

Supported by the NIH and the Veterans Administration.

NOTE ADDED IN PROOF: The single column benzene extraction method described here sometimes produces extracts that show inflated values in the assay. We have recently eliminated this problem by incorporating additional extraction steps using a second column and three ether-based reagents.

The new procedure involves attaching a second column containing unbonded silicic acid gel (Bio-Rad) below the original column of diatomaceous earth. The sample is prepared as usual and, after equilibrating on the first column for 30 min, is extracted with 30 ml benzene under vacuum. The first column, which retains the aqueous and hydrophilic phases of the matrix, is discarded. The benzene is also discarded, as the vitamin D metabolites are adsorbed onto the silicic acid in the second column. A series of solvents is then used to selectively remove the different metabolites, based on their polarity. The relatively nonpolar lipids and nonhydroxylated vitamin D_3 are eluted first using 4.5 ml 1 : 1 hexane : ether. Diethyl ether alone (6 ml) is used to remove the 25-hydroxy and much of the 24,25-dihydroxy forms. Finally, 7.5 ml 1 : 19 methanol : ether elutes the more polar 1,25-$(OH)_2D_3$. Only this final eluent is collected and evaporated for use in the assay. The cytoreceptor assay itself remains unchanged.

[6] S. C. Manolagas, R. E. Reitz, R. Horst, J. Haddad, and L. J. Deftos, *Lancet* **1**, 191 (1983).

[22] Monoclonal Antibodies as Probes in the Characterization of 1,25-Dihydroxyvitamin D$_3$ Receptors

By J. WESLEY PIKE and MARK R. HAUSSLER

The recent acquisition of both polyclonal and monoclonal antibodies to a variety of steroid receptor proteins has brought with it an entirely new approach to the study of receptor biology.[1-6] The number of applications of these immunological probes is enormous and includes immunocytochemical analysis, detection under both native and denaturing conditions, characterization, isolation, examination of synthesis and metabolism, and even the isolation of genes from which receptors are derived. All immunological techniques are based on high affinity, stable interactions between the antibody and receptor molecule, and are generally independent of highly unstable receptor–hormone interactions.

We have recently generated several rat spleen–mouse myeloma hybridomas which secrete monoclonal antibodies to the chick intestinal 1,25-dihydroxyvitamin D$_3$ [1,25-(OH)$_2$D$_3$] receptor,[7-9] a polypeptide known to mediate the action of this hormone at the level of the genome. These antibodies, which cross-react with all tested mammalian forms of the 1,25-(OH)$_2$D$_3$ receptor, have been used in a variety of experiments aimed at elucidating the physicochemical nature as well as biology of this specific protein. This chapter describes the methodology associated with the application of these monoclonal antibodies toward the detection and characterization of the 1,25-(OH)$_2$D$_3$ receptor.

[1] G. L. Greene, F. W. Fitch, and E. V. Jensen, Proc. Natl. Acad. Sci. U.S.A. 77, 157 (1980).

[2] G. L. Greene, C. Nolan, J. P. Engler, and E. V. Jensen, Proc. Natl. Acad. Sci. U.S.A. 77, 5115 (1980).

[3] B. Moncharmont, J.-L. Su, and I. Parikh, Biochemistry 21, 6916 (1982).

[4] F. Logeat, M. T. V. Hai, A. Fournier, P. Legrain, G. Buttin, and E. Milgrom, Proc. Natl. Acad. Sci. U.S.A. 80, 6456 (1983).

[5] D. P. Edwards, N. L. Weigel, W. T. Schrader, B. W. O'Malley, and W. L. McGuire, Biochemistry 23, 4427 (1984).

[6] M. V. Govindan and H. Gronemeyer, J. Biol. Chem. 259, 12915 (1984).

[7] J. W. Pike, C. A. Donaldson, S. L. Marion, and M. R. Haussler, Proc. Natl. Acad. Sci. U.S.A. 79, 7719 (1982).

[8] J. W. Pike, S. L. Marion, C. A. Donaldson, and M. R. Haussler, J. Biol. Chem. 258, 1289 (1983).

[9] J. W. Pike, J. Biol. Chem. 259, 1167 (1984).

METHODS IN ENZYMOLOGY, VOL. 123

Methodology

Preparation of Monoclonal Antibodies. Rat spleen–mouse myeloma antireceptor-secreting hybridoma cell lines SP2/0-4A5, SP2/0-9A7, and SP2/0-8D3 are grown in suspension culture in 1-liter glass spinner flasks in Iscove's modified Dulbecco's medium supplemented with 10% fetal bovine serum. Antibodies are harvested from the medium of 4- to 6-day cultures and precipitated twice with phosphate-buffered 40% saturated ammonium sulfate. Immunoglobulins are redissolved in phosphate-buffered saline (PBS) and stored at $-20°$ in 0.02% sodium azide. Serum-free preparations of monoclonal antibodies are obtained by removing the hybridoma cells from 4-day cultures of serum-containing media and permitting them to secrete antibody for an additional 4 days in serum-free medium. Purification of antibodies derived from SP2/0-9A7 and SP/0-4A5 (9A7γ 2b and 4A5γ 2a, respectively) is achieved by DEAE-cellulose chromatography. They are stored at $4°$ in PBS with 0.02% sodium azide at a concentration of 1–2 mg/ml.

Preparation of 1,25-$(OH)_2D_3$ Receptors. Chicken 1,25-$(OH)_2D_3$ receptors are obtained from the intestinal mucosa of 4-week-old Rhode Island Red chicks fed a diet deficient in vitamin D_3 from the day of hatch. This tissue, as well as kidney and liver, is homogenized in a glass Teflon tissue grinder in 5–10 volumes (w/v) of KETD-0.3 buffer (0.01 M Tris–HCl, pH 7.4, 1 mM EDTA, 0.3 M KCl, and 5 mM dithiothreitol), and the buffer soluble fraction of the tissue obtained after ultracentrifugation at 165,000 g for 45 min at $4°$. Tissues obtained from other species are prepared in identical fashion.

1,25-$(OH)_2D_3$ receptors are also obtained from extracts of cultured mammalian cell lines. Lines such as mouse 3T6 fibroblast, rat osteosarcoma (ROS 17/2.8), rat pituitary (GH$_3$), pig kidney (LLC-PK$_1$), human breast (MCF-7), human leukemia (HL-60), and human intestine (407) are grown under standard cell culture conditions and harvested utilizing trypsin. The cells are suspended in ice cold KETD-0.3 buffer and lysed by homogenization in a Dounce homogenizer at $25 × 10^6$ cells/ml. The buffer-soluble cell extract is then obtained by ultracentrifugation as above.

Hormone Labeling of 1,25-$(OH)_2D_3$ Receptors. Formation of 1,25-$(OH)_2D_3$ receptor complexes from both tissue as well as cultured cell extracts is achieved by incubating the cytosol with concentrations of 1,25-$(OH)_2[^3H]D_3$ (120–170 Ci/mmol) ranging from 1 to 4 nM in a final volume containing 10% absolute ethanol. Incubation periods can be as short as 1 hr or extended to 24 hr at $4°$. Prolonged incubation leads to proteolytic cleavage and eventual inactivation of 1,25-$(OH)_2D_3$ receptors. Quantitation of the concentration of receptor–hormone complexes is achieved by DEAE filter adsorption, dextran-coated charcoal removal of free hormone, or hydroxylapatite assay as described previously.[7–9] 1,25-(OH)$_2$D$_3$-receptor complexes may also be created by incubating intact cell suspensions with 1,25-OH[^{3}H]D$_3$ for 1 hr at 37° and then preparing cellular extracts in KETD-0.3 buffer as previously described.[10]

Sedimentation Displacement Analysis. Sedimentation displacement analysis is utilized to assess the presence and relative titer of specific antireceptor antibodies in the serum of immunized animals and to assess the secretion and general class (IgG or IgM) of monoclonal antibodies produced by selected hybridomas. Further, it can also be used to evaluate the qualitative reactivity of antiserum or monoclonal antibodies to 1,25-(OH)$_2$D$_3$ receptors derived from cell, tissue, or species sources other than those employed to immunize the original animal. This approach is thus useful in determining the relative immunochemical similarities of various 1,25-(OH)$_2$D$_3$ receptor species. Preformed 1,25-(OH)$_2$[^{3}H]D$_3$–receptor complexes (0.05–0.1 pmol) derived from chick intestinal cytosol or cytosols obtained from other tissues or cultured cell lines are incubated with preimmune or immune serum, hybridoma medium, or purified monoclonal antibody for periods of 1–4 hr at 4° in KETD-0.3 buffer. Following incubation, the samples (0.2–0.4 ml) are layered onto 4.8-ml gradients of 10–30% sucrose prepared in KETD-0.3 buffer and centrifuged at 2° in a Beckman SW50.1 rotor at 265,000 g for 16 hr. After fractionation of the gradient, 1,25-(OH)$_2$[^{3}H]D$_3$ is quantitated by liquid scintillation counting. Gradient pellets are solubilized in 0.1 N NaOH prior to quantitation as above. Qualitative reactivity of the serum or hybridoma medium reveals a displacement of the 1,25-(OH)$_2$[^{3}H]D$_3$–receptor complex from its native position at 3.7 S to a position consistent with a receptor-immunoglobulin complex. If receptors have been incubated with positive antiserum, it is likely that the native complex will be displaced to heavier sedimenting units indicative of the polyclonal nature of the antiserum. This can be seen clearly in Fig. 1A where chick intestinal cytosol receptor, which normally sediments at 3.7 S in 10–30% sucrose gradients, is displaced to the bottom of the gradient tube. Sedimentation of receptor in the presence of monoclonal antibodies of the IgG and IgM class will reveal migration of the 1,25-(OH)$_2$[^{3}H]D$_3$ at 7–8 S and >13 S, respectively.[8] As seen in Fig. 2, the 1,25-(OH)$_2$D$_3$ receptor from rat osteosarcoma cells (ROS 17/2.8) is displaced to the 7–8 S position with monoclonal antibody 4A5γ whereas with antiserum both a 7–8 S peak and a larger receptor–immunoglobulin

[10] J. W. Pike and M. R. Haussler, *J. Biol. Chem.* **258**, 8554 (1983).

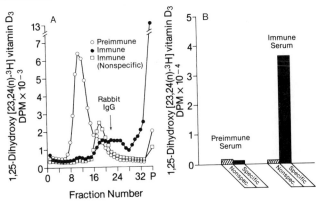

FIG. 1. Immunoreactivity of chick intestinal 1,25-(OH)$_2$D$_3$ receptors with rat antiserum. (A) 1,25-(OH)$_2$[^{3}H]D$_3$–receptor complex (0.1 pmol) was incubated with preimmune ($\bigcirc$) or immune ($\bullet$) rat serum for 4 hr at 2°. Following incubation, the mixtures were sedimented through 10–30% sucrose gradients prepared in KETD buffer containing 0.3 M KCl. Nonspecific binding ($\square$) was determined by forming the 1,25-(OH)$_2$D$_3$–receptor complex in the presence of 100-fold excess unlabeled hormone. (B) 1,25-(OH)$_2$[^{3}H]D$_3$–receptor complex (0.15 pmol) was incubated with either preimmune or immune rat serum for 4 hr at 2°, and then immunoprecipitated as described in the text. Nonspecific binding was determined as in A and subtracted from total binding to yield specific immunoprecipitable 1,25-(OH)$_2$[^{3}H]D$_3$ receptor.

complex at 13 S are observed. More precise estimations of the molecular mass of ternary 1,25-(OH)$_2$D$_3$–receptor–antibody complexes can be obtained via gel filtration chromatography.

Immunoprecipitation Analysis. Immunoprecipitation represents an extremely valuable technique whereby immunoglobulin-bound 1,25-(OH)$_2$[^{3}H]D$_3$ receptors can be quantitated and interactions between 1,25-(OH)$_2$[^{3}H]D$_3$ receptors precisely assessed. Whereas sedimentation displacement analysis has limited usefulness with respect to the processing of individual samples, immunoprecipitation is amenable to multiple assays, such as those necessary in radioimmunoassay or hybridoma screenings. Furthermore, immunoprecipitation is also indispensible in the enrichment of receptors which have been metabolically labeled by incubation of cultured cells with [^{35}S]methionine or from cell-free translation of isolated cellular poly(A)$^+$ RNA.

Immunoprecipitation of 1,25-(OH)$_2$[^{3}H]D$_3$–receptor complexes bound to monoclonal antibodies such as 9A7γ, 4A5γ, 8D3μ, or antiserum is achieved by the addition of carrier rat serum and anti-rat Ig antiserum, followed by overnight incubation at 4°. The immunoglobulin pellet obtained after centrifugation at 4000 g for 10 min is washed once with 1 ml

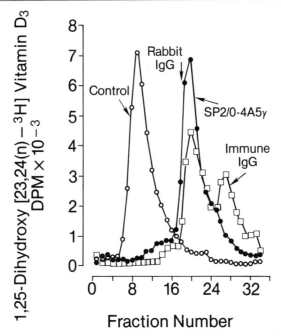

FIG. 2. Immunoreactivity of the rat osteosarcoma cell (ROS 17/2.8) 1,25-(OH)₂D₃ receptor by sedimentation displacement analysis. ROS 17/2.8 1,25-(OH)₂[³H]D₃–receptor complex (0.1 pmol) was incubated with either preimmune (○) or immune (□) serum and monoclonal antibody 4A5γ (●) for 4 hr at 2° and then sedimented on 10–30% sucrose gradients as described in the text.

of KETD-0.3 buffer, dissolved in 0.1 N NaOH, and quantitated via liquid scintillation spectrometry. An example of the results of immunoprecipitation of chick 1,25-(OH)₂D₃ receptor with antiserum and anti-rat Ig antibodies is observed in Fig. 1B. Less expensive methodology can be employed using either protein A coupled to agarose or formalin-fixed *Staphylococcus aureus* cells. Since 4A5γ and 9A7γ are rat antibodies of IgG class 2a and 2b, respectively, neither binds to protein A. However, an effective immunoglobulin bridge can be created by preincubating either protein A–agarose complex or *S. aureus* cells with rabbit or goat anti-rat IgG antiserum as described previously. Subsequent incubation of either of these reagents with 1,25-(OH)₂[³H]D₃–receptor–monoclonal antibody complexes for 1–4 hr at 4° followed by centrifugation as above will yield immunoprecipitated 1,25-(OH)₂D₃ receptor. Efficient washing of these precipitates is essential if background 1,25-(OH)₂[³H]D₃ is to be eliminated. Depletion of this radioactivity is achieved by washing in KETD-0.3 buffer which contains 0.5% Triton X-100 or Tween 20.

Immunoprecipitation of 1,25-$(OH)_2D_3$ receptors can also be achieved by the addition of monoclonal antibody 9A7γ or 4A5γ directly coupled to Sepharose. This reagent is particularly useful in binding metabolically labeled 1,25-$(OH)_2D_3$ receptors due to the extremely high affinity nature of 9A7γ for receptor (10^{-11} M) which permits the extensive washing procedures necessary to eliminate background. 1,25-$(OH)_2D_3$ receptors in mouse 3T6 fibroblasts are prepared by incubating confluent monolayer cultures (approximately 10^7 cells) in Dulbecco's modified Eagle's medium deficient in methionine but supplemented with 2% dialyzed calf serum and 0.5 mCi of [^{35}S]methionine (1400 Ci/mmol) for 1–8 hr at 37°. At the end of incubation, the cells are washed with PBS, removed from the plates in KETD-0.3 lysis buffer containing 0.1% Triton X-100, and the lysed cell extract obtained after centrifugation at 165,000 g for 30 min at 4°. Overnight incubation of aliquots of the lysate with 9A7γ-Sepharose is sufficient to achieve coupling of receptor, whereupon the Sepharose is washed sequentially with two 1-ml aliquots of KETD-0.3 buffer, six 1-ml aliquot of KETD buffer containing 0.5 M KCl, 1% NP-40, 1% sodium deoxycholate, 0.1% SDS, 0.5% Tween 20, and two 1-ml aliquots of KETD buffer without KCl. Subsequent SDS–polyacrylamide gel electrophoresis of precipitated proteins followed by fluorography reveals a 1,25-$(OH)_2D_3$ receptor species at approximately 54,000 Da which is typical of mouse 3T6 fibroblast receptor. Immunoprecipitation of internally [^{35}S]methionine-labeled proteins produced by cell-free translation of isolated intestinal poly(A)$^+$ RNA is also achieved by similar methodology.[11]

Radioligand Immunoassay (RLIA). Research aimed at quantitating both occupied and unoccupied 1,25-$(OH)_2D_3$ receptors during hormonal treatment of animals is essential in understanding the relationship between serum hormone levels and biological response. These types of studies can be accomplished with both hormone exchange assays[12–14] and with a specific radioligand immunoassay for 1,25-$(OH)_2D_3$ receptors.[15] The immunoassay, which recognizes the receptor molecule independent of hormonal status of the receptor or its hormone-binding capabilities, is essential in the characterization of unoccupied 1,25-$(OH)_2D_3$ receptors as well as in detecting receptor forms which are biologically inactive with

[11] W. A. Bornstein, J. W. Pike, M. R. Haussler, and H. M. Kronenburg, *Calcif. Tissue Int.* **36,** 510 (1984).

[12] W. Hunziker, M. R. Walters, and A. W. Norman, *J. Biol. Chem.* **255,** 9534 (1980).

[13] W. Hunziker, M. R. Walters, J. E. Bishop, and A. W. Norman, *J. Clin. Invest.* **69,** 826 (1982).

[14] E. R. Massaro, R. U. Simpson, and H. F. DeLuca, *Proc. Natl. Acad. Sci. U.S.A.* **80,** 2549 (1983).

[15] S. Dokoh, M. R. Haussler, and J. W. Pike, *Biochem. J.* **221,** 129 (1984).

respect to hormone-binding function. The radioligand imunoassay for receptors is a competition assay in which the standard curve is created by competing chick intestinal cytosolic 1,25-(OH)$_2$[^{3}H]D$_3$ receptor (16 fmol) with increasing concentrations of radioinert 1,25-(OH)$_2$D$_3$ receptor complex (0–240 fmol) for the 9A7γ monoclonal antibody. Samples, prepared in identical buffer, are likewise competed with the radiolabeled chick receptor for the antibody. Incubations are carried out in triplicate in 0.25 ml of KETD-0.3 buffer containing 1% fetal bovine serum until maximum binding is achieved (16–18 hr at 4°) whereupon the receptor–antibody complexes are precipitated with *S. aureus* cells which were precoated with rabbit anti-rat Ig as described above. The *S. aureus* cell pellets are washed twice in KETD buffer containing 0.5 M KCl and 0.5% Tween 20, the radiolabeled 1,25-(OH)$_2$[^{3}H]D$_3$ extracted for 1 hr with 1 ml of acetone, and then quantitated. This assay is useful in the quantitation of both chick as well as mammalian 1,25-(OH)$_2$D$_3$ receptors since the 9A7γ antibody displays similar if not identical equilibrium dissociation constants for all 1,25-(OH)$_2$D$_3$ receptors examined.[16] The assay is sensitive to 2 fmol receptor/tube and demonstrates intra- and interassay variations of 7 and 12%, respectively. Finally, the antibody recognizes both occupied and unoccupied 1,25-(OH)$_2$D$_3$ receptors with equivalent affinity, permitting the quantitation of total receptor concentration under physiologic conditions.

Immunoblot Assay. The superior resolving ability of Laemmli SDS–polyacrylamide gel electrophoresis has made this technique the method of choice for characterizing the molecular weights of all polypeptide species. 1,25-(OH)$_2$D$_3$ receptors, however, exist in such trace concentrations in cellular extracts that they cannot be detected directly by such techniques as silver staining nor can they be detected by radioactive hormone since the ligand is dissociated upon denaturation. Nevertheless, the 1,25-(OH)$_2$D$_3$ receptor can be detected in nanogram amounts in crude chick intestinal cytosol by means of immunoblot methodology.[17] Moreover, through simple enrichment procedures 1,25-(OH)$_2$D$_3$ receptors can also be detected and characterized in a host of cultured cells where the receptor specific activity is less than 10% that of chick intestine. This procedure is essential in the precise assessment of the molecular weights of both avian and mammalian forms of the 1,25-(OH)$_2$D$_3$ receptor and in characterizing both endogenously and exogenously created proteolytic cleavage products of the receptor. Finally, immunoblot methodology is essential in

---

[16] J. W. Pike, *Proc. 65th Annu. Meet., Am. Endocr. Soc.* p. 25 (1983).
[17] J. W. Pike and M. R. Haussler, *J. Biol. Chem.* (submitted for publication).

confirming that a particular protein species purified to homogeneity is in fact the 1,25-$(OH)_2D_3$ receptor.

Immunological detection of the 1,25-$(OH)_2D_3$ receptor is carried out following the resolution of crude cellular extracts, enriched protein preparations, or purified polypeptides on SDS–PAGE and their "Western" transfer to nitrocellulose. The nitrocellulose sheet is incubated sequentially in PBS containing 3% bovine serum albumin (3 hr at 25°), PBS containing 1% BSA and 2–6 μg purified 9A7γ monoclonal antibody/ml (16 hr at 4°), PBS containing 3% BSA (10 min at 25°), and then radioiodinated affinity purified rabbit anti-9A7 (10^5 cpm/ml) (2 hr at 25°). The nitrocellulose sheet is washed with 5–50 ml portions of TBS buffer (0.05 M Tris–HCl, pH 7.5, 0.2 M NaCl) containing 0.05% Tween 20 over a 2-hr period immediately after incubation with primary antibody and after incubation with radioiodinated anti-9A7γ. The nitrocellulose membrane is dried and then autoradiographed for periods ranging from 24 to 96 hr at $-70°$ using a Dupont Cronex Hiplus intensifying screen and Kodak Omat XAR or XRP film.

Concentrations of 1,25-$(OH)_2D_3$ receptor must be at least 0.002% of total soluble protein (~20 ng/mg protein) to permit detection by immunoblot assay with the 9A7γ antibody. This concentration of receptor is achievable only in cytosols of chick or rat intestinal mucosa prepared in KETD-0.3 buffer. Other tissues, such as kidney, as well as all cultured cell lines examined display receptor concentrations as low as 1–2% of this value. Thus, enrichment procedures are necessary for these proteins to be detected and characterized. 1,25-$(OH)_2D_3$ receptors can be enriched sufficiently by preparing the nuclear or chromatin fraction of receptor-poor tissues in KETD buffer without KCl and then analyzing the proteins extracted from this fraction with KETD-0.3 buffer. However, the most efficient method for increasing the relative concentration of 1,25-$(OH)_2D_3$ receptors is DNA cellulose chromatography of tissue or cultured cell extracts prepared in KETD-0.3 buffer. The resulting peak of 1,25-$(OH)_2D_3$ receptor activity eluting between 0.2 and 0.3 M KCl is purified 100- to 500-fold over cytosol, and trichloroacetic acid precipitation of these receptor-containing fractions yield receptor in sufficient concentration for immunoblot detection. This procedure allows the detection of 1,25-$(OH)_2D_3$ receptors in concentrations as low as 500 copies/cell. An example of this type of protocol is observed in Fig. 3, where 1,25-$(OH)_2D_3$ receptors obtained from the rat osteosarcoma cell line ROS 17/2.8 are enriched by DNA cellulose chromatography, and the resulting peak fractions electrophoresed and then immunoblotted, clearly revealing a protein of approximate 54,000 Da (inset). Employing this type of approach, 1,25-$(OH)_2D_3$ receptors can be detected and characterized for all chick, mouse,

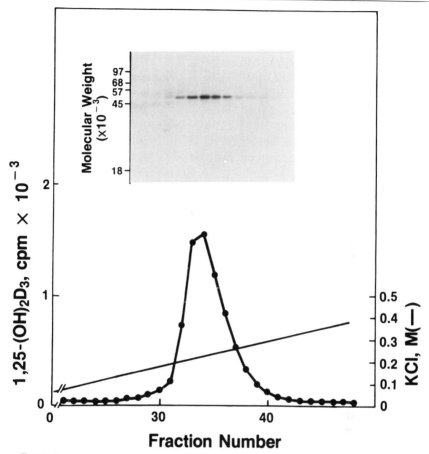

FIG. 3. Immunologic characterization of the 1,25-(OH)₂D₃ receptor in a rat osteosarcoma cell line (ROS 17/2.8). A cell suspension of ROS 17/2.8 cells (3×10^8) was incubated with 2 nM 1,25-(OH)₂D₃ and the labeled receptor extracted from the nucleus as described in ref 10. Receptor was chromatographed on a column of DNA-cellulose and eluted during a KCl gradient as depicted. Inset: Aliquots of the individual chromatographic fractions were precipitated with 6% trichloroacetic acid, electrophoresed in 0.1% SDS, and immunoblotted as outlined in the text. Molecular weight protein standards are indicated on the axis.

rat, pig, monkey, or human tissues or cell lines in which receptors are of interest.

Experimental Observations

The above described immunological methodology, as well as procedures not considered here, have been extremely useful in examining both the physicochemical nature of 1,25-(OH)$_2$D$_3$ receptors and certain features of the protein's cellular biology. The following represent several ,interesting observations we have made concerning 1,25-(OH)$_2$D$_3$ receptors using these techniques.

Immunochemical Similarity. Table I summarizes the qualitative results we have obtained utilizing sedimentation displacement analysis as well as immunoprecipitation analysis to characterize the reactivity of antibodies 4A5γ, 9A7γ, and 8D3μ to 1,25-(OH)$_2$D$_3$ receptors from a variety of avian and mammalian sources. It is clear from these data that while the antibodies do not react with glucocorticoid or estrogen receptors, they are

TABLE I
CROSS-REACTIVITY OF MONOCLONAL ANTIBODY TO CHICK
1,25-(OH)$_2$D$_3$ RECEPTORS

	Monoclonal antibody		
Receptor protein	4A5γ	9A7γ	8D3μ
1,25-(OH)$_2$D$_3$ receptors			
Intestine			
Fish (*M. salmoides*)	±[a]	±	−
Frog (*R. catesbiana*)	+	+	−
Chick	+	+	+
Rat	+	+	−
Human (407)[b]	+	+	−
Kidney			
Chick	+	+	+
Rat	+	+	−
Pig (LLC-PK$_1$)[b]	+	+	−
Bone			
Rat (ROS 17/2.8)[b]	+	+	−
Fibroblast			
Mouse (3T6)[b]	+	+	−
Pituitary			
Rat (GH$_3$)[b]	+	+	−
Breast			
Human (MCF-7)[b]	+	+	−
Glucocorticoid receptors			
Liver			
Rat	−	−	−
Estrogen receptors			
Breast			
Human (MCF-7)[b]	−	−	−

[a] Signifies limited cross-reactivity when assessed in comparison to chick receptor.
[b] Indicates that source is a cultured cell line.

generally cross-reactive with all mammalian forms of the 1,25-(OH)$_2$D$_3$. The single exception is the reaction of the 8D3 antibody (IgM), which is specific for the avian receptor and shows no cross-reactivity with mammalian receptor protein. Further work utilizing immunoprecipitation techniques has shown that the equilibrium dissociation constants of the 9A7γ antibody for avian as well as mammalian receptor is similar and in the range of 1–2 × 10^{-11} M.[15,16] Thus, there exists a high degree of amino acid sequence conservation in at least two epitopes on 1,25-(OH)$_2$D$_3$ receptors, although the single epitope to which 8D3μ was created is unique to chick receptor protein.

Biology and Characterization of Normal 1,25-(OH)$_2$D$_3$ Receptors and Detection of Biological Variants of Human Receptors by Radioimmunoassay. The RLIA has been useful in combination with ligand-binding assay in determining the level of occupied and unoccupied 1,25-(OH)$_2$D$_3$ receptors in chick intestine as a function of the hormonal status of chicks.[15] RLIA is used to determine the total concentration of receptor, 1,25-(OH)$_2$[^{3}H]D$_3$ binding is used to assess the unoccupied level of receptor, and the difference represents an assessment of receptor which is bound to endogenous hormone. Results show that in rachitic chicks, less than 13% of the 1,25-(OH)$_2$D$_3$ receptor exists in the occupied state, whereas normal chicks and chicks treated with 2 nmol of 1,25-(OH)$_2$D$_3$ (48 and 24 hr prior to sacrifice) display 20 and 56% respectively, of the 1,25-(OH)$_2$D$_3$ receptor in the occupied form. These data, coupled with evaluation of serum levels of 1,25-(OH)$_2$D$_3$, suggest that occupied receptors exist in simple equilibrium with the prevailing level of circulating hormone.

RLIA has been useful in characterizing the sedimentation coefficient of the unoccupied 1,25-(OH)$_2$D$_3$ receptor and in evaluating the ability of the occupied receptor to bind to immobilized DNA.[15] Immunoassay of fractions obtained after sedimentation of receptor through 10–30% sucrose gradients prepared in KETD-0.3 buffer reveal an S value of 3.7. Moreover, similar assays of chromatographic fractions derived from DNA-cellulose columns demonstrate that the 1,25-(OH)$_2$D$_3$ receptor binds to DNA and is eluted between 0.2 and 0.3 M KCl. Thus, the immunoassay is useful in evaluating physicochemical as well as putative functional properties of native 1,25-(OH)$_2$D$_3$ receptors.

Of particular interest is the use of the immunoassay to detect human 1,25-(OH)$_2$D$_3$ receptors which are biologically inactive in fibroblasts derived from patients with vitamin D$_3$-dependent rickets, type II.[18] This

---

[18] J. W. Pike, S. Dokoh, M. R. Haussler, U. A. Liberman, S. J. Marx, and C. Eil, *Science* **224**, 879 (1984).

rare, heritable human syndrome is characterized by hypocalcemia, secondary hyperparathyroidism, and rickets, all of which persist despite high circulating levels of 1,25-$(OH)_2D_3$. This constellation of features results from peripheral target organ resistance to vitamin D_3 and is a disease which arises almost exclusively from defects in either the receptor's interaction with 1,25-$(OH)_2D_3$ or its nuclear site of action. One particularly abundant cellular phenotype in these patients is one in which no detectable hormone binding can be demonstrated.[19] The RLIA was utilized to evaluate whether fibroblasts from these patients were unable to synthesize the 1,25-$(OH)_2D_3$ receptor or expressed proteins incapable of hormone binding. The latter proved true, providing initial evidence that deficiencies in hormone binding associated with inherited tissue resistance to 1,25-$(OH)_2D_3$ probably arise from structural abnormalities in the 1,25-$(OH)_2D_3$ receptor molecule, and are not due to defective receptor synthesis.

Immunoblot Characterization. Immunoblot methodology has been extremely useful recently in precisely characterizing the molecular weights of 1,25-$(OH)_2D_3$ receptors from both avian and mammalian sources.[17,20] Purification of the 1,25-$(OH)_2D_3$ receptor from chick intestinal mucosa has consistently yielded a series of Coomassie-stained protein species on SDS–polyacrylamide gels. However, unequivocal demonstration that one or more of these species was the native 1,25-$(OH)_2D_3$ receptor could not be made due to rapid loss of hormonal ligand under either native or denaturing conditions. Nevertheless, immunoblot of 10 ng of purified material yielded a pair of signals at 58,000 and 60,000 Da which exactly comigrate with Coomassie stainable protein species. Moreover, these species can be identified in immunoblots of crude chick intestinal cytosol, although the larger species at 60,000 Da is most prevalent. It is clear from these results that the native receptor species is the 60K protein, but that endogenous proteolytic action rapidly cleaves the native monomeric form to a series of polypeptides of lesser molecular weight. This species is likewise observed in cellular extracts of chick kidney.[17]

1,25-$(OH)_2D_3$ receptor from mammalian sources have also been characterized by immunoblot techniques.[20] Cell lines from mouse, rat, pig, monkey, and human have been examined, and receptors evaluated after enrichment techniques as described above and illustrated in Fig. 3. Table II summarizes the estimated molecular mass of each of these receptor proteins. Clearly, all mammalian forms have a lower molecular mass, with the human receptor displaying a molecular mass of 52,000–53,000

[19] U. A. Liberman, C. Eil, and S. J. Marx, *J. Clin. Invest.* **71,** 192 (1983).
[20] J. W. Pike and M. R. Haussler, in preparation.

TABLE II
COMPARISON OF THE MOLECULAR MASS OF 1,25-(OH)$_2$D$_3$
RECEPTORS FROM AVIAN AND MAMMALIAN SOURCES

Receptor source[a]	Molecular Mass (Da)
Chick intestine	60,000
Chick kidney	60,000
Mouse fibroblast (3T6)	54,000–56,000
Rat osteosarcoma (ROS 17/2.8)	54,000
Pig kidney (LLC-PK$_1$)	54,000
Human intestine (407)	51,000–53,000
Human leukemia (HL-60)	51,000–53,000

[a] The designated cultured cell line is indicated in parentheses.

Da. While it cannot be ruled out that proteolytic cleavage has yielded these smaller species, no evidence exists at present that these forms are derived from larger molecular weight precursors. The demonstrated smaller protein, however, is consistent with the early finding that mammalian receptors have generally displayed smaller sedimentation coefficients than avian receptors.

Section IV

Vitamin E Group, Tocopherols

[23] An HPLC Method for the Simultaneous Determination of Retinol and α-Tocopherol in Plasma or Serum[1,2]

By G. L. CATIGNANI

Nearly all recently published methods for determination of retinol and α-tocopherol and their derivatives, metabolites, isomers, etc. employ some form of HPLC. Increasingly these procedures are being used in both clinical and research laboratory settings. Numerous HPLC assays for both retinol and α-tocopherol have been reported. Several simultaneous assays for both vitamins are available.[3-6]

The method described here offers the researcher or clinician a rapid, sensitive, simple, specific, precise, and nondestructive microprocedure for the simultaneous determination of retinol and α-tocopherol in plasma or serum. The method is applicable to samples obtained from humans or laboratory animals.

Reagents

HPLC grade hexane and methanol. Reagent grade diethyl ether and absolute ethanol.

Chromatographically pure standards: α-tocopherol, α-tocopheryl acetate, retinol, and retinyl acetate.

Stock standard solutions: α-tocopherol and α-tocopherol acetate 5 g/liter, retinol, and retinyl acetate 100 mg/liter in ethanol.

Working standard solutions: dilute each of above standard solutions 100-fold with ethanol.

Remarks. Store vitamin stock solutions at −20° in actinic glassware. Prepare bimonthly as necessary. Make working solution fresh as needed. The following absorptivities ($A_{1\,cm}^{1\%}$ in ethanol) should be used to verify

[1] Paper No. 9584 of the Journal Series of the North Carolina Agricultural Research Services, Raleigh, NC 27695-7601. The use of trade names in this publication does not imply endorsement of the products by the North Carolina Agricultural Service.

[2] Supported in part by a grant from the North Carolina Department of Justice.

[3] J. G. Bieri, T. J. Tolliver, and G. L. Catignani, *Am. J. Clin. Nutr.* **32**, 2143 (1979).

[4] G. L. Catignani and J. G. Bieri, *Clin. Chem. (Winston-Salem, N.C.)* **29**, 708 (1983); *in* "Selected Methods of Clinical Chemistry" (G. R. Cooper, ed.), Vol. 10, p. 230. American Association for Clinical Chemistry, 1983.

[5] A. P. DeLeenheer, V. De Bevere, M. G. M. DeRuyter, and A. E. Claeys, *J. Chromatogr.* **162**, 408 (1979).

[6] W. J. Driskell, J. W. Neese, C. C. Bryant, and M. M. Bashor, *J. Chromatogr.* **231**, 439 (1982).

concentrations: α-tocopherol 75.8 at 292 mm, α-tocopheryl acetate 43.6 at 285 mm, retinol 1780 at 325 mm, and retinyl acetate 1510 at 328 mm. Should absorbance or peak height (area) change or additional peaks or shoulders appear, new standard solutions should be prepared.

HPLC Apparatus

An isocratic HPLC system capable of reproducing the conditions listed below including a UV detector and a 10-mV, 10-cm recorder, intergrating recorder or data handling system.

Materials and Miscellaneous Equipment

Vortex mixer, 0.45-μm pore filters (for filtering solvents), Lang-Levy pipettes, laboratory centrifuge, 6 × 50-mm disposable glass test tubes, 3- or 5-ml conical centrifuge tubes, and a 100-μl injection syringe.

Conditions

Column, 3.9 mm i.d. × 30 cm, C_{18} reverse phase, 10 μm particle size; mobile phase, 95 : 5 methanol–water; flow rate 2.5 ml/min, detection, UV 280 nm; sensitivity, 0.01 AUFS, and chart speed 1 cm/min.

Procedures

Sample Preparation

Pipette 50 μl each of retinyl acetate and α-tocopheryl acetate working standards into a 6 × 50-mm disposable glass test tube. Add 100 μl of sample and vortex mix vigorously for 10 sec. Add 100 μl of hexane and vortex mix intermittently and vigorously for 45 sec. Centrifuge at 800 g for 5 min. Transfer 75 μl of the hexane layer to a 3-ml conical centrifuge tube. Evaporate the hexane under a stream of air or nitrogen. Tubes may be placed in a 60° water bath to speed evaporation. Dissolve the lipid residue in 25 μl of diethyl ether. With gentle mixing, add 75 μl of methanol. Using a 10 μl flush of methanol behind the sample, inject 90 μl of the solution into the chromatograph.

Remarks. Step requiring hexane should be carried out in a hood. Alternatively, heptane may be used for extraction. Either plasma or serum can be used as sample. Specimens collected after an overnight fast are preferred. Direct exposure to natural illumination should be avoided. Previous evaluators of this method[4] noted that retinol and α-tocopherol in serum are stable to repeated freezing (−20°) and thawing (17 cycles over a period of 5 weeks). They also note considerable variability in stability among different serum samples, but most remain stable for 1 day at 25°, 4 weeks at 4°, and 1 year at −20° or −70°. If stored for longer periods of time, it is extremely important to add the hexane immediately after the ethanol or to add ascorbic acid (1 g/liter) to the ethanol. Otherwise, most of the retinol, tocopherol, and retinyl acetate are destroyed.

Sample sizes of 100–400 μl do not affect the linearity of the assay, provided the proportion of ethanol to plasma in the initial precipitation of proteins is not changed. Injection volumes of 30–90 μl do not affect linearity of the assay.

Standard Curves and Calculation

A typical standard curve for peak height ratio vs weight ratio for retinol and α-tocopherol is shown in Fig. 1. A constant amount of the acetate form of each vitamin was combined with variable amounts of the corresponding alcohol form of each vitamin to give solution with a 3-fold range of weight ratio.

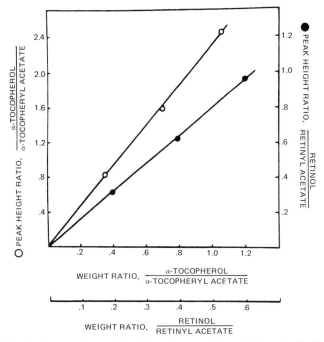

FIG. 1. Standard curves for peak height ratio vs weight ratio for retinol/retinyl acetate (●) and α-tocopherol/α-tocopheryl acetate (○). (Reprinted with permission from *Clinical Chemistry, American Association for Clinical Chemistry,* and *American Journal of Clinical Nutrition,* American Society for Clinical Nutrition.)

Peak height ratios of samples are converted to known quantities of retinol and α-tocopherol from the standard curve as follows:

$$\frac{\text{vitamin peak height of sample}}{\text{vitamin acetate peak height (internal std)}} = R$$

$$\frac{R}{\text{slope of std. curve}} \times \frac{\text{amount of added internal std.}}{\text{volume of sample size}} = \begin{array}{l}\text{vitamin concn in}\\\text{the injected}\\\text{sample}\end{array}$$

The concentration of both vitamins is generally reported in μg/dl after appropriate correction for sample size and dilution.

For laboratories using intergrating recorders or data handling systems standard curves may also be constructed for peak area ratio vs weight ratio.

Remarks. Where possible the peak area ratio method should be used since this method eliminates errors (slight changes in flow rates or retention times) which would affect peak height but not area. The analytical

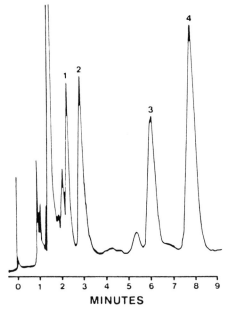

FIG. 2. Chromatogram of normal plasma with internal standards of retinyl acetate and α-tocopheryl acetate added: Peak 1, retinol; 2, retinyl acetate; 3, α-tocopherol; 4, α-tocopheryl acetate. (Reprinted with permission from *Clinical Chemistry, American Association for Clinical Chemistry,* and the *American Journal of Clinical Nutrition,* American Society for Clinical Nutrition.)

validity of the assay is not affected if either α-tocopherol acetate or retinyl acetate is used alone for the internal standard for both vitamins. Most intergrating recorders and some data handling systems only allow one peak to be designated as internal standard. If retinyl acetate is used the total run time is reduced by 2 min.[6]

It is necessary to use α-tocopheryl acetate as internal standard when assaying samples taken from plastic blood collection bags. Some phthalate esters (used as plasticizers) are extracted by the plasma and either cochromatograph or interfere with quantification of retinyl acetate. In addition, tocol has been used successfully as a single internal standard in a simultaneous assay.[5]

A typical chromatogram of a normal plasma sample with internal standards added is shown in Fig. 2. It has been demonstrated[3] that no interfering peaks are present in plasma with retention times of the two internal standards. In addition using plasma from a vitamin A deficient rat and a β-lipoproteinemic human, it has been shown that no peaks occur in plasma which interfere with retinol and tocopherol, respectively.[3]

The analytical validity of this method has been established.[4] Intralaboratory within day and day to day precision, interlaboratory comparisons of three different common plasma pools over a 5-week period, analytical recoveries, and comparison of this method to the colorimetric methods for determination of retinol and α-tocopherol were reported.

Section V

Vitamin K Group

[24] Assay of Coumarin Antagonists of Vitamin K in Blood by High-Performance Liquid Chromatography

By Martin J. Shearer

Most of the vitamin K antagonists currently used in medicine as oral anticoagulant drugs and in pest control as rodenticides belong to a group of compounds with a 3-substituted 4-hydroxycoumarin structure. A second group with a similar action, but a more limited commercial role, are the 2-substituted 1,3-indanediones; their assay being rarely required will not be considered further in this chapter.

4-Hydroxycoumarin anticoagulants are not direct antagonists of vitamin K but act indirectly seemingly by inhibiting one or more of the microsomal enzymes which reduce vitamin K 2,3-epoxide to vitamin K and vitamin K to vitamin K hydroquinone. These reactions normally operate in concert, as part of a metabolic cycle, to supply the reduced vitamin K needed as a cofactor by a carboxylase which converts specific glutamate residues of vitamin K-dependent proteins to γ-carboxyglutamate residues.[1]

As moderately polar compounds with both UV-absorbing and fluorescent properties, 4-hydroxycoumarin antagonists are well suited to analysis by high-performance liquid chromatography (HPLC) and have been successfully assayed in plasma by both reverse-phase[2–7] and normal-phase methods.[8–12] Such systems are also capable of resolving a number of similar coumarin anticoagulants[9,13] and so may form the basis for screening procedures. Since no single chromatographic system has been

[1] J. W. Suttie, in "Nutrition in Hematology" (J. Lindenbaum, ed.), p. 245. Churchill-Livingstone, Edinburgh and London, 1983.

[2] T. D. Bjornsson, T. F. Blaschke, and P. J. Meffin, J. Pharm. Sci. 66, 142 (1977).

[3] L. T. Wong, G. Solomonraj, and B. H. Thomas, J. Chromatogr. 135, 149 (1977).

[4] M. J. Fasco, L. J. Piper, and L. S. Kaminsky, J. Chromatogr. 131, 365 (1977).

[5] M. J. Fasco, M. J. Cashin, and L. S. Kaminsky, J. Liq. Chromatogr. 2, 565 (1979).

[6] L. T. Wong and G. Solomonraj, J. Chromatogr. 163, 103 (1979).

[7] J. X. de Vries, J. Harenberg, E. Walter, R. Zimmermann, and M. Simon, J. Chromatogr. 231, 83 (1982).

[8] R. A. O'Reilly and C. H. Motley, Fed. Proc., Fed. Am. Soc. Exp. Biol. 35, 756 (1976).

[9] J. H. M. van den Berg, J. P. M. Wielders, and P. J. H. Scheeren, J. Chromatogr. 144, 266 (1977).

[10] S. H. Lee, L. R. Field, W. N. Howald, and W. F. Trager, Anal. Chem. 53, 467 (1981).

[11] R. A. R. Tasker and K. Nakatsu, J. Chromatogr. 228, 346 (1982).

[12] M. J. Shearer, Adv. Chromatogr. 21, 243 (1983).

[13] R. Vanhaelen-Fastré and M. Vanhaelen, J. Chromatogr. 129, 397 (1976).

shown to resolve all currently available drugs and rodenticides, a combination of normal-phase and reverse-phase methods may be necessary for the identification of a particular antagonist.

In pharmacological studies it is often necessary to measure blood levels of metabolites in addition to the administered anticoagulant. The concurrent analysis of anticoagulants and their polar metabolites by HPLC is often difficult. In reverse-phase systems the retention time of parent drugs such as warfarin is very long compared to their polar, hydroxylated metabolites and their analysis in a single run is best achieved by gradient elution methods.[5] Greater flexibility is obtained with normal-phase partition HPLC since selectivity is more easily influenced by the mobile phase composition. Polar-bonded phases have been used for the normal-phase separation of warfarin from its major human metabolites with isocratic elution.[10,12] The order of elution of compounds on polar-bonded phases may be difficult to predict; in one method[10] the elution order was exactly opposite to that of the reverse-phase method of Fasco et al.[4] while in another method,[12] warfarin behaved as a more polar compound than two of its metabolites.

Compounds with a 4-hydroxycoumarin ring structure have weakly acidic properties (pK_a warfarin is 5.0) due to their conjugated hydroxycarbonyl moiety. Therefore to obtain retention and good peak shapes with normal- or reverse-phase HPLC it is necessary to suppress ionization by acidification of the mobile phase. For normal-phase HPLC with nonaqueous mobile phases the addition of as little as 0.1% acetic acid is effective although we have found further benefits for the separation of warfarin metabolites when the concentration is increased to about 1%. Similar concentrations of acetic acid are also effective in suppressing ionization in the semiaqueous mobile phases used in reverse-phase HPLC. Alternatively, the pH of the aqueous component of the mobile phase may be precisely controlled with a buffer. Fasco et al.[4] found that the resolution of warfarin metabolites by reverse-phase HPLC is critically dependent on pH making it necessary to buffer the pH of the mobile phase to within 0.1 of a pH unit.

For routine assays of most 4-hydroxycoumarin anticoagulants in blood, UV detection is suitable. Although the wavelength maximum is usually around 280 nm, it is often an advantage to monitor the elution of peaks at a higher wavelength (about 305 nm) since this minimizes interference from coextracted UV-absorbing compounds in plasma. To detect low plasma levels of anticoagulant drugs and their metabolites, fluorescence detection is both more sensitive and selective.[10] One problem, however, is that the native fluorescence of many 4-hydroxycoumarins is severely quenched by the acidic mobile phases commonly used to suppress ionization. This may be overcome by the postcolumn manipulation of pH[10,14] or by the use of ion-pair chromatography using a cationic counter-ion.[15]

Procedures for Plasma Assays of Warfarin with UV Detection

There are three main areas in which an assay of 4-hydroxycoumarin antagonists in blood is most often needed: in therapeutics to monitor compliance in anticoagulant therapy, in toxicology to investigate possible poisoning with vitamin K antagonists, and in pharmacological studies to monitor blood levels of drugs and their metabolites.

With their wide variety of chemical structures it is not within the scope of this chapter to give precise experimental details for the analysis of all 4-hydroxycoumarins. Instead, procedures will be given for the assay in plasma of a typical and widely used clinical anticoagulant, warfarin, by both reverse-phase and normal-phase methods. Both these methods are readily adaptable to the analysis of other anticoagulants with minimal changes in methodology. The reverse-phase method is a modification of the method of Bjornsson et al.[2] with p-chlorowarfarin as an internal standard. The normal-phase method uses a cyanopropyl-bonded support with another clinical anticoagulant, acenocoumarin, as an internal standard. The latter method was developed in our laboratory as a more flexible procedure for the rapid analysis of small volumes of plasma. This method also gives a good separation of warfarin metabolites and may be used to screen for other common 4-hydroxycoumarins in plasma.

Equipment

Both reverse-phase and normal-phase methods may be carried out with a standard HPLC pump and a UV photometer, preferably a variable wavelength instrument able to operate at the optimal wavelength of 305 nm.

Sample extracts are injected via a syringe-loading injection valve (model 7125 from Rheodyne Ins., Cotati, CA) fitted with a 100 μl loop.

Reverse-Phase Method

Reagents

Warfarin [3-(α-acetonylbenzyl)-4-hydroxycoumarin] and p-chloro-warfarin [3-(α-acetonyl-p-chlorobenzyl)-4-hydroxycoumarin] from Sigma Chemical Co.

[14] K. Hunter, J. Chromatogr. 270, 267 (1983).
[15] K. Hunter, J. Chromatogr. 270, 277 (1983).

Sodium phosphate buffer, 0.025 M, pH 4.0. A solution of $NaH_2PO_4 \cdot 2H_2O$ (analytical grade) containing 3.9 g/liter is prepared in glass distilled water and the pH adjusted exactly to 4.0 with a 5% solution of H_3PO_4.[16]

Hydrochloric acid, 5.0 M.

Solvents. 1,2-Dichloroethane, analytical grade from BDH Chemicals Ltd., Poole, England. Dichloromethane and methanol, HPLC grade from Rathburn Chemicals Ltd., Walkerburn, Scotland.

Preparation of Standard Solutions. Separate stock solutions of warfarin and p-chlorowarfarin containing 1 g/liter are prepared in 1,2-dichloroethane and stored in the dark at 4°.

An internal standard solution containing 40 mg/liter of p-chlorowarfarin in 1,2-dichloroethane is prepared by dilution of the stock solution.

A series of calibration standards in 1,2-dichloroethane containing 2, 10, 20, 30, 40, 60, and 80 mg/liter of warfarin together with 40 mg/liter of p-chlorowarfarin are prepared by serial addition and dilution of the stock solutions.

When stored in the dark at 4° all the above solutions are stable for at least 3 months.

Extraction of Plasma. Usually 1 ml of plasma is extracted; where necessary smaller volumes may be extracted with appropriate adjustments being made in the proportion of extract injected onto the HPLC column (see under Sample Injection). The plasma is pipetted into a glass stoppered, centrifuge tube and 50 μl of the internal standard solution (2 μg p-chlorowarfarin) is accurately added. The solution is acidified with 0.5 ml 5 M HCl and extracted with 3 ml dichloromethane by vigorous agitation on a vortex-mixer. Upon centrifuging in a bench-top centrifuge (2000 g) two phases separate, an upper aqueous phase and, beneath a middle layer of precipitated proteins, a lower dichloromethane phase containing extracted warfarin, the internal standard and other lipids. The upper aqueous-phase is removed with a Pasteur pipette and discarded. The lower dichloromethane phase is transferred in the same way to a tapered, glass tube (care being taken to minimize transfer of protein), and the solvent removed under a stream of nitrogen at 60°.

HPLC Conditions

Column packing, Hypersil ODS (5 μm) (Shandon Southern Ltd.)
Column dimensions, 100 × 5 mm (i.d.)
Mobile phase, 35% 0.025 M phosphate buffer, pH 4.0 in methanol
Flow rate, 1 ml/min

[16] P. J. Twitchett and A. C. Moffat, *J. Chromatogr.* **111**, 149 (1975).

Temperature, ambient

Detection, UV, 305 nm

Sample Injection. The dichloromethane extract of 1 ml of plasma is redissolved in 100 μl of mobile phase. The contents are vortex mixed, centrifuged briefly, and 5–10 μl of the supernatant is injected onto the column. When extracting smaller volumes of plasma the proportion of the dichloromethane extract injected onto the column should be increased accordingly.

Calibration. Aliquots of 1 ml of a plasma pool from subjects not taking warfarin are spiked with 50 μl of each of the calibration standards (2–80 mg of warfarin and 40 mg of *p*-chlorowarfarin per liter) so that each tube contains from 0.1 to 4.0 μg of warfarin and 2 μg of *p*-chlorowarfarin. The contents of each tube are extracted as described above for plasma samples, the extracts redissolved in 100 μl of mobile phase and 5–10 μl of each injected onto the column. From the resulting chromatograms (Fig. 1)

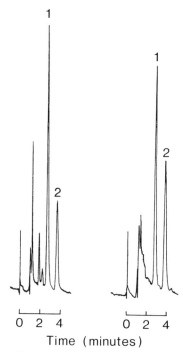

Fig. 1. Chromatograms illustrating the assay of warfarin in plasma by reverse-phase HPLC and showing the separation of warfarin (peak 1) from the internal standard, *p*-chlorowarfarin (peak 2) for dichloromethane extracts of plasma from a patient taking warfarin (left) and a spiked plasma standard containing 1.0 μg/ml of warfarin and 2.0 μg/ml of *p*-chlorowarfarin (right). Procedures and chromatographic conditions are those described in the text for the reverse-phase method.

the peak heights of warfarin and *p*-chlorowarfarin are measured and a calibration graph constructed in which the peak height ratios of warfarin to internal standard are plotted against the equivalent weight ratios. The graph is linear and passes through the origin (Fig. 2). An identical calibration graph is obtained when standard solutions are injected directly onto the column (i.e., without addition to and extraction from plasma).

Calculation. A typical chromatogram of a plasma extract is shown in Fig. 1. From the peak height ratio of warfarin to the internal standard given by the sample, an equivalent weight ratio may be obtained by reference to the calibration graph (Fig. 2). Multiplication of this weight ratio by the weight of internal standard originally added gives the amount of warfarin in the volume of plasma extracted.

Precision and Recovery. The within-run precision and recovery of the above reversed-phase method have been established in our laboratory by replicate analyses of four plasma pools to which warfarin (in 0.02 *M* NaOH) had been added so that the final plasma concentrations ranged

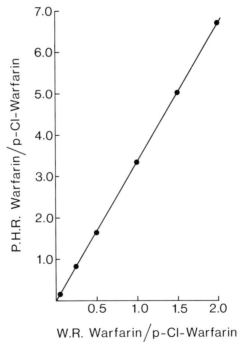

FIG. 2. Calibration graph for the assay of warfarin in plasma by the reverse-phase method. The calibration of peak height ratios (P.H.R.) against weight ratios (W.R.) of warfarin/*p*-chlorowarfarin is shown for a plasma warfarin range of 0.1–4.0 μg/ml (W.R. 0.05–2.0) when 1 ml plasma is extracted.

TABLE I
WITHIN-RUN PRECISION AND RECOVERY OF REVERSE-PHASE
METHOD FOR PLASMA WARFARIN[a]

Plasma warfarin concentration (μg/ml)		Coefficient of variation (%)	Mean recovery (%)
Prepared	Measured		
0.1	0.09	3.3	90.0
0.5	0.45	3.4	90.0
1.0	0.98	3.5	98.0
2.0	1.88	2.3	94.0

[a] Five replicate analyses at each plasma concentration.

from 0.1 to 2.0 μg/ml. As shown in Table I the precision (2–4%) and recovery (90–98%) were similar at all plasma levels within this range.

Normal-Phase Method

Reagents

Warfarin [3-(α-acetonylbenzyl)-4-hydroxycoumarin] from Sigma Chemical Co.

Acenocoumarin or nicoumalone [3-(α-acetonyl-*p*-nitrobenzyl)-4-hydroxycoumarin] from Geigy Pharmaceuticals.

Solvents. Methyl *t*-butyl ether, hexane, propan-2-ol, and dichloromethane, all HPLC grade from Rathburn Chemicals Ltd., Walkerburn, Scotland.

Acids. Hydrochloric acid, 5.0 *M* and glacial acetic acid (analytical or aldehyde-free grade) from BDH Chemicals Ltd., Poole, England.

Preparation of Standard Solutions. Separate stock solution of warfarin and acenocoumarin with concentrations of 1 and 0.5 g/liter, respectively, are prepared in 1,2-dichloroethane and stored in the dark at 4°. From these stocks are accurately prepared dilute stock solutions of warfarin and acenocoumarin containing 100 and 20 mg/liter, respectively.

The working solution of the internal standard has a concentration of 2 mg/liter of acenocoumarin in methyl *t*-butyl ether. To prepare this, 5 ml of the dilute stock solution of acenocoumarin (20 mg/liter) is accurately transferred to a 50-ml volumetric flask, the dichloroethane removed under a stream of nitrogen at 60°, and the acenocoumarin redissolved in 50 ml methyl *t*-butyl ether.

A series of calibration standards in 1,2-dichloroethane containing 2, 10, 20, 30, 40, 60, and 80 mg/liter of warfarin together with 20 mg/liter of acenocoumarin are prepared from their respective concentrated stock solutions. For use in the calibration of the method (by direct injection onto columns or the addition to and extraction from plasma), aliquots of these standards are evaporated to dryness and a 10-fold dilution made with methyl *t*-butyl ether.

For testing the accuracy of the calibration, solutions of warfarin containing from 1.0 to 10.0 μg/ml in 0.01 M NaOH are prepared by evaporation and dilution of appropriate volumes of the 100 mg per liter warfarin stock.

With the exception of solutions in methyl *t*-butyl ether all the above stock and standard solutions are kept in the dark at 4° and are stable for several months. The working internal standard solution in methyl *t*-butyl ether is stored in a dark cupboard. If kept tightly stoppered this solution is stable for several weeks. However, to guard against the possible evaporation of ether from the working internal solution, periodic tests are carried out in which the solutions of warfarin in 0.01 M NaOH with accurately known concentrations are taken through the assay procedure in place of plasma, and the calibration of the assay checked.

Extraction of Plasma. Plasma (100 μl) is pipetted into a micro, capped, centrifuge tube (polypropylene, 1.5-ml capacity) and 100 μl of the internal standard solution (0.2 μg acenocoumarin) is accurately added. Further volumes of 100 μl methyl *t*-butyl ether and 50 μl 5 M HCl are added and the contents mixed briefly but vigorously with a vortex mixer. After centrifuging for 5 min on a microcentrifuge, the contents separate into an upper ether phase and a lower aqueous-phase with a protein layer sandwiched between them.

HPLC Conditions

Column packing, Spherisorb nitrile (5 μm) (Phase Separations Ltd.)
Column dimensions, 250 × 5 mm (i.d.)
Mobile phase, 82% hexane, 12% propan-2-ol, 5% dichloromethane, and 1% acetic acid
Flow rate, 2 ml/min
Temperature, ambient
Detection, UV, 305 nm

Sample Injection. Volumes varying from 30 to 70 μl of the upper methyl *t*-butyl ether phase are injected directly onto the column.

Calibration. The method is initially calibrated with plasma as follows. Aliquots of 100 μl of plasma (containing no warfarin) are spiked with 100 μl of each of the calibration standards in methyl *t*-butyl ether (0.2–8 mg of warfarin and 2.0 mg of acenocoumarin per liter) so that each tube contains from 0.02 to 0.8 μg of warfarin and 0.2 μg of acenocoumarin. After addition of 100 μl methyl *t*-butyl ether and 50 μl 5 *M* HCl, the tubes are extracted as described above for plasma samples and 50 μl of the ether extract is injected onto the column. From the resulting chromatograms (Fig. 3), a calibration graph is constructed of peak height ratios of warfarin/acenocoumarin against their equivalent weight ratios. The graph is linear and passes through the origin (Fig. 4). Since an identical graph is obtained with pure standard solutions, in routine use the extraction step may be omitted and the calibration graph obtained by the direct injection of standards.

Calculation. The plasma concentration of warfarin is calculated by reference to the peak height ratio of the sample (Fig. 3) and the calibration graph (Fig. 4) in exactly the same way as described previously for the reverse-phase method.

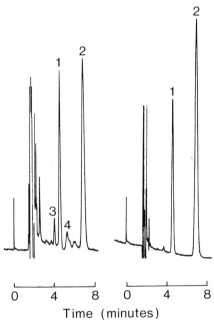

Time (minutes)

FIG. 3. Chromatograms illustrating the assay of warfarin in plasma by normal-phase HPLC and showing the separation of warfarin (peak 1) from the internal standard, acenocoumarin (peak 2) for *t*-butyl ether extracts of plasma from a patient taking warfarin (left) and a spiked plasma standard containing 1.0 μg/ml of warfarin and 2.0 μg/ml of acenocoumarin. The small peaks denoted 3 and 4 show the elution of warfarin alcohol and 7-hydroxywarfarin metabolites, respectively. Procedures and chromatographic procedures are those described in the text for the normal-phase method.

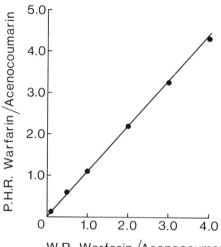

W.R. Warfarin/Acenocoumarin

Fig. 4. Calibration graph for the assay of warfarin in plasma by the normal-phase method. The calibration of peak height ratios (P.H.R.) against weight ratios (W.R.) of warfarin/acenocoumarin is shown for a plasma warfarin range of 0.2–8.0 μg/ml (W.R. 0.1–4.0) when 100 μl plasma is extracted.

Precision and Recovery. The within-run precision of the normal-phase method with plasma pools spiked with warfarin to concentrations of 0.5–4.0 μg/ml ranged from 1 to 5% and the recovery of warfarin ranged from 97 to 106% (Table II). In a "blind" analysis of 10 plasma samples from patients (warfarin concentrations 0.5–4.5 μg/ml) the between-run precision for five replicate analyses carried out over a period of 3 months ranged from 3 to 9% and averaged 6%.

TABLE II

WITHIN-RUN PRECISION AND RECOVERY OF NORMAL-PHASE
METHOD FOR PLASMA WARFARIN[a]

Plasma warfarin concentration (μg/ml)		Coefficient of variation (%)	Mean recovery (%)
Prepared	Measured		
0.5	0.53	5.1	106.0
1.0	0.97	3.4	97.0
2.0	2.05	2.7	102.5
4.0	3.98	1.3	99.5

[a] Five replicate analyses at each plasma concentration.

Comments

Both the normal- and reverse-phase methods described above are capable of the rapid, selective, and reproducible analysis of warfarin in plasma samples.

In the reverse-phase method, warfarin metabolites, being more polar than warfarin, elute before the parent drug. Although they do not interfere with the analysis of warfarin, their retention is too short to allow their individual measurement. The normal-phase method, however, conveniently resolves warfarin from its major metabolites, namely, warfarin alcohols (both pairs of diastereoisomers eluting as a single peak), 6-, 7-, 8-, and 4'-hydroxywarfarin, and would allow these to be measured concurrently if present in sufficient concentrations. Even when as little as 100 μl of plasma is extracted, peaks corresponding to warfarin alcohols and 7-hydroxywarfarin are regularly detected in plasma from patients on warfarin therapy and occasionally a small peak is seen which corresponds to 6-hydroxywarfarin. If there is a need for the precise measurement of these metabolites, a further increase in sensitivity would be desirable either by increasing the volume of plasma extracted or by using a more sensitive detection method such as fluorimetry.[10]

With photometric detection at 305 nm the sensitivity of both methods (for equivalent volumes of plasma injected) was similar and less than 0.1 μg/ml. This is below the plasma level at which warfarin would be expected to exert a pharmacological effect and is adequate for most clinical and toxicological purposes. Using the reverse-phase assay the plasma concentrations in 56 patients, whose anticoagulant control with warfarin was both stable and in the therapeutic range, ranged from 0.7 to 3.2 μg/ml (mean 1.6 μg/ml). These values are almost identical to those obtained by others[17] using a combined thin-layer chromatographic and fluorometric assay which though time-consuming is highly specific.

Our modifications to the reverse-phase method of Bjornsson et al.[2] included using a phosphate buffer instead of acetic acid to maintain a low pH, a different packing material, and a shorter column length. These modifications did not alter the excellent precision of the method (Table I) which was very similar to that obtained by Bjornsson et al.[2] If preferred, the methanol component of the reverse-phase mobile phase may be replaced with a similar concentration of acetonitrile which will give a higher column efficiency at a lower pressure. Against these advantages, ace-

---

[17] A. Breckenridge and M. L'E. Orme, in "Biological Effects of Drugs in Relation to their Plasma Concentrations" (D. S. Davies and B. N. C. Prichard, eds.), p. 145. Macmillan, New York, 1973.

tonitrile is more expensive and has a greater inhalational toxicity than methanol.

From the practical point of view we prefer the design of the normal-phase method because of the facility in being able to make direct injections of the extracting solvent onto the HPLC column. In this respect the solvent methyl t-butyl ether is an excellent extracting solvent and compared to other ethers has a favorable boiling point and a low risk from peroxide formation. Methyl t-butyl ether is also compatible with the mobile phase of the normal-phase system and relatively large volumes (up to 70 μl) may be injected without significant loss in column efficiency. In marked contrast, the injection of organic solvents onto reverse-phase columns which are being eluted with a semiaqueous mobile phase invariably leads to severe band spreading resulting in poor peak shapes and resolution. In their reverse-phase method for warfarin, Fasco et al.[4] avoided an evaporation step by the partition of plasma between saturated sodium borate and chloroform followed by the direct injection of the aqueous phase. Later, they reported[5] that this method resulted in increased column pressures and a shortened column life due to the incomplete removal of protein and lipid components from the injected supernatant. We also experienced problems with the stability of our reverse-phase columns which in the short term we also attributed to the accumulation of insoluble material originating from the sample at the top of the column. Another problem in maintaining column performance is the slow dissolution of the silica matrix with mobile phases containing water. To some extent these problems with reverse-phase columns may be alleviated by using a guard column to protect the analytical column or by filtration of the sample. Such precautions were not necessary with the normal-phase method and the cyano-bonded columns have shown excellent long-term stability and reproducibility. Although the column in current use in our laboratory is used only intermittently, it has shown no significant deterioration over a 3-year period. When not in use the column may be left without harm in situ with the mobile phase.

Apart from its use for the assay of warfarin in plasma, the normal-phase method is also capable of resolving other clinical oral anticoagulants. With no modification to the mobile phase composition, the method may be used for the assay of acenocoumarin (sintrom, used as the internal standard for the warfarin assay), tromexan (ethyl biscoumacetate), and phenprocoumon (marcoumar). Bishydroxycoumarin (dicumarol) elutes near the solvent front but could be assayed with slight adjustments to make the mobile phase less polar.

[25] Assay of K Vitamins in Tissues by High-Performance Liquid Chromatography with Special Reference to Ultraviolet Detection

By MARTIN J. SHEARER

Of all the fat-soluble vitamins, the K vitamins have proved one of the most difficult to measure in biological tissues by physicochemical methods. Tissue concentrations are generally low even in organisms able to synthesize them. Problems concerning their isolation and measurement are compounded by the existence in many tissues of several molecular forms with biological activity. Again, their instability toward alkali precludes the use of saponification to remove bulk lipids, a procedure that has facilitated the isolation of other fat-soluble vitamins.

In cells, K vitamins are largely concentrated in membranous fractions. Phylloquinone (vitamin K_1) synthesized by plants is mainly located in green areas and concentrated in the chloroplast lamellae whereas menaquinones (vitamins K_2) are synthesized by bacteria and are located in the plasma membrane.[1] In animal cells, the vitamin is found in various cellular membranes especially those of the microsomes and mitochondria.[2]

The main features of early physicochemical methods for the assay of phylloquinone in plants were the separation of the vitamin by paper, column, or thin-layer chromatography followed by quantitation by colorimetry,[3] UV spectroscopy,[4] or simple visual comparison with known amounts of phylloquinone on the chromatograms.[5] Such methods have not always proved reliable owing to limitations of sensitivity, difficulties in correcting for losses, and problems of chromatographic resolution such as that in resolving phylloquinone from plastoquinone.[1]

Until the late 1960s the isolation of K vitamins from animal tissues was an intractable problem beyond the resolving power of traditional chromatographic methods. Even with the introduction of more sophisticated reverse-phase[6] and silver ion[7] chromatographic procedures, small

---

[1] J. F. Pennock, *Vitam. Horm. (N.Y.)* **24**, 307 (1966).

[2] J. W. Suttie, *in* "The Fat-Soluble Vitamins" (H. F. DeLuca, ed.), p. 211. Plenum, New York, 1978.

[3] H. K. Lichtenthaler, *Planta* **57**, 731 (1962).

[4] H. K. Lichtenthaler and M. Calvin, *Biochim. Biophys. Acta* **79**, 30 (1964).

[5] K. Egger, *Planta* **64**, 41 (1965).

[6] J. T. Matschiner and W. V. Taggart, *Anal. Biochem.* **18**, 88 (1967).

[7] J. T. Matschiner and J. M. Amelotti, *J. Lipid Res.* **9**, 176 (1968).

METHODS IN ENZYMOLOGY, VOL. 123

amounts of lipid contaminants often persisted throughout the purification procedure.[8]

The procedures, both spectroscopic and chromatographic, that were available for the analysis of K vitamins in 1970 have been reviewed in a previous volume of this series.[9] The recent growth of high-performance liquid chromatography (HPLC) has brought two major advantages to vitamin K analysis. First, advances in column technology have given an extra dimension to resolution and, second, developments in detectors have provided the means for sensitive on-line detection; in the beginning with UV detection and lately with even more sensitive detection based on amperometric or fluorescence measurements.

To date, there is no single assay which can simultaneously measure all the molecular forms of vitamin K in a given tissue. This chapter will mainly describe procedures which have been developed in our laboratory for the HPLC assay of phylloquinone in plant and animal tissues using UV detection. The assay of phylloquinone using electrochemical detection and the analysis of bacterial menaquinones is described elsewhere in this volume. To date, there has been little work on the analysis of menaquinones in animal tissues.

Assay Design

To assay K vitamins in biological tissues by HPLC, it is usually necessary to employ a multistage procedure. For example, the assay of phylloquinone with UV detection can have the following stages: (1) extraction of tissues, (2) preliminary purification of lipid extracts, and (3) high-performance liquid chromatography: (a) semipreparative adsorption HPLC and (b) analytical reverse-phase HPLC. The factors that influence both the overall design of the assay and the choice of procedures for individual stages will now be considered.

Extraction of Tissues

The extraction of K vitamins from tissues may be successfully accomplished by a number of methods; that chosen will depend on the nature of the tissue, ease of manipulation, and individual preference.

To extract plant and liver tissues, we use a method in which tissues are ground in a mortar and extracted into acetone.[3] Apart from its inherent

---

[8] J. T. Matschiner, "The Biochemistry, Assay and Nutritional Value of Vitamin K and Related Compounds," p. 21. Association of Vitamin Chemists, Chicago, Illinois, 1971.
[9] P. J. Dunphy and A. F. Brodie, this series, Vol. 18, p. 407.

simplicity the method has the advantage that after adding water and hexane to the acetone extract, the nonpolar K vitamins partition entirely in the upper hexane phase. Although the acetone extracts are routinely shaken on a mechanical shaker, additional disruption procedures, such as ultrasonic disintegration or refluxing with hot acetone in a Soxhlet extractor, do not increase extraction efficiencies. Extraction with alternative solvents such as ethanol–diethyl ether (3 : 1, v/v) or hot isopropanol in conjunction with acetone[10] did not increase the amounts of phylloquinone extracted from green leafy tissues. Although the extraction of phylloquinone from plant tissues has been reported to be difficult,[11] we find acetone extraction is both reproducible and exhaustive with only 1%, or less, of further extractable phylloquinone remaining in the residue. Freeze-drying of tissues before extraction is not recommended since this resulted in losses in the recovery of the vitamin from plant tissues.[12]

To extract liquid tissues, such as milk, the method of Folch with chloroform–methanol (2 : 1, v/v) has been successfully used.[13] Its disadvantage is that excessive amounts of polar lipid are also extracted, especially phospholipids. A milder procedure which has been extensively used to extract phylloquinone from serum or plasma[14–16] is to extract with hexane after flocculation of proteins with ethanol. To ensure good recovery of the vitamin the ratio of ethanol to serum should be 2 : 1 (v/v).[16]

Preliminary Purification of Lipid Extracts

Depending on the sensitivity of detection of the analytical HPLC stage which in turn governs the amount of starting material to be extracted, a preliminary purification step may be necessary to remove excessive amounts of polar lipids.[10,17] In our laboratory we originally used conventional column chromatography on silica gel[13,18] but this step has now been made easier by the commercial availability of Sep-Pak silica cartridges

[10] J. N. Thompson, G. Hatina, and W. B. Maxwell, *NBS Spec. Publ. (U.S.)* **519**, 279 (1979).

[11] L. P. Kegel and F. L. Crane, *Nature (London)* **194**, 1282 (1962).

[12] Y. Haroon, Ph.D. Thesis, London University, U.K. (1981).

[13] Y. Haroon, M. J. Shearer, S. Rahim, W. G. Gunn, G. McEnery, and P. Barkhan, *J. Nutr.* **112**, 1105 (1982).

[14] M. F. Lefevere, A. P. De Leenheer, A. E. Claeys, I. V. Claeys, and H. Steyaert, *J. Lipid Res.* **23**, 1068 (1982).

[15] M. J. Shearer, S. Rahim, P. Barkhan, and L. Stimmler, *Lancet* **2**, 460 (1982).

[16] T. Ueno and J. W. Suttie, *Anal. Biochem.* **133**, 62 (1983).

[17] M. J. Shearer, *Adv. Chromatogr.* **21**, 243 (1983).

[18] M. J. Shearer, V. Allan, Y. Haroon, and P. Barkhan, *in* "Vitamin K Metabolism and Vitamin K-Dependent Proteins" (J. W. Suttie, ed.), p. 317. Univ. Park Press, Baltimore, Maryland, 1980.

(Waters Associates) which have been designed for the rapid clean-up of samples.

High-Performance Liquid Chromatography (HPLC)

The theory and application of HPLC to the analysis of K vitamins and their analogs have recently been reviewed.[17] K vitamins are nonpolar molecules and they are ideally suited to both adsorption and reverse-phase modes of HPLC. Adsorption HPLC is mainly, but not exclusively, used for separations according to gross differences in the polarity of lipids (e.g., separation of lipid classes) while reverse-phase HPLC is used to separate lipids with similar polarities and/or structures (e.g., separation of homologous series).

For the analysis of plant tissues or blood plasma containing pharmacological concentrations of phylloquinone, an assay based on a single HPLC stage may be sufficient using either adsorption or reverse-phase systems.[19,20]

To measure endogenous concentrations of phylloquinone in foods and animal tissues, including blood plasma, a procedure with two sequential stages of HPLC is often necessary to enhance the resolution of the vitamin from interfering lipids. This is particularly true of assays based on UV detection which is not very selective.[10,14,17] But two HPLC stages have also been needed for assays based on electrochemical detection.[16,21] In all these methods adsorption HPLC has been used as a semipreparative purification stage from which a phylloquinone-containing fraction is collected and analyzed by a second reverse-phase stage. With electrochemical detection a semipreparative stage is desirable not only to increase selectivity by removing electroactive interferences but also to reduce the amount of lipid. This is necessary because the solubility of nonpolar lipids in the semiaqueous mobile phases needed for amperometric detection is low and this severely restricts the limit of sensitivity by reducing the amount of lipid that can be injected onto the column. With UV detection this problem may be partially overcome by using a highly retentive octadecylsilane (ODS)-bonded support with a completely nonaqueous mobile phase in which phylloquinone and coextracted lipids are freely soluble.[17] Other advantages of nonaqueous mobile phases are their low viscosity which need a lower pumping pressure, give higher column efficiencies, and confer long-term stability to the column bed.[17]

[19] H. K. Lichtenthaler and U. Prenzel, J. Chromatogr. 135, 493 (1977).
[20] A. C. Wilson and B. K. Park, J. Chromatogr. 277, 292 (1983).
[21] J. P. Hart, M. J. Shearer, P. T. McCarthy, and S. Rahim, Analyst 109, 477 (1984).

Detection

Assays of phylloquinone by HPLC have been described with UV absorption[10,13–15,17] and electrochemical[16,21,22] and fluorometric detection.[23,24] Published assays using photometric detectors have had a limit of detection of 500–1000 pg while even lower amounts can be detected by electrochemical and fluorometric assays. While UV detection is the least sensitive and selective, photometric detectors are widely available and their stability and ease of operation are generally better than for electrochemical detectors. Naturally occurring K vitamins do not exhibit native fluorescence and fluorometric assays are based on the fluorescence exhibited by the quinol form or other products of photodecomposition.[23] Fluorometric assays therefore require a postcolumn derivatization step either with a photochemical reactor[23] or by chemical[23] or electrochemical[24] reduction. The potential advantage of fluorometric assays is that their sensitivity is coupled to a much greater selectivity than either UV or electrochemical detection. Problems of detector drift found with assays based on reductive electrochemical detection[16,21] may be resolved by using a dual electrode detector in the redox mode.[22,24]

Quantitation

Because of the multiple chromatographic steps, many assays for phylloquinone have incorporated an internal standard, either a compound structurally related to vitamin K[17] or tritium-labeled phylloquinone of high specific activity.[14,16] Phylloquinone 2,3-epoxide and *cis*-2-chlorophylloquinone can be used as internal standards with UV detection and 2′,3′-dihydrophylloquinone with electrochemical detection.

Ideally, for assays which have two stages of HPLC, the internal standard should elute with exactly the same retention as phylloquinone in the semipreparative stage. In practice, this has been difficult to achieve other than with phylloquinone 2,3-epoxide which has the disadvantage of being a natural metabolite of phylloquinone. On the other hand, significant tissue concentrations of the epoxide relative to phylloquinone occur only in animals treated with coumarin anticoagulants. In our experience phylloquinone 2,3-epoxide has been a most valuable and reliable internal standard for the assay by UV detection of phylloquinone in foods and plant

[22] Y. Haroon, C. A. W. Schubert, and P. V. Hauschka, *J. Chromatogr. Sci.* **22**, 89 (1984).

[23] M. F. Lefevere, R. W. Frei, A. H. M. T. Scholten, and U. A. T. Brinkman, *Chromatographia* **15**, 459 (1982).

[24] J. P. Langenberg and U. R. Tjaden, *J. Chromatogr.* **305**, 61 (1984).

tissues as well as for animal tissues such as blood and livers. Positive advantages of phylloquinone epoxide as an internal standard compared to other compounds we have tried are its chromatographic properties, ease of synthesis, and long-term stability on storage.

Validation

The validity of an HPLC assay may be examined from several standpoints such as peak identity, precision, and recovery. UV spectroscopy and/or mass spectrometry have been used to examine peak identity in the photometric HPLC assay of plant tissues, milks, animal livers, and blood plasma. It is usually necessary to collect the putative phylloquinone peak from many injections and pool the fractions for further analysis. To obtain UV absorption spectra of collected fractions we have used a procedure which may be carried out on-line with HPLC using a detector capable of stopped-flow wavelength scanning. This is a useful technique since the UV absorption spectrum of K vitamins is very characteristic and the

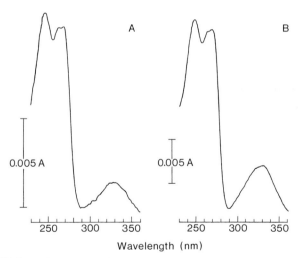

Wavelength (nm)

FIG. 1. UV absorption spectra of authentic phylloquinone (K_1) (A) and a pooled phylloquinone fraction collected from the final analytical stage of HPLC during the assay of rabbit liver (B). The spectra were obtained on-line with HPLC by stopped-flow scanning of eluting peaks as described by Krstulovic *et al.*[25] Column: Partisil-5 silica (Whatman). Mobile phase: 0.6% acetonitrile in hexane. Detector: Schoeffel model SF 770 spectrophotometric detector equipped with SFA 339 wavelength drive and MM 700 memory module. Absorbance range: 0.02 *A* full scale (authentic K_1) and 0.04 *A* full scale (liver K_1).

PRECISION OF HPLC ASSAY FOR PHYLLOQUINONE USING UV DETECTION
AND PHYLLOQUINONE 2,3-EPOXIDE AS INTERNAL STANDARD

Tissue or food	Mean phylloquinone content (ng/g or ng/ml)	Number of determinations	Within-run coefficient of variation (%)
Kale	7290 (ng/g)	20	4.7[a]
Green beans	460	10	5.2[a]
Infant formula	262	8	6.9[a]
Infant formula	25	6	2.5[a]
Human liver	4.8	7	22.9[b]
Human milk	2.5 (ng/ml)	60	8.6[c]
Cow's milk	4.9	24	8.6[c]
Human plasma	232	8	3.0[b]
	72	8	4.0[b]
	13	7	7.4[b]
	0.42	6	11.0[b]

[a] Calculated from the variation between the values of duplicate analyses for the same sample performed over several weeks.

[b] Calculated from replicate analyses for the same sample performed in the same run.

[c] Calculated from the variation between the values of duplicate analyses for different samples performed over several weeks.

spectrum of a chromatographic peak can be obtained with good resolution at an absorbance range of 0.02 A full scale (Fig. 1).[25]

The precision of the HPLC assay using UV detection and phylloquinone 2,3-epoxide as an internal standard has been determined for tissues of both vegetable and animal origin (see the table). The assay of leaf tissues from plants gave a higher coefficient of variation than was expected from their high concentrations of phylloquinone. Most of this error was found to lie outside the chromatographic procedures and could be attributed to difficulties in obtaining uniform tissue samples or to extraction differences. For example, although the overall error for the assay of kale was 5%, the coefficient of variation for replicate analyses of lipid extracts was only 1–2%. One of the most difficult tissues to analyze is human liver; the low concentrations are difficult to resolve from interfering lipids and this is reflected by the relatively poor coefficient of variation

[25] A. M. Krstulovic, R. A. Hartwick, P. R. Brown, and K. Lohse, *J. Chromatogr.* **158**, 365 (1978).

which is usually more than double that for other tissues with comparable concentrations.

Procedures for Assay of Phylloquinone with UV Detection

Details of the procedures used in our laboratory for the assay of phylloquinone in various tissues are described below. Taken together, the methods described here constitute an assay procedure with UV detection. On the other hand individual procedures such as those for the extraction, preliminary purification, and semipreparative HPLC have also been used to prepare samples for the assay of phylloquinone by electrochemical detection.[21]

General Precautions

Stability. K vitamins are sensitive to light and alkaline conditions but are reasonably stable toward heat, oxygen, and acidic conditions. It is important that all manipulations be carried out away from direct sunlight and under subdued artificial light. Solvents may be removed at temperatures of 60° either under reduced pressure in a rotary evaporator or in a water bath under a stream of nitrogen.

Storage. Phylloquinone in whole tissues, in dry lipid extracts, or in isolated fractions may be safely stored in the dark at −20°. Human milk stored under these conditions showed no significant difference in phylloquinone content after 5 months.[13]

Solvents. HPLC grade solvents are used for all chromatographic procedures. HPLC grade hexane and methanol are also used for tissue extractions. Other solvents used for extractions such as acetone, chloroform, and ethanol are of analytical grade and are used without further purification.

Contamination. With the improved sensitivity of HPLC detectors, care is essential to avoid contamination of samples from extraneous sources. Possible sources of contamination are rubber fittings, lubricants (e.g., from rotary evaporators) and vitamin K standards in HPLC (e.g., from syringes or valve injectors). All glass tubes and flasks are washed with acetone immediately before use.

Extraction Methods

Extraction with Acetone (Example: Plant and Liver Tissues). Tissue samples, weighing 5–50 g (fresh weight), are chopped into small fragments and ground thoroughly with fine quartz granules to a paste-like consistency with a pestle and mortar. Acetone (about 50 ml) is added and the tissue is ground further. The contents of the mortar are quantitatively transferred with acetone (total volume about 150 ml) to a 250-ml capacity glass-stoppered conical flask and agitated on a mechanical shaker for 30 min. The acetone extract is filtered under reduced pressure through a Büchner funnel with a sintered glass filter (porosity 3) and the residue on the filter washed with a small portion of acetone (50–100 ml). The acetone filtrate is transferred to a separating funnel (1-liter capacity) and an approximately equal volume of hexane added (using a small portion of this to rinse the filter flask). Distilled water (total volume about 250 ml) is then added slowly down the side of the separating funnel and the contents left to stand (about 10 min) until two phases separate: an upper hexane phase containing vitamin K and other lipids and a lower aqueous–acetone phase. The lower phase is run off and discarded. The upper hexane phase is washed with a further volume of water and the lower aqueous phase again discarded. The hexane extract is transferred to a flask and dried over anhydrous sodium sulfate. The sodium sulfate is removed by filtration through a Büchner funnel (the filter being rinsed with a small volume of hexane) and the extract taken to dryness in a rotary evaporator. The lipid residue is transferred with hexane to a weighing bottle and the hexane removed under a stream of nitrogen until a constant lipid weight is attained.

Extraction with Hexane (Example: Blood): Separated serum or plasma (5-ml aliquots) is pipetted into a glass stoppered centrifuge tube (35-ml capacity) and 2 volumes of ethanol and 4 volumes of hexane added. The mixture is vigorously mixed (by hand and on a vortex mixer) and centrifuged to separate an upper hexane phase from a lower aqueous–ethanolic phase and precipitated proteins. The upper hexane layer is transferred to a clean tube and the lower phase reextracted with a further four volumes of hexane. The two hexane extracts are pooled and evaporated to dryness under a stream of nitrogen.

Extraction with Chloroform–Methanol (Example: Milk and Milk Formulas). Volumes of 5–20 ml milk or reconstituted milk formulas are extracted directly in a glass-stoppered conical flask with 20 volumes of a chloroform–methanol mixture (2 : 1, v/v).[26] After mixing on a mechanical shaker for 30 min, precipitated proteins are removed by centrifugation or filtration under reduced pressure through a glass microfibre (GF/A filters from Whatman Ltd., Maidstone, U.K.). Further small portions of the chloroform–methanol mixture are used to rinse the flask and the filter (where used), and the washings pooled with the extract. To the chloroform–methanol extract is added 0.2 its volume of water. Upon centrifug-

[26] J. Folch, M. Lees, and G. H. Sloane Stanley, *J. Biol. Chem.* **226,** 497 (1957).

ing or standing, two phases separate: an upper phase which is discarded and a lower chloroform phase containing vitamin K and other lipids which is evaporated to dryness in a rotary evaporator or under nitrogen.

Preliminary Purification of Lipid Extracts

Column Chromatography (Gravity Flow). Conventional column chromatography of lipid extracts is performed on slurry-packed columns of kieselgel 60, particle size 0.2–0.5 mm (E. Merck, A. G., Darmstadt, Germany). The silica is activated at 110° for 1 hr and packed into glass columns (10–15 mm diameter fitted with glass stopcocks). Each 1 g of lipid is chromatographed on 20 g silica and each fraction eluted with 200 ml solvent. Lipid extracts are quantitatively applied onto columns in 2–5 ml hexane. The columns are developed by stepwise elution, first with hexane to elute a hydrocarbon fraction and then with 3% diethyl ether in hexane to elute a vitamin K-containing fraction. The latter fraction is collected and taken to dryness in a rotary evaporator.

Sep-Pak Silica Cartridges. Sep-Pak silica cartridges (Waters Associates, Northwich, U.K.) are prepacked cartridges of silica designed for rapid sample preparation and through which solvents may be pumped by attachment to the Luer end fitting of a glass syringe.

Each silica cartridge is first washed with 10–20 hexane. The lipid extract is dissolved in 2 ml hexane, introduced into the glass syringe, and pumped through the cartridge to concentrate the sample at the head of the cartridge. The tube which contained the lipid is rinsed with a further 2 ml hexane and this loaded onto the cartridge in the same way. After loading, the cartridge is eluted with 10 ml hexane to elute a hydrocarbon fraction (discarded) followed by 10 ml of 3% diethyl ether in hexane to elute a fraction containing phylloquinone. The eluate is collected and the solvent removed under a stream of nitrogen.

We have found that up to 100 mg of a lipid extract from liver may be processed without overloading the cartridge and consequent losses of phylloquinone.

HPLC Equipment

For the two-stage HPLC assay, it is convenient to have two chromatographs dedicated to the adsorption and reverse-phase stages, respectively. If only one chromatograph is available, care should be taken when changing from reverse-phase to adsorption columns to eliminate all traces of polar solvents from the system. Contamination of silica columns with only trace amounts of methanol will result in column deactivation with consequent loss of retention for phylloquinone.

Pumps. For the semipreparative stage of HPLC we use a standard, twin-piston, reciprocating HPLC pump. For the analytical stage we find a constant pressure (pneumatic amplifier) pump particularly advantageous in providing the pulseless flow required at maximum sensitivities. Some modern reciprocating pumps also meet these requirements or the problem may be overcome by the addition of a pulse damper.

UV Detectors. UV detection in the final analytical stage should be carried out with a high-performance instrument with an absorbance range down to 0.005 A full scale and preferably with a variable wavelength facility. Although the semipreparative HPLC stage is also monitored by a UV detector, sensitivity requirements are not critical and fixed wavelength detectors (e.g., 254 nm) are also suitable.

Sample Injector. Samples are dissolved in mobile phase and up to 70 μl injected via a syringe-loading injection valve (model 7125 from Rheodyne Inc., Cotati, CA) fitted with a 100-μl loop.

Semipreparative Adsorption HPLC

HPLC Conditions

Column packing, Partisil 5 silica (5 μm) (Whatman Inc.)
Column dimensions, 250 × 5 mm (i.d.)
Mobile phase, 15–22% dichloromethane (50% water-saturated) in hexane
Flow rate, 1 ml/min
Temperature, ambient
Detection, UV, 270 nm (or 250 nm)
Sample injection, 70 μl in mobile phase (100-μl loop)
Retention time, 8–10 min (phylloquinone)

Preparation of 50% Water-Saturated Dichloromethane. We find 50% water-saturated dichloromethane a good moderator for adsorption HPLC of K vitamins.[17]

A stock solution of dichloromethane is kept in a flask under a layer of water. Immediately before use, the flask is shaken well, poured into a separating funnel, and the dichloromethane run off into a clean flask (100% water-saturated dichloromethane). Equal volumes of this solution and dry dichloromethane (HPLC grade is usually sufficiently dry but may be dried further with calcium chloride) are mixed to produce 50% water-saturated dichloromethane. Since the degree of water saturation declines on storage (due to adsorption by glass), for reproducible retention times the 50% water-saturated dichloromethane should be prepared on the day of use.

Column Equilibration. The operating conditions are first standardized by allowing the mobile phase to flow through the column until the silica and mobile phase have equilibrated. The rate of equilibration is often slow and should be tested at periodic intervals by injecting a standard solution of phylloquinone and the internal standard until constant retention is achieved. The precise proportion of 50% water-saturated dichloromethane in hexane is adjusted so that the retention of phylloquinone is in the working range (usually 8–10 min).

Sample Injection. The valve injector is first thoroughly washed with mobile phase to prevent contamination of the sample with standards. The sample, in a tube with a tapered end, is injected after the addition of a suitable volume of mobile phase (70 μl if the total sample is to be injected) and brief vortex mixing to dissolve the lipid.

Collection of Phylloquinone Fraction. The fraction of eluate which encloses the retention of both phylloquinone and the internal standard is collected into a clean, tapered, glass tube and evaporated to dryness under nitrogen. The collection of this fraction is the most crucial and difficult manipulation of the assay. Depending on the tissue analyzed, if the collection window is too wide, the phylloquinone and internal standard peaks may be masked by interfering peaks in the final stage of the assay. If the collection window is too narrow there may be a danger of a differential loss of phylloquinone and the internal standard, especially if these do not entirely coelute.

In the assay of some plant tissues and of pharmacological levels of phylloquinone in plasma, the peak corresponding to phylloquinone can be clearly seen at the semipreparative stage with a UV detector (Fig. 2). This allows precise collection of the phylloquinone-containing fraction. For most tissues, however, the phylloquinone peak is masked and collection of the fraction must be carried out by reference to the retention of vitamin K standards.

A useful guide to the collection of the phylloquinone-containing fraction is the "fingerprint" pattern given by other UV-absorbing lipids to which the elution of phylloquinone can be related (Fig. 2). This is especially useful when the retention time of phylloquinone is slowly changing, as may occur during the course of a day.

Although naturally occurring phylloquinone is all trans, pharmacological preparations contain *cis*-phylloquinone which is separated from, and elutes before, the trans-isomer on silica columns.[17] Therefore when measuring total phylloquinone of synthetic origin in tissues, the collection window must be increased to include *cis*-phylloquinone.

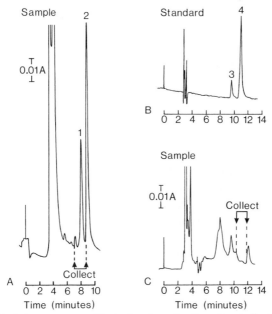

FIG. 2. Typical chromatograms illustrating the collection of a phylloquinone-containing fraction from the semipreparative HPLC stage using a silica column and phylloquinone epoxide as an internal standard. Chromatographic conditions are as described in the text. Two examples are shown: (A) a plant extract (dwarf green beans) where the peak containing phylloquinone and phylloquinone epoxide (peak 1) is clearly detectable in the sample and resolved from plastoquinone (peak 2), and (C) a plasma extract in which endogenous *trans*-phylloquinone is not detectable in the sample. In the latter case the fraction is collected by reference to the retention time of a standard mixture of phylloquinone and phylloquinone epoxide (B) in which peak 3 contains *cis*-phylloquinone (not included in collection) and peak 4 contains *trans*-phylloquinone and *cis*- and *trans*-phylloquinone epoxide.

Choice of Silica Support. Partisil 5 (Whatman Ltd., Maidstone, U.K.) was chosen as the support for the semipreparative stage because, with 50% water-saturated dichloromethane as moderator, *trans*-phylloquinone coeluted with phylloquinone 2,3-epoxide, the compound we most frequently use as an internal standard. This property is pertinent because it reduces the risk of errors caused by the incomplete recovery of either the vitamin or the internal standard when collecting fractions from silica columns.

Analytical Reverse-Phase HPLC

HPLC Conditions

Column packing, Hypersil ODS (5 μm) (Shandon Southern Ltd.)
Column dimensions, 250 × 5 mm (i.d.)

Mobile phase, 15% dichloromethane in methanol
Flow rate, 1 ml/min
Temperature, ambient
Detection, UV, 270 nm
Sample injection, 10–70 μl
Retention time, 7–8 min (phylloquinone epoxide), 9–10 min (phylloquinone)

Sample Injection. The fraction collected from the semipreparative stage is redissolved in mobile phase (70–100 μl) and, after brief vortex mixing, the total or an aliquot is injected onto the column.

Resolution and Detection of Phylloquinone. Examples of final chromatograms with phylloquinone 2,3-epoxide as an internal standard are shown in Fig. 3. The appearance and complexity of these chromatograms vary widely with different tissues and the concentration of phylloquinone. The identity of the phylloquinone and internal standard peaks in the sample are checked by reference to the retention times of standards. If the

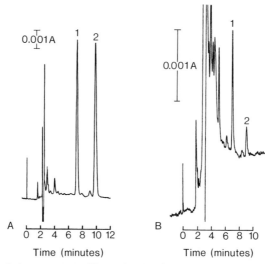

FIG. 3. Typical chromatograms illustrating the final analytical stage of the assay using reverse-phase HPLC, UV detection, and phylloquinone epoxide as internal standard. The chromatograms show the resolution of phylloquinone epoxide (peak 1) and phylloquinone (peak 2) for a plant tissue (dwarf green beans) (A) having a phylloquinone concentration of 0.46 μg/g and for an endogenous concentration of 0.5 ng/ml in blood plasma (B). Detection is at 270 nm with full-scale absorbance ranges of 0.2 and 0.005 A for plant and plasma analyses, respectively. Other chromatographic conditions are as described in the text.

chromatograms are complex, the phylloquinone peak may be identified by comparing chromatograms of the sample alone with that given when sample and phylloquinone standard are injected together.

After the elution of phylloquinone it is advisable to wait for several minutes before reinjecting since many tissues contain UV-absorbing peaks with long retention times which would otherwise interfere with subsequent analyses.

Choice of Reverse-Phase Support. For reasons already discussed we prefer to use a reverse-phase support which is highly retentive and can be used with nonaqueous mobile phases. For a given packing the retention increases with the chain length, surface coverage, and percentage loading of the bonded stationary phase (carbon loading). Two packings with suitable retention characteristics for nonaqueous chromatography of vitamin K are Hypersil ODS (Shandon Southern Ltd., Runcorn, U.K.) and Zorbax ODS (Du Pont Ltd., Stevenage, U.K.). Both have been used for the assay just described. Zorbax ODS is more retentive than Hypersil ODS and with methanol/dichloromethane mixtures requires about double the concentration of dichloromethane to elute phylloquinone with a similar retention.

Because of the difficulty in resolving phylloquinone in some tissues it is recommended that columns are maintained at high efficiencies ($8-12 \times 10^3$ theoretical plates) for maximum resolution and sensivity.

Selectivity of Mobile Phases. With both Hypersil ODS or Zorbax ODS columns we have found that the most satisfactory mobile phases for the separation of phylloquinone and phylloquinone epoxide from interfering lipids are mixtures of methanol and dichloromethane. Although substitution of methanol with acetonitrile gives higher efficiencies and a higher selectivity for the separation of some vitamin K compounds,[17] the selectivity for plasma and liver analyses is inferior to that found with methanol/dichloromethane mixtures.

Quantitation Procedure for Phylloquinone in Tissues

The amount of phylloquinone in tissue samples is calculated by reference to the chosen internal standard by the method of peak height ratios.

Procedure with 2,3-Phylloquinone Epoxide as Internal Standard

Vitamin K Compounds

Phylloquinone (Sigma Chemical Co.). This is purified by HPLC as necessary.

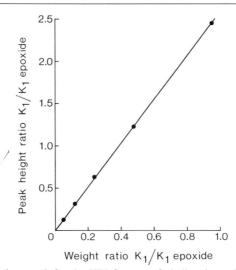

FIG. 4. Calibration graph for the HPLC assay of phylloquinone in tissues with phylloquinone epoxide as an internal standard. Chromatographic conditions are those described in the text for analytical, reverse-phase HPLC.

Phylloquinone 2,3-epoxide. This is synthesized by the method of Tishler *et al.*[27] and if necessary purified from residual phylloquinone by HPLC.

Standard Solutions. Stock solutions of phylloquinone and phylloquinone 2,3-epoxide are prepared in ethanol and stored at $-20°$. A convenient stock solution of phylloquinone has a concentration of about 20 μg/ml which can be accurately measured and periodically tested by UV absorption spectroscopy. For phylloquinone epoxide a convenient stock solution is 100 μg/ml (determined by weighing) and which may be accurately diluted to concentrations suitable for the addition to tissues containing different concentrations of phylloquinone. For the assay of low levels of phylloquinone (e.g., plasma or liver) an internal standard solution of 1 μg/ml is suitable. To obtain approximately equal peak height ratios in the final analytical stage of the HPLC the amount of phylloquinone epoxide added as internal standard should be about twice the total amount of phylloquinone in the sample analyzed.

Calibration. A series of phylloquinone/phylloquinone epoxide standards are prepared by pipetting varying volumes of a solution of phylloquinone of accurately known concentration and a constant volume of a

[27] M. Tishler, L. F. Fieser, and N. L. Wendler, *J. Am. Chem. Soc.* **62**, 2866 (1940).

standard solution of phylloquinone epoxide into volumetric flasks and making up to the mark with ethanol. Injection of these solutions on the reverse-phase system enables a calibration graph to be constructed in which peak height ratios of phylloquinone to phylloquinone epoxide are linearly related to their weight ratios (Fig. 4). For maximum accuracy the internal standard solution added to the sample at the beginning of the assay should be the same solution (or an accurately made dilution) as that used to obtain the calibration graph.

Calculation. From the chromatogram of the sample analyzed, the peak heights of phylloquinone and phylloquinone epoxide are measured. The peak height ratio of phylloquinone to phylloquinone epoxide given by the sample may then be converted to an equivalent weight ratio by reference to the calibration graph. Multiplication of this weight ratio by the weight of internal standard originally added to the sample gives the total weight of phylloquinone present in the amount of tissue processed.

[26] Analysis of Bacterial Menaquinone Mixtures by Reverse-Phase High-Performance Liquid Chromatography

By JIN TAMAOKA

There is a wide variety in length and degree of saturation of the isoprenoid side chain of menaquinone extracted from bacteria. Molecular species of menaquinones can be identified by mass spectrometry or chromatography. A number of analyses of menaquinone by high-performance liquid chromatography (HPLC) have been reported.[1-5] I describe a reverse-phase HPLC method for isolation and identification of menaquinones.

The menaquinones are referred to as MK-n, where n is the number of isoprene units in the side chain. If the isoprenoid side chains are partially saturated, they are referred to as MK-n(H$_m$), where m is the number of hydrogen atoms which saturate the side chain.

[1] D. O. Mack, *J. Liq. Chromatogr.* **3,** 1005 (1980).

[2] Y. Haroon, M. J. Shearer, and P. Barkhan, *J. Chromatogr.* **206,** 333 (1981).

[3] R. M. Kroppenstedt, *J. Liq. Chromatogr.* **5,** 2359 (1982).

[4] M. D. Collins, *J. Appl. Bacteriol.* **52,** 457 (1982).

[5] J. Tamaoka, Y. Katayama-Fujimura, and H. Kuraishi, *J. Appl. Bacteriol.* **54,** 31 (1983).

Preparation of Sample for HPLC Determination

Extract menaquinone mixture from 20 mg of lyophilized cells with 10 ml of chloroform–methanol (2 : 1, v/v). Although the content of mena-quinone in bacteria is variable, 20 mg of lyophilized cells is enough for HPLC determination in many cases. Recently, sequential extraction of isoprenoid quinones and polar lipids from bacterial cells have been re-ported.[6] Filter the extract with a filter paper and dry *in vacuo*. Dissolve the residue in a small quantity of acetone and purify quinones by thin-layer chromatography (TLC),[7] using Merck kieselgel 60 GF_{254} or an F_{254} plate and a developing mixture of petroleum ether (bp 60–80°)–diethyl ether (85 : 15, v/v). Use vitamin K_1 as a standard which has an R_f of about 0.7. Elute the menaquinone mixture from the silica gel with acetone. Filter the solution through a 0.45-μm pore filter. Store at −20°.

HPLC System

Various components for an HPLC system are commercially available. We have analyzed menaquinone mixtures with ease using the system described below.

Pump

Either a single-plunger or double-plunger type pump give good results. A gradient elution is not needed for routine determination of mena-quinone mixtures.

Column

A column packed with octadecylsilane (ODS)-bonded silica such as Zorbax ODS, μBondapak C_{18}, and Nucleosil C_{18} is suitable. The length of the column should be 25 cm or longer, because for good separation of menaquinones the number of theoretical plate must be at least 10,000. Recently, a column packed with fine particles (5 or 3 μm) became avail-able. That column's number of theoretical plate will be 10,000 at 15 cm long.

[6] D. E. Minnikin, A. G. O'Donnell, M. Goodfellow, G. Alderson, M. Athalye, A. Schaal, and J. H. Parlett, *J. Microbiol. Methods* **2**, 233 (1984).

[7] M. D. Collins, T. Pirouz, M. Goodfellow, and D. E. Minnikin, *J. Gen. Microbiol.* **100**, 221 (1977).

Detector

The use of a variable wavelength type UV detector is preferable as one can optimize the monitoring wavelength and measure the UV spectrum of a peak of interest. To obtain the UV spectrum, stop the flow of eluent at the top of the peak. Then operate manually a grating monochromator (wavelength dial) and record the detector response. Although this is a rough spectrum, one can check for artifact peaks.

Data Analyzer

The author has used Chromatopac C-R1A, C-R2AX (Shimadzu), and Data Module Model 730 (Waters) to measure elution times and peak areas. These instruments are essential for quantitative analysis of menaquinone composition.

Identification of Menaquinones

Identification by HPLC is based on the elution time, and use of at least one standard menaquinone is necessary. Figure 1 shows a HPLC chromatogram and Fig. 2 shows the relationship between common logarithm

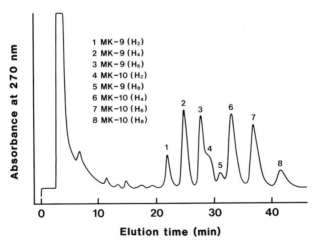

FIG. 1. HPLC chromatogram of the menaquinone mixture from *Nocardiopsis dassonvillei* JCM 3237. Conditions: pump, LC-5A (Shimadzu); column, Zorbax ODS (Du Pont) 250 × 4.6 mm; eluent, methanol–isopropyl ether (4 : 1, v/v); flow rate, 1 ml/min; column temperature, 45°; sample volume, 10 μl; detector, SPD-2A (Shimadzu) at 270 nm.

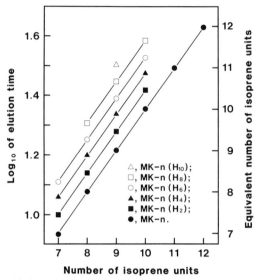

FIG. 2. Relationship between elution time and number of isoprene units of menaquinone. Conditions of HPLC are the same as for Fig. 1.

of the elution time [not the measured time but the adjusted time (measured time − elution time of acetone)] and the number of isoprene units of equally hydrogenated menaquinones. The relation is almost linear and the lines are almost parallel. Therefore, the author proposes the following equation for routine determination:

$$\text{log of elution time of MK-}n(\text{H}_m) = A + B \times n + C \times m \qquad (1)$$

where A is -0.012, B is 0.137, and C is 0.029 in the system. MK-7 to MK-12 and MK-9(H_2) to MK-9(H_8) may be enough to determine parameters A, B, and C in other HPLC systems. The value of an equivalent number of isoprene units (ENIU)[5] is the transform of the logarithm as shown in Fig. 2. This value is more useful than a simple logarithm and the equation is as follows:

$$\text{ENIU of MK-}n(\text{H}_m) = n + D \times m \qquad (2)$$

where D is 0.212 in the system. Differences between the measured times and the calculated times were, with few exceptions, less than 2% of time or less than 36 sec for the elution time of 30 min.

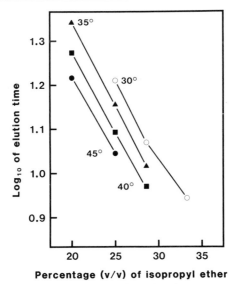

FIG. 3. Relationship between elution time of MK-9 and percentage of isopropyl ether in eluent mixture (methanol–isopropyl ether). Conditions of HPLC are the same as for Fig. 1 except for column temperature and composition of the eluent.

Preparation of Standard Menaquinones

Availability of commercial menaquinones, especially partially hydrogenated menaquinones, is limited. The review by Collins and Jones[8] is the best reference for selecting microorganisms when preparing a standard mixture of menaquinones. Extract the menaquinone mixture from the selected microorganism. Purify menaquinones by HPLC and collect the eluate where the detector indicates the peak. Commercially available isopropyl ether often contains antioxidants but these compounds will not cause serious difficulties in HPLC or mass spectrometry. Dry the eluate *in vacuo* and dissolve it in a small quantity of acetone. Analyze this sample by mass spectrometry. In the HPLC system, the elution times of MK-$(n + 1)$ and MK-$n(H_4)$ or MK-$n(H_6)$ sometimes show little difference. Confirmation of purity by mass spectrometry is therefore essential in preparing standard menaquinones.

[8] M. D. Collins and D. Jones, *Microbiol. Rev.* **45,** 316 (1981).

Composition of Eluent and Column Temperature

Figure 3 shows the relationship between elution time of MK-9 and eluent composition or column temperature. A high percentage of isopropyl ether as well as a high temperature of the column decrease the elution time. In some cases, the elution times of MK-$(n + 1)$ and MK-$n(H_4)$ show little difference. A change in eluent composition can also affect elution time. Temperature affects the elution times of all menaquinones equally and has little effect on the separation of peaks.

Section VI

Carnitine

[27] Radioisotopic Assay of Acetylcarnitine and Acetyl-CoA

By Shri V. Pande

$$\text{Acetyl}(-)\text{-carnitine} + \text{CoA} \xrightarrow{\overset{\text{carnitine}}{\underset{}{\text{acetyltransferase}}}} \text{acetyl-CoA} \qquad (1)$$

$$\text{Acetyl-CoA} + [^{14}\text{C}]\text{oxaloacetate} \xrightarrow{\overset{\text{citrate}}{\underset{}{\text{synthase}}}} [^{14}\text{C}]\text{citrate} \qquad (2)$$

$$[^{14}\text{C}]\text{Oxaloacetate} \xrightarrow[\text{+ glutamate}]{\overset{\text{aspartate}}{\underset{}{\text{aminotransferase}}}} [^{14}\text{C}]\text{aspartate} \qquad (3)$$

Principle

Acetyl-CoA is stoichiometrically converted to $[^{14}\text{C}]$citrate using citrate synthase (EC 4.1.3.7) and excess of freshly generated $[\text{U-}^{14}\text{C}]$oxaloacetate of high specific radioactivity [Eq. (2)]. The leftover $[^{14}\text{C}]$oxaloacetate is then converted back to $[^{14}\text{C}]$aspartate by adding glutamate and aspartate aminotransferase (glutamate-oxaloacetate transaminase, EC 2.6.1.1.) [Eq. (3)]. The $[^{14}\text{C}]$aspartate is then removed from the reaction mixture by the addition of a cation exchanger and the radioactivity of the supernatant due to the $[^{14}\text{C}]$citrate formed is measured.[1] Because of the high affinity of the citrate synthase for acetyl-CoA and oxaloacetate and of favorable equilibrium of the reaction toward citrate formation, near quantitative conversion of picomoles of acetyl-CoA to citrate is obtained in a short time even in the presence of only a modest excess of $[^{14}\text{C}]$oxaloacetate. On inclusion of CoASH and carnitine acetyltransferase (EC 2.3.1.7) the same procedure enables the estimation of acetylcarnitine [Eq. (1)].

Reagents

Salt-Depleted[2] *Enzymes.* The commercially obtained ammonium sulfate suspensions of the enzymes are freed from excess salts as follows:

---

[1] S. V. Pande and M. N. Caramancion, *Anal. Biochem.* **112**, 30 (1981).

[2] Being simpler this procedure is preferred over dialysis[1]; the results obtained using the salt-depleted enzyme were found to be indistinguishable from those obtained using the dialyzed enzymes.

275 μl of citrate synthase (10 mg/ml; 110 units/mg; preparations that contain citrate are usable only after the removal of citrate[3]), 400 μl of carnitine acetyltransferase (5 mg/ml, 80 units/mg), and 300 μl of glutamate-oxaloacetate transaminase (10 mg/ml, 200 units/mg), taken in separate tubes, are centrifuged at 12,000 g for 10 min at 0°, and the supernatant is aspirated away. Each tube then receives 1 ml of 20 mM potassium phosphate of pH 7.4 to dissolve the precipitated enzyme. The resulting solutions provide enzymes for 200 assay tubes; stored at −20°, these enzymes remain usable for at least a month.

Glutamate Plus Glutamate Oxaloacetate Transaminase Mix. For 40 assay tubes 1 ml of chilled 160 mM potassium L-glutamate is mixed with 200 μl of the salt-depleted glutamate oxaloacetate transaminase and is kept in an ice bath. This mixture, when stored at −20°, remains usable for at least a week.

Freshly Reduced Coenzyme A for Acetylcarnitine Assay. Reagent for 40 assay tubes is obtained by mixing 40 μl of 2 mM CoA, 40 μl of 100 mM potassium pyrophosphate, pH 9.0, and 320 μl of 4 mM freshly dissolved dithiothreitol. This mixture is left at 25° for 30 min, then 800 μl of 0.5 M potassium phosphate, pH 7.4, is added and the tube is transferred to an ice bath.

[14C]Oxaloacetate–Enzyme Premix for Acetylcarnitine Assay. This reagent is prepared fresh each day and should be used immediately after its preparation. Reagent for 40 assay tubes is prepared by mixing 4.5 μl of 1 M potassium phosphate, pH 7.4, 20 μl of 20 mM α-ketoglutarate, 4 μl of 40 mM ethylenediaminetetraacetic acid, pH 7.4, 18 μl of 220 μM [U-14C]aspartate, 32 μl of H_2O, and 1.5 μl of the salt-depleted glutamate-oxaloacetate transaminase. The tube is left at 25° for 10 min, then 20 μl of 1 M perchloric acid is added and a further 15 min incubation at 25° is allowed. The tube then receives with mixing and, in the order given, 40 μl of chilled 0.6 M KOH and 60 μl of 67 mM ethylene diamine tetraacetate, pH 7.4, 1.2 ml of freshly reduced CoASH, 200 μl of salt-depleted carnitine acetyltransferase, and 200 μl of salt-depleted citrate synthase.

[14C]Oxaloacetate–Citrate Synthase–N-Ethylmaleimide Premix for Acetyl-CoA Assay. This is prepared exactly as described above for the [14C]oxaloacetate–enzyme premix for acetylcarnitine assay except that CoASH and carnitine acetyltransferase are replaced by 1400 μl of freshly dissolved N-ethylmaleimide (2 mg) in 300 mM potassium phosphate of pH 7.4.

---

[3] For this, dialyze against 500 volumes of 20 mM potassium phosphate of pH 7.4 at 0° with fresh phosphate buffer being replaced at 30 min intervals three times.

Acetyl(−)-carnitine, 1 μM, pH 7.0.
Acetyl-CoA, 1 μM, pH 7.0.

Procedure for Acetylcarnitine Assay. Reaction tubes, 13 × 100 mm, kept in ice, receive the following in the order given: sample to be assayed (unknown or standards containing 1–70 pmol of acetylcarnitine), water to bring the volume to 155 μl (water only in controls), then 45 μl of freshly prepared [^{14}C]oxaloacetate–enzyme premix. This last reagent is added to successive tubes at every 30 sec intervals, the contents are mixed, and the tubes are left in a 25° water bath. Twenty minutes later, 30 μl of glutamate–glutamate-oxaloacetate transaminase mixture is added with mixing to the successive assay tubes at 30 sec intervals, and the tubes are returned to the 25° bath for another 20 min. Each tube then receives AG 50W-X8 (H$^+$ form, 200–400 mesh) and with mixing 0.6 ml of water from a repeating dispenser. For resin addition, a plastic 1-ml insulin syringe with 0.6 mm inner diameter (Becton, Dickinson and Co., Rutherford, NJ) is cut at the 0 mark and the resin is filled to the volume mark corresponding to 0.5 ml through the cut end by pressing the inverted syringe in a 100-ml beaker containing the resin with plunger held at the 0.5 ml mark. By depressing the plunger the resin is added to the assay tubes. This technique allows both resin and water to be added almost simultaneously to successive tubes at every 30 sec intervals. The tubes are shaken mechanically for about 10 min at room temperature, allowed to stand undisturbed for at least 2 min, then 0.3 ml of the supernatant is withdrawn and is thoroughly mixed with a suitable volume of appropriate scintillant (e.g., 2.4 ml of Ready Solv EP of Beckman) for radioactivity measurement.

Procedure for Acetyl-CoA Assay. This is carried out exactly as described for acetylcarnitine above except that the [^{14}C]oxaloacetate–citrate synthase–N-ethylmaleimide premix is used here instead of the one described as [^{14}C]oxaloacetate–enzyme premix for acetylcarnitine assay.

Application to Tissue Extracts. Appropriate further processing of the neutralized perchloric acid extracts of tissues is necessary as described below to exclude interference from endogenous substances. Thus, for acetylcarnitine assay, a portion of the neutralized perchloric acid extract is freed of any endogenous oxaloacetate, acetyl-CoA, and other anions by using AG 2-X8 (Cl$^-$ form, 200–400 mesh) resin. For example, for rat liver, 300 μl of the neutralized extract corresponding to about 8 mg wet tissue is mixed with about 120 mg of the resin, the tubes are shaken for 10 min at room temperature, briefly centrifuged, and then portions (2–30 μl) of the supernatant are withdrawn for acetylcarnitine assay.

For acetyl-CoA assay, any endogenous oxaloacetate is eliminated by its selective decomposition. For example, for rat liver, 400 μl of the neutralized perchloric acid extract corresponding to about 10 mg wet tissue is mixed with 20 μl of 400 mM potassium acetate buffer, pH 4.8, plus 20 μl of 1 mM CuSO$_4$, the tubes are incubated at 25° for 30 min and then portions of this mixture (up to 100 μl) are withdrawn for acetyl-CoA assay.

Comments

To minimize complications from the spontaneous decarboxylation of oxaloacetate in solutions, the [^{14}C]oxaloacetate is generated last and the oxaloacetate–enzyme premixes are used immediately after their preparation. Because of the large dilution of tissue extracts needed for the present assays owing to their high sensitivity, it is unlikely that interfering substances of tissue origin would invalidate the assay. Nevertheless, appreciable quantities of at least the following substances in the unknown samples are likely to interfere with the assay: (1) polyvalent cations, certain amines, and amino acids; these promote the decarboxylation of oxaloacetate[4,5]; (2) high concentrations of imidazole and sucrose; these can compete with oxaloacetate for accepting the acetyl group of acetyl-CoA nonenzymatically[6]; and (3) Tris and >1 mM concentrations of various thiols during acetylcarnitine assay, because these substances start serving as acetyl group acceptor in the presence of carnitine acetyltransferase. Where such interference is suspected, additional controls should be set up to ensure that the values of acetyl-CoA and acetylcarnitine being picked up in the unknown samples are reliable. The unknown samples should not contain high concentrations of analogs of oxaloacetate or acetyl-CoA that can impede or prevent the stoichiometric conversion of these two substrates to citrate. For example, estimation of acetyl-CoA, with 100 pmol [^{14}C]oxaloacetate present per assay tube is not possible when the sample being analyzed contributes $\geq$10 nmol of malonate, because at this large molar excess, malonate competes sufficiently with oxaloacetate to cause a marked interference.[7] The concentrations of propionyl-CoA and propionylcarnitine normally encountered in tissues extracts are not high enough to interfere, and the same applies to endogenous citrate, CoA, carnitine, α-ketoglutarate, aspartate, and glutamate. The present procedure, because of its high sensitivity, allows acetylcarnitine estimation conveniently with very low sample sizes of tissue extracts; for example extract corresponding to 1 mg of liver and 2 μl of serum suffices.

[4] H. A. Krebs, *Biochem. J.* **36,** 303 (1942).
[5] J. F. Speck, *J. Biol. Chem.* **178,** 315 (1949).
[6] C. Hebb, S. P. Mann, and J. Mead, *Biochem. Pharmacol.* **24,** 1007 (1975).
[7] T. S. Lee and S. V. Pande, unpublished (1982).

Alternative Procedures

With samples containing nanomole quantities, both acetyl-CoA and acetylcarnitine can be assayed by following changes in NADH concentration spectrophotometrically or fluorimetrically by coupling of the citrate synthase reaction to that of malate dehydrogenase.[8,9] Acetylcarnitine together with certain other short chain acylcarnitines in picomole range can be assayed by allowing equilibration of radioactive (−)-carnitine into the acylcarnitine pool using carnitine acetyltransferase, followed by high-performance liquid chromatographic or thin-layer chromatographic separation of the acylcarnitines.[10] For acetyl-CoA, methods of comparable sensitivity have been described based on the enzymatic recycling of the CoA moiety using citrate synthase plus phosphate acetyltransferase (EC 2.3.1.8).[11,12] Because of the high enzymatic amplification in the latter types of assays, the possibility of an otherwise slower spurious reaction being picked up needs being considered. For example, we found that femtomole quantities of acetyl-CoA and/or CoA could be estimated by the present procedure of acetyl-CoA analysis simply by supplementing the assay incubation mixtures either with 1 mM acetylcarnitine plus carnitine acetyltransferase or with 1 mM acetyl phosphate plus phosphate acetyltransferase while excluding N-ethylmaleimide, but the procedures proved unreliable because the various thiols, at the mM concentrations needed to keep the CoA reduced during these analyses, then mimicked CoA to a noticeable degree.[7]

[8] J. R. Williamson and B. E. Corkey, this series, Vol. 13, p. 434.
[9] D. J. Pearson, P. K. Tubbs, and J. F. A. Chase, in "Method en der Enzymatischen Analyse" (H.-U. Bergmeyer, ed.), 3rd ed., Vol. 4, p. 1758. Verlag Chemie, Weinheim, 1974.
[10] J. Kerner and L. L. Bieber, Anal. Biochem. 134, 459 (1983).
[11] T. Kato, Anal. Biochem. 66, 372 (1975).
[12] D. B. McDougal and R. V. Dargar, Anal. Biochem. 97, 103 (1979).

[28] Short-Chain Acylcarnitines: Identification and Quantitation

By L. L. Bieber and J. Kerner

Although acetylcarnitine is the predominant acylcarnitine in most tissues,[1,2] the occurrence of other short-chain acylcarnitines in biological systems has been recognized for more than a decade. Early studies demonstrated that mitochondria have the capacity to form branched-chain acylcarnitines and propionylcarnitine depending on the substrate.[3,4] Investigations in this area have been limited due to the lack of adequate methodology. With the development of a gas chromatographic technique, it was shown that mammalian tissues[5,6] and human urine[7] contain small amounts of short-chain, aliphatic acylcarnitines. However, some samples such as sows colostrum[8] contain large quantities of specific short-chain acylcarnitines and propionylcarnitine was the major urinary acylcarnitine in a patient with an undiagnosed metabolic disease.[9] Such studies indicated carnitine may act as an acyl sink or may be involved in selected dextoxification mechanisms.[9] More recent studies support such a role for carnitine with the demonstration that large amounts of urinary propionylcarnitine are associated with a human propionic acidemia,[10] and other urinary acylcarnitines are associated with some metabolic diseases.[11] Further evidence has come from studies which show formation of pivaloylcarnitine due to a pivaloyl[1'-¹⁴C]oxyethyl-(S)-3-(3,4-dihydroxyphenyl)-2-methylalaninate load.[12]

With the finding that urinary acylcarnitines may be an indicator of specific metabolic blocks of acyl-CoA metabolism, the need for rapid and

[1] D. L. Pearson and P. K. Tubbs, *Biochem. J.* **105,** 953 (1967).

[2] A. M. Snoswell and P. R. Koundakjian, *Biochem. J.* **127,** 133 (1972).

[3] T. Bohmer and J. Bremer, *Biochim. Biophys. Acta* **152,** 559 (1968).

[4] H. E. Solberg and J. Bremer, *Biochim. Biophys. Acta* **222,** 372 (1970).

[5] L. L. Bieber and Y. R. Choi, *Proc. Natl. Acad. Sci. U.S.A.* **74,** 2795 (1977).

[6] Y. Choi, P. Fogle, P. Clarke, and L. L. Bieber, *J. Biol. Chem.* **252,** 7930 (1977).

[7] K. J. Valkner and L. L. Bieber, *Biochem. Med.* **28,** 197 (1982).

[8] J. Kerner, J. A. Froseth, E. R. Miller, and L. L. Bieber, *J. Nutr.* **114,** 854 (1984).

[9] L. L. Bieber, R. Emaus, K. Valkner, and S. Farrell, *Fed. Proc., Fed. Am. Soc. Exp. Biol.* **41,** 2858 (1982).

[10] C. R. Roe and T. P. Bohan, *Lancet* June 19, 1411 (1982).

[11] C. R. Roe, C. L. Hoppel, T. E. Stacey, R. A. Chalmers, B. M. Tracey, and D. S. Millington, *Arch. Dis. Child.* **58,** 916 (1983).

[12] S. Vickers, C. A. H. Duncan, J. L. Smith, R. W. Walker, H. Flynn, and B. H. Arison, *Fed. Proc., Fed. Am. Soc. Exp. Biol.* **43,** Pap. No. 322 (1984).

more precise measurements of acylcarnitines has been recognized. Recently, two new approaches for identification and separation of short-chain acylcarnitines have been reported: (1) an enzymatic exchange of radioactive carnitine into an acylcarnitine pool, which provides a very sensitive covalent tage that can be used both for detection and quantitation[13] and (2) the use of fast atom bombardment (FAB) to identify and potentially quantitate acylcarnitines.[14]

Isolation of a Short-Chain Acylcarnitine Fraction

Most methods, require partial purification of the acylcarnitine fraction prior to analyses. The isolation of a short-chain acylcarnitine fraction of sufficient purity to permit saponification and subsequent identification of the acyl residues by gas chromatography is described in a previous volume.[15] This procedure[15] has some limitations. (1) It is time consuming and must be modified for at least one biological specimen of importance, human urine, due to the occurrence of large amounts of unusual medium-chain acylcarnitines in some samples which can be partly retarded by both Dowex-1 and Dowex-50 columns. (2) When the samples are saponified and the volatile fatty acids analyzed by gas chromatography, contamination by volatile acids such as acetic and propionic can occur due to contamination of ammonium hydroxide or other alkaline solutions by vapors from chromatographic solvents. (3) Highly purified ether must also be used due to trace materials which have retention times similar to some volatile fatty acids. (4) After concentration of samples, the solubility of the longer chain acylcarnitines and some fatty acids in acidic solutions is limited. This limited solubility can cause selective loss of both acylcarnitines and volatile fatty acids.

We have recently used ion-paper precipitation to concentrate the acylcarnitines from urine. The ion-pair precipitation is an excellent procedure for enriching the acylcarnitine fraction and eliminating water-soluble materials, but the precipitation is not complete for free carnitine and the shorter chain acylcarnitines.

With the radioexchange HPLC assay and the fast atom bombardment method, samples which are relatively impure can be used. Neutralized perchloric acid extracts of tissues have been used for the radioexchange assay and Dowex-1 column chromatography for the fast atom bombardment procedure. However, Dowex resins can partly retard the more hy-

[13] J. Kerner and L. L. Bieber, *Anal. Biochem.* **134,** 459 (1983).
[14] D. S. Millington, C. R. Roe, and D. A. Maltby, *Biomed. Mass Spectrom.* **11,** 236 (1984).
[15] L. L. Bieber and L. M. Lewin, this series, Vol. 72, part D, p. 276.

RECOVERY OF URINARY
CARNITINE/ACYLCARNITINES FROM DOWEX-1[a]

Sample	Eluate (%)
Free carnitine	97.6 ± 0.6 ($n = 5$)
Acylcarnitine	81.4 ± 2.6 ($n = 5$)
Total carnitine	84.3 ± 2.2 ($n = 6$)

[a] Urine samples were treated with C : M 3/2 as described in the text. The residue was dissolved in distilled water and adjusted to a known volume. Aliquots were passed through Dowex-1, Cl⁻, X8, 100–200 mesh (5 × 0.5 cm) columns and the columns washed with 3 volumes of distilled water. Aliquots of the chloroform/methanol extracted urine were analyzed for free and acylcarnitine before and after chromatography on Dowex-1. Aliquots were also subjected to the radioisotopic exchange procedure and the total carnitine (acyl + free) determined.

drophobic acylcarnitines due to hydrophobic interaction with the polystyrene matrix. This is illustrated in the table, which shows loss of some urinary acylcarnitines on Dowex-1. Similar losses occur with Dowex-50. The recovery of free carnitine and the shorter-chain acylcarnitines, i.e., C_2 to C_5 is excellent, but recovery of higher molecular weight, longer chain acylcarnitines is not complete. For any isolation method the recovery of the acylcarnitines in each step should be determined.

Determination of Free Carnitine, Total Carnitine, and Acylcarnitine Amounts

The distribution of total tissue carnitine into free carnitine, short-chain acylcarnitine, and long-chain acylcarnitine fractions has also been described in a previous volume.[15] The designation of a fraction as long-chain and short-chain acylcarnitine is arbitrary. It does not account for the medium-chain acylcarnitines and the partial solubility of some acylcarnitines which can partition both into the long-chain acylcarnitine fraction and also into the short-chain acylcarnitine fraction. Similarly inadequate washing of the protein precipitate containing the long-chain acylcarnitine fraction can cause significant quantitation errors.[16]

[16] R. C. Fishlock, L. L. Bieber, and A. M. Snoswell, *Clin. Chem. (Winston-Salem, N.C.)* **30**, 316 (1984).

Identification of Short-Chain Acylcarnitines

Radioexchange HPLC

The lack of a sensitive method for detection of individual acylcarnitines on chromatographs or in column effluents has hampered the use of thin-layer, paper, and column chromatographic techniques with biological samples that contain considerable acetylcarnitine and smaller quantities of other water-soluble acylcarnitines. This limitation has partly been overcome by enzymatically exchanging high specific activity radioactive carnitine into aliphatic acylcarnitines.[17]

When radioactive carnitine is exchanged to isotopic equilibrium and the acylcarnitines separated by HPLC, picomole quantities of individual acylcarnitines can be detected and quantitated. This method has been used to detect short-chain acylcarnitines on thin-layer chromatograms, paper chromatograms, and column effluents; however, the preferred separation method is HPLC coupled to an instream continuous flow through a beta counter as illustrated previously.[8] The acylcarnitines in the sample must be substrates for the use of carnitine acetyltransferase or carnitine acyltransferase(s). If samples contain medium-chain or longer chain acylcarnitines, then enzymes such as carnitine octanoyltransferase are recommended. The latter has a higher catalytic capacity for medium-chain and banched-chain acylcarnitines. The enzyme or enzymes must be *totally free* of acylcarnitine and acyl-CoA hydrolytic activity.

Gas Chromatographic–Mass Spectral Analyses

Individual short-chain acylcarnitines in a $HClO_4$ extracts of tissues have been identified by isolating the acylcarnitine fraction using a series of column chromatographic steps and then separating and quantitating the fatty acids.[15] This approach when coupled to an instream mass spectrometer has permitted unequivocal identification of a number of short-chain acylcarnitines which occur in tissues such as liver and heart.[5,6] Several short-chain aliphatic, branched-chain and short-chain unsaturated fatty acyl residues have been identified and other longer chain (less-volatile) derivatives have been detected particularly in human urine. This approach gives data similar to the radioexchange HPLC technique, when precautions as described in the previous sections are taken. A comparison of the radioexchange HPLC and gas chromatographic procedure using a human urine specimen which contains considerable amounts of medium-

[17] J. Kerner and L. L. Bieber, *Anal. Biochem.* **134,** 459 (1983).

chain acylcarnitines is given in Fig. 1. The gas chromatographic technique has the advantage of greater resolving power and direct coupling to mass spectrometry, but it is an indirect method. For example, 4-carbon and 5-carbon acid isomers are resolved by gas chromatography, but to date, we have not completely separated the respective acylcarnitine isomers using HPLC.

Fast Atom Bombardment

Recently, a different approach has been applied to water-soluble acyl-carnitines using fast atom bombardment, in which the mass of the molecular ion of individual molecules can be determined. This technique has been used to identify and to quantitate acetylcarnitine and propionylcarnitine in human urine and to detect the occurrence of other water-soluble acylcarnitines.[14] Higher molecular weight acylcarnitines have been detected in human urine by both the gas chromatographic technique and the radioexchange HPLC procedure as shown in Fig. 1. Determining the absolute identity of individual (medium-chain) acylcarnitines using fast atom bombardment is difficult due to lack of structural details. For example, we have isolated and purified acylcarnitine derivatives with retention times of 33.4, 34.7, and 35.6 min (see Fig. 1C) from urine using ion-pair precipitation, silicic acid column chromatography, and HPLC techniques and then assaying using fast atom bombardment, gas chromatography, and electron impact (EI) mass spectrometry of the acyl residues. Both fast atom bombardment and radioexchange coupled to HPLC separation indicated the presence of medium-chain acylcarnitines. Fast atom bombardment demonstrated the occurrence of C_8 and C_9 acylcarnitines, but the positions of double bonds, branching, etc. were not apparent.

Quantitation

The total short-chain acylcarnitines present in tissues can be determined as indicated above. Usually the tissue carnitine is partitioned between an acid-soluble and acid-insoluble fraction.

Different approaches have been used for the quantitation of individual water-soluble acylcarnitines.

Specific Enzymatic Analyses

Acetylcarnitine can be accurately quantitated using a coupled enzyme assay in which the acetylcarnitine is converted to acetyl-CoA and the acetyl-CoA quantitated enzymatically with citrate synthase and malate

[17a] L. L. Bieber and J. Kerner, *Biochem. Prep.*, in press, (1986).

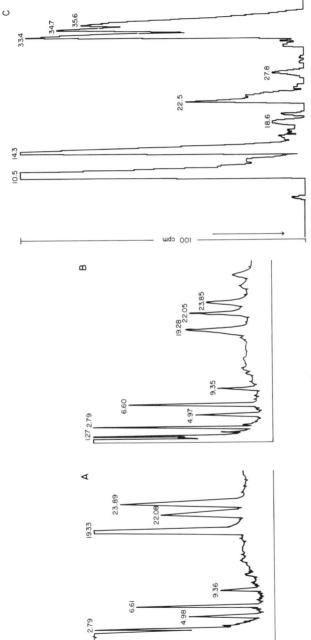

FIG. 1. Comparison of gas chromatographic and HPLC separations of the medium-chain acylcarnitines from human urine. (A) The gas chromatographic procedure was used with diethyl ether as the solvent; (B) metaphosphoric acid was the free fatty acid solvent; (C) the identical acylcarnitine fraction was subjected to the radioisotopic exchange procedure described in the text and subjected to HPLC as described in Refs. 8 and 17. (A and C reproduced from Bieber and Kerner,[17a] with permission of Dekker, New York. Copyright © 1986.)

dehydrogenase.[18] This method is satisfactory for measurement of small amounts of acetylcarnitine, but it is not applicable to the measurement of other acylcarnitines. We use this procedure to check the acetylcarnitine determinations done by either the gas chromatographic technique or the radioexchange procedure.

Gas Chromatographic Analyses

The short-chain acylcarnitine fraction is isolated as described in another volume[15] with the precautions mentioned above. A known amount of specific acylcarnitine such as valerylcarnitine can be added as an internal standard. Although with tissue samples such as liver, heart, and colostrum, the absolute amounts of acylcarnitines, determined by gas chromatography and radioexchange HPLC, are very similar, recoveries from human urine are more variable. Since the details of the gas chromatographic technique have been published previously,[15] they will not be given here with the exception of the following modifications.

Step 2 [Vol. 72 (p. 283)]: 200–400 nmol valerylcarnitine is used as the internal standard. Crotonylcarnitine can be an adequate internal standard, but the double bond is readily air oxidized.

Step 6 [Vol. 72 (p. 284)]: 3 column volumes of distilled H_2O are used to elute the Dowex-1 column.

Step 7 [Vol. 72 (p. 284)]: 1.2 N NH_4OH in ethyl alcohol is used rather than 0.3 N NH_4OH. The saponification time is 1 hr rather than 30 min and the pH is adjusted to 9.0 with 1 N HCl prior to evaporation to dryness. The dry samples can be stored at this stage or dissolved in various solvents such as metaphosphoric acid or acidic ether for gas chromatographic analyses.

Step 9 [Vol. 72 (p. 285)]: the residue from step 7 is dissolved in 300 μl H_2O and saturated with NaCl, acidified to 0.2 N with HCl, and then extracted 4× with 150 μl of ether. For samples in which the volume of ether needs to be reduced, 25 μl of 1.0 N KOH is added and the ether evaporated with N_2; 10 μl 6 N HCl is added, followed by the appropriate volume of diethyl ether. For samples which contain medium-chain acylcarnitines, use of ether rather than metaphosphoric acid is recommended because of the limited solubility of medium-chain acids in metaphosphoric acid. This is illustrated in Fig. 1 where identical aliquots of the fatty acid from the acylcarnitines of the same urine were extracted with ether (Fig. 1A) or dissolved in metaphosphoric acid (Fig. 1B). The amounts of the shorter chain acids, propionic (2.79), isobutyric (4.98), and internal standard (6.6) were essentially identical, but the longer chain C_8 and C_9 deriv-

---

[18] D. J. Pearson and P. K. Tubbs, *Nature (London)* **202**, 91 (1964).

atives with retention times of 19.3, 22.0, and 23.85 min are not quantitatively dissolved in metaphosphoric acid, causing a large underestimation of these acylcarnitines.

Radioisotopic Exchange HPLC

Since acylcarnitines are products or substrates for the reversible carnitine acyltransferases and there are a family of carnitine acyltranferases with different substrate specificities, it is possible to put a quantitative "handle" onto individual acylcarnitines by exchanging labeled L-carnitine into an acylcarnitine pool using purified carnitine acyltransferases.[8,17] The short-chain acylcarnitines and acyl-CoAs have different K_m values for the enzyme, thus for quantitative purposes this technique requires sufficient reaction time to allow attainment of isotopic equilibrium for the *slowest exchanging acylcarnitine.* We have found (unpublished data) that for substrates such as isovalerylcarnitine, the carnitine octanoyltransferase purified from liver peroxisomes is a better catalyst for exchanging carnitine into the medium-chain acylcarnitine fraction. However, this latter enzyme is not commercially available. Purification details of carnitine octanoyltransferase are described in another section of this volume. The methodology described below is for commercial carnitine acetyltransferase, but it works well with carnitine octanoyltransferase as well as mixtures of the two enzymes. The exchange assay is a rapid and accurate technique for quantitating small amounts of acylcarnitines[8] in biological samples such as those obtained from liver, heart, and blood.

Quantitation by Mass Spectrometry

As indicated above, fast atom bombardment mass spectrometry has recently been used to identify and quantitate the amount of acetylcarnitine and propionylcarnitine in urine from human subjects.[11,14] From the data given, it is difficult to determine how accurately the minor components which occur in urine and other biological specimens can be measured. Regardless, this technique has great promise for identification and quantitation of individual acylcarnitines, especially as more reference acylcarnitines become available and the techniques are refined.

Procedures

Initial Purification from Tissues

Freeze clamped tissue is extracted twice with 5 volumes of ice-cold perchloric acid (0.6 M) by mechanical disruption (Polytron). The homogenates are centrifuged at 16,000 g for 10 min at 0°, the supernatant fluids combined, and 50 μl of 0.5 M potassium phosphate buffer, pH 7.0, per ml of supernatant fluid is added. The pH is adjusted to 6.5 with 6 N KOH, let stand for 30 min, and the potassium perchlorate precipitate removed by centrifugation. One aliquot of the neutralized supernatant fluid (0.5–1.0 ml) is put on a Dowex 1-X8, 100–200 mesh (Cl^-)column (5 × 0.5 cm) and the acylcarnitines eluted with three column volumes of distilled water. This step is necessary to remove any CoA esters which can produce an overestimation of specific acylcarnitines when the radioexchange HPLC technique is used.

Free and total saponifiable carnitine are measure before and after Dowex-1 column chromatography. If carnitine is retained by the Dowex column such as described for some urine specimens, then the presence of medium-chain acylcarnitines should be suspected. We are currently attempting to use DEAE-cellulose as a column support for removing the acyl-CoA fraction. Aliquots containing 0.2–5 nmol of total carnitine are used for the enzymatic exchange reaction as described below.

Urine

Some urines contain large quantities of anions and other substances which can affect carnitine analyses. We have found it necessary to pretreat urine. Initially a Dowex-1 step was used, but this reduced the recovery of medium-chain acylcarnitines. More recently we have used a chloroform/methanol treatment which removes much of the salts.

A 0.2-ml aliquot of urine is combined with 19 volumes of chloroform/methanol (3 : 2, v/v) and the precipitated material removed by centrifugation. The residue is extracted with chloroform/methanol mixture as before and evaporated. The dry or (sometimes) slightly oily residue is taken up in 1.0 ml of distilled water. Following measurement of free and total saponifiable carnitine, aliquots containing 0.2–5.0 nmol of total carnitine (preferable 1.0 nmol) are used for the radioisotopic exchange reaction. Some urine specimens contain large amounts of three acylcarnitines, tentatively identified as 2-methyloctanoyl-, 2-methyl-$\Delta^?$-octanoyl-, and $\Delta^?$-octenonylcarnitine which exchange very slowly when carnitine acyltransferase is the catalyst (see Fig. 1). The unsaturated acylcarnitines are readily oxidized so precautions must also be taken to minimize oxidation.

Exchange Assay Conditions

To remove $(NH_4)_2SO_4$, commercial carnitine acyltransferase (CAT) is dialyzed against 50 mM potassium phosphate buffer, pH 7.4, containing 1 mM EDTA and 1 mM DTT (DL-dithiothreitol) using a microdialysis system. The dialyzed enzyme in stable for at least 8 days at 0–4° and it is free of any detectable acyl-CoA hydrolase activity.

Exchange Assay. Combine 5 μl 250 mM K$_2$HPO$_4$, pH 7.4, 5 μl 50 μM CoASH in 10 mM dithiothreitol (DTT), 5 μl L-[^{3}H]- or L-[^{14}C]carnitine containing 0.2–0.3 μCi, 5–30 μl sample containing 0.2–5.0 nmol total carnitine, 1.5 U dialyzed CAT or 1.5 U CAT and 1.5 U COT, and water to make final volume of 50 μl. Incubate 1 hr at 30° and then add 5 μl of 30 mM N-ethylmaleimide (NEM) and incubate 30 min. NEM is added to convert acyl-CoA derivatives to acylcarnitines.

If radioactive DL-carnitine is available, the L-isomer can be obtained by enzymatic acetylation and subsequent separation by column chromatography.[19] We recommend use of the L-isomer rather than the DL mixture.

Separation of Radioactive Acylcarnitines

Paper and Thin-Layer Chromatography. Aliquots of the reaction mixtures can be spotted directly onto paper chromatograms and thin-layer chromatograms. Chromatographic solvent systems have been described for separation of the shorter chain acylcarnitines,[20,21] but solvent systems for adequate separation of straight-chain, branched-chain, and unsaturated acylcarnitines such as valeryl- and isovalerylcarnitine have not been developed.

Reference acylcarnitines can be synthesized by chemical methods[22] or by enzymatic methods using the appropriate acyl-CoA derivatives, L-carnitine and carnitine acetyltransferase as described for Fig. 2.

HPLC. The incubation mixture (50 μl) described previously is acidified with 10 μl of 6% HClO$_4$ and, after standing on ice for 10 min the precipitated protein is removed by centrifugation for 1 min in an Eppendorf microcentrifuge. The supernatant fluid is treated with a predetermined amount of 1 N KOH to bring the pH to 6.5–7.0 and left on ice for 15 min. After centrifugation the supernatant fluid is filtered through a 0.2-μm pore diameter nylon filter by a brief centrifugation using centrifugal filter tubes.

For Fig. 1C, 20 μl of sample was applied to Partisil 10 ODS-3 (250 × 4.6 mm) column and the column developed using 5 mM butanesulfonic acid, adjusted to pH 3.4 with acetic acid (solvent A) and 100% methanol (solvent B) as described.[8,17] The solvent composition and gradients can be varied depending on the acylcarnitines in the mixture. For example acylcarnitines from C$_2$ to C$_9$ can be separated on a C$_{18}$ reverse-phase column

[19] R. R. Ramsay and P. K. Tubbs, *FEBS Lett.* **54**, 21 (1975).
[20] H. E. Solberg and J. Bremer, *Biochim. Biophys. Acta* **222**, 372 (1970).
[21] L. M. Lewin and L. L. Bieber, *Anal. Biochem.* **96**, 322 (1979).
[22] T. Bohmer and J. Bremer, *Biochim. Biophys. Acta* **152**, 559 (1968).

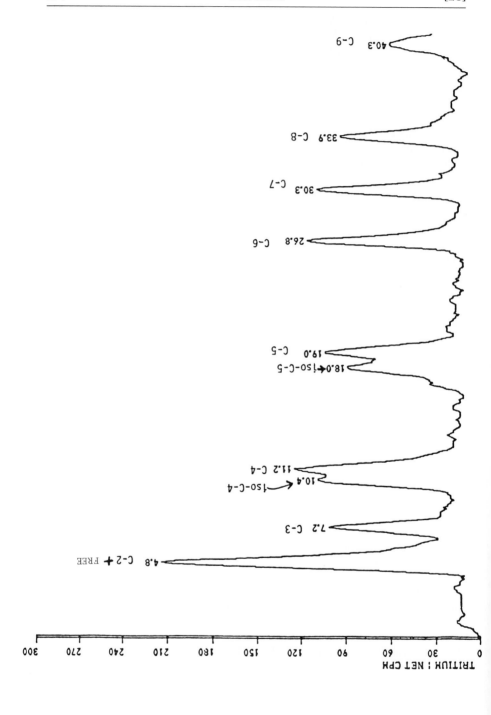

using the phosphate/dibutylamine/hyptanesulfonic acid–methanol gradient given in the legend of Fig. 2.[23]

After isotopic equilibrium is attained, the dpm in individual acylcarnitines are proportional to the amounts of the individual acylcarnitines in the mixture. The amount of the individual acylcarnitines can be calculated from the specific radioactivity of the initial total carnitine pool, if the specific activity of the carnitine fraction has been determined or if the absolute amount of one of the acylcarnitines such as acetylcarnitine is known. Acetylcarnitine can be quantitated rapidly using the enzymatic assay described above. Alternatively a known quantity of an exchangeable acylcarnitine can be added as an internal standard. The total amount

[23] S. Clausen, O. Olsen, and H. Sorensen, *J. Chromatogr.* **260**, 193 (1983).

FIG. 2. HPLC separation of C_2 to C_9 acylcarnitines. CAT was dialyzed to remove $(NH_4)_2SO_4$ and acyl-L-[*methyl*-³H]carnitine esters were enzymatically prepared. The assay mixture contained, in a final volume of 0.25 ml, 10 mM potassium phosphate buffer, pH 7.4, 1.0 mM NEM, 125 nmol acyl-CoA, 1.0 nmol L-[*methyl*-³H]carnitine (2 Ci/mmol), 15 nmol L-carnitine, and 1–2 U of CAT. After 2 hr at room temperature, 1 ml of methanol was added, the mixtures were allowed to sit at $-20°$ for 15 min, and then centrifuged for 2 min in an Eppendorf microcentrifuge. Following room temperature rotary evaporation the acylcarnitines were dissolved in 1 ml water and then filtered through a 0.2-μm pore diameter regenerated cellulose filter by brief centrifugation using centrifugal filter tubes.

For preparation of reference acylcarnitines, excess acyl-CoA was used to ensure complete conversion of labeled carnitine to acylcarnitines. A combined mixture of radioactive acylcarnitines was separated by HPLC. The HPLC chromatographic equipment consisted of Waters Model 6000A pumps controlled by a Waters Model 720 microprocessor and a Waters U6-K sample injector. The column was a Waters μ-Bondapak C_{18} (3.9 mm × 30 cm). The gradient system described previously[23] was used at pH 2.5. Solvent A (0.02 M sodium phosphate buffer, 0.02 M dibutylamine, and 0.02 M sodium heptanesulfonic acid in 50% acetonitrile) was filtered prior to addition of acetonitrile through a 0.45-μm pore diameter tortuous pore disk membrane. Solvent B (0.01 M sodium phosphate buffer, 0.01 M dibutylamine, and 0.01 M sodium heptanesulfonic acid) was also filtered. The mobile phase consisted of solvent AM (2A : 8B) and solvent BM (7A : 3B). For developing, washing, and reequilibrating the column the following program was used: 0–6 min, linear increase from 0 to 10% solvent BM; 6–9 min, linear increase from 10 to 22% solvent BM; 9–11 min, linear increase from 22 to 25% solvent BM; 11–16 min, linear increase from 25 to 50% solvent BM; 16–21 min, linear increase from 50 to 100% solvent BM. Solvent BM (100%) was maintained for 23 min followed by a linear decrease of solvent BM to 0% in 20 min and then held for 10 min. The flow rate was 1.0 ml/min. When the composition of solvent AM is changed to (0.5A : 9.5B) better resolution of free carnitine from acetylcarnitine and C_4 from iso-C_4 is achieved. Complete separation of free carnitine from acetylcarnitine can be obtained using the chromatographic conditions previously described.[13] The radioactivity in the effluent was detected by a FLO-ONE Model IC radioactivity flow detector using Flow Scint II nongelling scintillator (Radiomatic Instruments and Chemical Co., Inc., Tampa, FL) and recorded by a computer controlled printer with data reduction capabilities. The numbers above the peaks indicate the retention times in minutes.

of acylcarnitines can be determined from the total dpm in each fraction as follows:

$$\frac{\text{dpm in A} \times \text{nmol B}}{\text{dpm in B}} = \text{nmol A}$$

where A is the acylcarnitine to be determined and B is the known amount of carnitine or acylcarnitine.

[29] Purification and Assay of Carnitine Acyltransferases

By L. L. BIEBER and C. FIOL

$$\text{Acetyl-CoA} + \text{L-carnitine} \underset{\text{CAT}}{\longleftrightarrow} \text{acetylcarnitine} + \text{CoASH}$$

$$\text{Octanoyl-CoA} + \text{L-carnitine} \underset{\text{COT}}{\longleftrightarrow} \text{octanoylcarnitine} + \text{CoASH}$$

$$\text{Palmitoyl-CoA} + \text{L-carnitine} \underset{\text{CPT}}{\longleftrightarrow} \text{palmitoylcarnitine} + \text{CoASH}$$

Short-chain (acetyl), medium-chain (octanoyl), and long-chain (palmitoyl) carnitine acyltransferases (CAT, COT, CPT) have been purified from different sources. Both soluble and membrane-bound carnitine acyltransferases occur in biological systems. In liver, soluble acetyl- and octanoyltransferases occur in peroxisomes,[1] membrane-bound acetyl- and palmitoyltransferases occur in mitochondria,[2] and membrane-bound acetyl- and octanoyltransferases are associated with microsomes.[3] Classical enzyme purification procedures have been used for the soluble acyltransferases,[4] while the use of solubilizing agents is required for the membrane-bound enzymes.[2]

Several different chromatographic sequences have been applied to the purification of carnitine acyltransferases including affinity and absorption gels, molecular sieves, and anion and cation exchangers. However for adequate reproducibility one has to carefully control the conditions used, particularly with the membrane bound enzymes. An initial step in the purification of CPT, a mitochondrial enzyme, requires solubilization of

[1] S. O. Farrell and L. L. Bieber, *Arch. Biochem. Biophys.* **22,** 123 (1983).

[2] P. R. H. Clarke and L. L. Bieber, *J. Biol. Chem.* **256,** 9861 (1981).

[3] M. A. K. Markwell, N. E. Tolbert, and L. L. Bieber, *Arch. Biochem. Biophys.* **176,** 479 (1976).

[4] S. Miyazawa, H. Ozasa, S. Furita, S. Osumi, and I. Hashimoto, *J. Biochem. (Tokyo)* **93,** 439 (1983).

the membrane to release >95% of the total activity. This can be accomplished using high salt and/or nonionic detergents such as Triton X-100, Tween 20, or octylglucoside. During the early stages of the purification, a high concentration of detergent should be maintained to prevent reaggregation of membrane fragments which would interfere with the separation of the individual proteins. Similarly the concentration of protein must be kept relatively low for the initial chromatographic steps. For CPT, affinity chromatography on Cibacron Blue Sepharose 4B has been used as the initial step.[1] It gave a good resolution of the acyltransferase activities with an overall 47-fold purification. However, because of the large amounts of protein in the initial homogenates, the stationary phase capacity is reduced after several uses which can be restored by treatment with pronase. Column supports, such as Blue Sepharose, vary in their binding capacity, depending on lot number and manufacturer. Detergent solubilized CPT does not bind to the commercially available Blue Sepharose from Pharmacia although this support is very effective in the purification of soluble mouse liver peroxisomal CAT and COT.[1] Nonionic detergents can interfere with Blue sepharose affinity chromatography by encapsulation of the dye into the micelle.[5] This limitation can be overcome by reducing the detergent concentration or increasing the capacity of the support phase. Two methods to achieve this are synthesis in the laboratory of a high capacity resin or its use in a later step of the purification procedure with lower detergent and protein concentrations. Herein we describe procedures used in this laboratory for purification of membrane-bound mitochondrial CAT and CPT and soluble peroxisomal CAT and COT. Alternative purification procedures of others have yielded partially purified enzyme,[6-8] and others have obtained similar enzymes showing a high degree of purity.[4]

Purification of CAT and COT from Peroxisomes

Preparation of Homogenates

Livers from animals treated with hypolipidemic drugs such as Wy-14,643 or livers from control mice that were starved for 24 hr are removed and rinsed with ice cold 0.25 M sucrose containing 1 mM sodium phosphate, pH 7.5. The livers are minced and suspended in 10 volumes w/v of

[5] J. B. Robinson, Jr., J. M. Strottmann, D. G. Wick, and E. Stellwagen, *Proc. Natl. Acad. Sci. U.S.A.* **77**, 5847 (1980).
[6] R. J. Tomec and C. L. Hoppel, *Arch. Biochem. Biophys.* **170**, 716 (1975).
[7] B. Kopec and I. B. Fritz, *Am. J. Biochem.* **49**, 941 (1971).
[8] J. D. Bergstrom and R. C. Reitz, *Arch. Biochem. Biophys.* **204**, 71 (1980).

the same buffer containing 0.1 mM PMSF and 5 mg/liter pepstatin and homogenized using a loose-fitting Teflon pestle and a glass Potter-Elve-jhem homogenizer. The homogenates are frozen at $-20°$ for at least 1 day.

Purification of COT

Homogenates are thawed and centrifuged at 500 g for 15 min and the supernatant fluid collected and centrifuged at 10,000 g for 15 min. The supernatant fluid is made 40% in ammonium sulfate, centrifuged, and the pellet discarded. The supernatant fluid is made 60% in ammonium sulfate, centrifuged, and the pellet suspended in 10 mM sodium pyrophosphate, 0.25 mM EDTA, 0.02% sodium azide, pH 7.5 (buffer A). The pellet should be assayed at this stage to determine the degree of acyltransferase precipitation. The pellet is dissolved in and dialyzed overnight against 20 volumes of buffer A and applied at a flow rate of 1 ml/min to a column (40 $\times$ 2.5 cm) of Cibacron Blue Sepharose CL-6B equilibrated with buffer A. After washing with 500 ml of the same buffer, a 500 ml linear gradient of 0–1 M KCl in buffer A is used to elute the enzyme.

Fractions containing COT activity are pooled, dialyzed overnight against 20 volumes of buffer A, and applied at a flow rate of 1 ml/min to a column (35 $\times$ 1.5 cm) of QAE Sephadex A-25 (Pharmacia) equilibrated with buffer A. COT activity passes through without binding. The void volume is dialyzed overnight against 20 volumes of 20 mM sodium phosphate, 0.25 mM EDTA, 0.02% sodium azide, pH 7.5 (buffer B) and applied at a flow rate of 1 ml/min to a column (20 $\times$ 2.5 cm) of hydroxylapatite equilibrated with buffer the same but without the EDTA. After the column is washed with 400 ml of buffer B, a 500 ml gradient of 20 mM sodium phosphate, 60 mM KCl, pH 7.5 to 900 mM sodium phosphate, 60 mM KCl, pH 7.5 is applied. Fractions containing COT are pooled and dialyzed overnight in 20 volumes of 20 mM sodium pyrophosphate, 0.02% sodium azide, pH 7.5 (buffer C) concentrated to 3 ml using an Amicon PM-10 filter, and applied at a flow rate of 1 ml/min to a column (95 $\times$ 2.5 cm) of Sephadex G-100 equilibrated with buffer C. COT is eluted with buffer C, the fractions pooled, and dialyzed overnight against 20 volumes of 2 mM sodium pyrophosphate, 0.002% sodium azide, pH 7.5, and applied at a flow rate of 1 ml/min to a column (2.5 $\times$ 8 cm) of QAE-Sephadex A-25 (Sigma) equilibrated with the same buffer. The column is washed with 200 ml of this buffer before a 200 ml linear gradient of 2–100 mM sodium pyrophosphate, pH 7.5 is used to elute the enzyme. Fractions containing COT are pooled and analyzed for homogeneity (see Table I). Similar procedures, but slightly different protocols, were used to isolate COT of high purity by another group of investigators.[4]

TABLE I
SUMMARY OF THE PURIFICATION OF CARNITINE
OCTANOYLTRANSFERASE FROM MOUSE LIVER PEROXISOMES[a]

Purification step	Specific activity		Percentage recovery
	Units/mg	Fold	
Crude homogenate	0.14	1.	100
250 g supernatant	0.15	1.1	96
15,000 g supernatant	0.16	1.2	80
40–60% $(NH_4)_2SO_4$	0.39	2.9	69
Cibacron Blue Cl-6B	5.69	42	40
QAE-Sephadex I	11.75	86	47
Hydroxylapatite	17.07	126	38
Sephadex G-100	39.53	291	32
QAE-Sephadex II	72.06	530	20

[a] One unit of activity is the amount of enzyme necessary to convert 1 μmol acyl-CoA to acylcarnitine in 1 min.

Purification of CAT

The first carnitine acyltransferase peak eluting from Cibacron Blue Sepharose column above is used for purification of peroxisomal CAT. The pooled fractions are dialyzed overnight in QAE buffer (25 mM sodium pyrophosphate, 0.25 mM EDTA, 0.02% sodium azide, pH 7.5) and applied at a flow rate of 1 ml/min to a column (35 × 1.5 cm) of QAE Sephadex A-25 (Pharmacia) that is equilibrated with the same buffer. CAT washes through without binding. The effluent is pooled, dialyzed overnight against 20 volumes of CM buffer (5 mM HEPES, 60 mM KCl, 0.25 mM EDTA, 0.02% sodium azide, pH 7.3) and applied to a column (35 × 1.5 cm) of CM Sephadex equilibrated with CM buffer. The column is washed with 300 ml of buffer and the enzyme eluted with a 400 ml linear gradient of 60–560 mM KCl in the CM buffer. Fractions containing CAT are pooled and dialyzed overnight in 20 volumes of buffer B, concentrated to 3 ml using an Amicon PM-10 filter, and applied at a flow rate of 1 ml/min to a column (95 × 2.5 cm) of Sephadex G-100 equilibrated with buffer B. CAT activity is eluted with buffer A and the fractions pooled and analyzed for purity. Membrane-bound CAT has been purified from mitochondria using detergent extractions[2] similar to the purification described for CPT. CAT has also been purified from both peroxisomes and mitochondria obtained from *C. Tropicalis*.[9] Again isolation procedures similar to those described herein were used.

[9] M. Ueda, A. Tamaka, and S. Fukui, *J. Biochem. (Tokyo)* **124,** 205 (1982).

Purification of Mitochondrial CPT and CAT

Mitochondrial suspensions are thawed, pooled, and added to one-half volume of 3 M KCl in 6% Triton X-100. The suspension is mixed and then homogenized with six passes of the Teflon-glass homogenizer at 15° and centrifuged for 90 min at 89,000 g. The floating lipid is removed by suction and the clear apricot-colored supernatant fluid is removed from the flocculant surface of the mitochondrial pellet. The soluble mitochondrial protein solution is then equilibrated with blue buffer (1.0% Triton X-100, 2.5 mM HEPES, 0.25 mM EDTA, 60 mM KCl, pH 7.5) by exhaustive dialysis and centrifuged at 15,000 g to remove protein precipitated by the decreased ionic strength. If only CPT is desired, the ionic strength can be lowered further by equilibrating against 20 mM KCl. This precipitates CAT. The ionic strength is increased by equilibrating against blue buffer in 300 mM KCl. The supernatant fluid containing both CPT and CAT is applied in 35 ml batches to a 40 × 5.0 cm column of Sephacryl S-300 equilibrated with blue buffer, at a flow rate of 75 ml/hr. Peak fractions containing CPT are pooled and dialyzed against QAE Buffer (0.5% Triton X-100, 5.0 mM Bis-Tris Propane buffer, 0.25 mM EDTA, 20 mM KCl, pH 9.7) and applied to a 30 × 4.1 cm column of QAE-Sephadex Q-25-120 (Pharmacia). The CPT activity which washes through the column is pooled concentrated using immersible CX-30 Millipore ultrafiltration unit with a nominal molecular weight limit of 30,000, dialyzed extensively against Blue Buffer (0.1% Triton X-100, 2.5 mM HEPES, 0.25 mM EDTA, 60 mM KCl, pH 7.6) and loaded into a 12 × 1.6 cm column of Cibacron Blue Sepharose.[10] Four bed volumes of buffer are used to wash off unbound protein before a 50-ml linear gradient of 60–869 mM KCl in Blue Buffer is used to elute CPT. This final step in the purification of CPT serves to further concentrate the enzyme and optionally to exchange Triton X-100 for octylglucoside. This detergent exchange can be accomplished by reducing the concentration of Triton X-100 in the equilibrating buffer to 0.002% and washing off unbound protein with one bed volume of Blue Buffer in 0.002% Triton X-100, followed by three bed bolumes of Blue Buffer in 25 mM octylglucoside (see Table II). If the Cibacron Blue Sepharose is not available it should be possible to substitute DEAE-cellulose chromatography as described Miyazawa et al.[4] or hydroxyapatite chromatography as described in Clarke and Bieber.[2] Peak fractions eluting from the Sephacryl S-300 column containing CAT can be pooled and CAT can be purified following the procedure described previously.[2]

---

[10] H. J. Bohme, G. Kopperschager, J. Schulz, and E. Hofman, *J. Chromatogr.* **69,** 209 (1972).

TABLE II
SUMMARY OF THE PURIFICATION OF CARNITINE
PALMITOYLTRANSFERASE FROM BEEF HEART MITOCHONDRIA[a]

Purification step	Specific activity (units/mg)	Percentage recovery	Fold
Thawed mitochondria	0.03	100	1
Solubilized mitochondria	0.04	92	1.3
Dialysis supernatant	0.065	85	2.1
Sephacryl S-300	0.60	75	20.0
QAE-Sephadex	1.5	60	50.0
Cibacron Blue Sepharose	40.0	44	1333

[a] One unit of activity is the amount of enzyme necessary to convert 1 μmol of acyl-CoA to acylcarnitine in 1 min.

Assays

General

Several different assays have been used for measurement of carnitine acyltransferases. The limitations of some assays were described previously.[11] Two general types of assays have been used for measuring carnitine acyltransferases: (1) continuous assays which measure directly the formation or disappearance of acyl-CoA derivatives at 232 nm or indirectly measure the release of CoASH by reaction with sulfhydryl reagents and (2) discontinuous (end point) assays for which three general approaches have been used: measurement of the formation of a specific radioactive acylcarnitine,[12] measurement of the formation of a hydroxamates, and measurement of the exchange of radioactive carnitine into the acylcarnitine fraction.[13]

Assays in Impure Preparations

For measurements of acyltransferase activities in impure preparations or tissue homogenates several factors must be considered:

1. Acyl-CoA hydrolase activity can be greater than carnitine acyltransferase especially in some tissue homogenates. For assays which quantitate acyl-CoA formation or its disappearance or those which de-

---

[11] L. L. Bieber and S. O. Farrell, in "The Enzymes" (P. D. Boyer, Ed.), 3rd Ed., Vol. 16, p. 627. Academic Press, New York, 1983.

[12] J. Bremer, *Biochim. Biophys. Acta* **655,** 628 (1981).

[13] S. A. Stakkestad and J. Bremer, *Biochim. Biophys. Acta* **711,** 90 (1982).

pend on maintenance of saturating amounts of acyl-CoA, the presence of acyl-CoA hydrolase can influence the data, especially for end point assays. The effect of hydrolase is often difficult to assess, i.e., with exchange assays and assays which measure acyl-CoA formation. When comparing one sample to another, differences in acyl-CoA hydrolase can alter the apparent amounts of carnitine acyltransferase.

2. Tissues such as liver contain a family of broad specificity carnitine acyltransferases, with overlapping substrate specificity and multiple organelle distributions as indicated above. Thus with homogenates or impure organelle preparations the measured transferase activity can represent the contribution of more than one enzyme. For example, COT from liver peroxisomes has considerable carnitine acyltransferase activity with palmitoyl-CoA,[14] although the K_m is considerably higher. If one is assessing total acyltransferase, the multiple enzymes and multiple organelle location and broad specificity may not present a difficulty; however, if, for example, one is attempting to measure total "real" CPT activity in liver, then the contribution of peroxisomal and microsomal COT enzymes to apparent CPT activity can be appreciable.

3. To minimize the detergent effects of the long-chain acyl-CoA substrates on the mitochondrial membrane, albumin is commonly used in the assays. Albumin changes the effective concentration of the acyl-CoA due to binding of acyl-CoAs to albumin. Therefore standardized assay conditions in which a fixed substrate/albumin ratio is used are necessary for comparative work and in studies of possible activators or inhibitors of the enzyme. The effect of these on the substrate and the agent binding to albumin must also be considered.

End-Point Assays

The use of end-point assays to measure acyltransferase activity has several limitations when proper controls are not performed. For example, the data may not be a true measure of the initial rate of the reaction if a time course has not been established. Enzymes like CPT show variable K_m values, for both carnitine and acyl-CoA substrates which depend on the tissue source and the experimental conditions. Lags in activity can be variable and nonlinear kinetics have been observed for CPT. Low estimates of activity can be obtained if lags occur or nonsaturating amounts of substrate are used. A good illustration of some of the effects mentioned above are shown in Fig. 1 of Ref. 15, where both lags and effects of experimental conditions are apparent for CPT.

[14] S. O. Farrell, C. J. Fiol, J. K. Reddy, and L. L. Bieber, *J. Biol. Chem.* **259**, 13089 (1985).
[15] E. D. Saggerson, *Biochem. J.* **202**, 397 (1982).

In summary, end-point assays should only be used with very controlled assay conditions. As a minimum time courses for each assay should be done when comparing one sample or experimental condition to another. Even then caution must be exercised with CPT, because of its capacity to show sigmoid kinetics.[16,17] For CAT and COT which exhibit Michaelis–Menten type kinetics, validity of end-point assays is easier to establish. When end-point assays are used, those which measure the amount of radioactive acylcarnitine formed are recommended.[18] With the introduction of HPLC separations of short-chain and medium-chain acylcarnitines quantitation of acylcarnitine formation can be both rapid and sensitive (see this volume [28]).

We recommend use of continuous initial rate assays for both crude preparations and purified enzymes, although assay difficulties can be encountered. Measurement of the appearance or disappearance of acyl-CoAs at 232 nm is difficult when nonpurified enzymes are used because of the large background absorbance. This assay is more suitable for purified enzymes. For CPT assays where Triton X-100 cannot be used, octylglucoside or Tween 20 is an adequate substitute. DTNB and DTBP can be used to monitor the release of CoASH with crude systems, but often very high blank corrections must be made due to acyl-CoA hydrolase activity. Some preparations contain up to 10 times more hydrolase than transferase. Thus, multiple assays per sample are necessary to obtain a reliable average of the transferase activity in question.

DTNB can inhibit carnitine acyltransferase activities; however, at pH 8.0, DTNB is not inhibitory in the presence of acyl-CoA when initial rate assays are performed and the enzyme is *not* preincubated with DTNB. DTBP has the advantage of being utilizable over a wider pH range. It has a higher extinction coefficient and is less inhibitory to CAT, and therefore is the preferred chromophoric agent.

Initial Rate Assay Protocol

Principle. The enzyme activity is measured in the forward direction by monitoring the initial rate of CoASH formation with DTBP at 323 nm at saturating levels of both substrates. The activity can be measured through a pH range of 6–8, but pH 8.0 is routinely used because of the lower K_m for the substrates. Hydrolase activity (the blank) is measured as the carnitine independent release of CoASH.

[16] C. J. Fiol and L. L. Bieber, *J. Biol. Chem.* **259**, 13084 (1985).
[17] L. L. Bieber and C. J. Fiol, *Curr. Top. Cell. Regul.* **24**, 111 (1984).
[18] J. Bremer, *Biochim. Biophys. Acta* **665**, 628 (1981).

Reagents

Bis-Tris Propane buffer, 0.74 M, pH 8.0 at 25° (for CAT assays, HEPES buffer is used instead of Tris)
Triton X-100, 1%
NaEDTA, 50 mM
Acyl-CoA, 2 mM
L-Carnitine (neutralized), 100 mM
DTBP in 0.01 M NaHCO$_3$, 2.5 mM

Procedure. Prepare a stock (A) reaction mix containing 2.0 ml of Bis-Tris Propane buffer, 0.65 ml of EDTA, 1.45 ml of L-carnitine, 2.90 ml of Triton X-100, and 1.0 ml of distilled water. Prepare a second stock (B) reaction mix, substituting water for L-carnitine to be used for hydrolase (blanks) assays. Refrigerate. On the day of the assay, prepare a reaction mix by adding 0.7 ml of stock solution A to 0.1 ml of the acyl-CoA, 0.15 ml of DTBP, and 0.05 ml of distilled water. Prepare a similar reaction mix using stock solution B. To a 200 μl cuvette, add 100 μl of reaction mix A or B and a 100 μl of sample plus distilled water. The reaction is started by the addition of sample to the cuvette, mixed rapidly, and monitored "immediately." A molar extinction coefficient of 19,800/cm is used. CPT activity is obtained by substracting the initial rate of formation of CoASH without L-carnitine (assay using solution B) for the initial rate when solution A is used.

[30] Determination of the Specific Activity of Long-Chain Acylcarnitine Esters

By RICHARD ODESSEY

Fatty acids are a major fuel for many types of tissue including vascular smooth muscle. However, the measurement of tissue fatty acid metabolism in these tissues using a radioactively labeled precursor is usually an underestimate since it does not include the oxidation of fatty acid derived from endogenous lipids. To overcome this problem measurement of the specific activity of the intracellular pool of fatty acid has been attempted. However, artifacts may arise due to fatty acid present in the extracellular space of the tissue or bound to the plasma membrane. This is especially significant when the incubation medium contains albumin which avidly binds fatty acids. In addition, long times are required to achieve isotopic equilibrium[1] and the true intracellular specific activity may still be uncertain due to compartmentation of the intracellular fatty acid pools.

Measurement of the specific activity of the cellular long-chain acylcarnitine pool avoids many of these problems. It is present only within the cell and is a direct precursor of fatty acid oxidation. Cytoplasmic and mitochondrial pools are in equilibrium.[2] The pool is very small and may equilibrate more rapidly with the external label than would the total fatty acid pool. When the measurement of long-chain acylcarnitine specific activity was attempted using previous methods, a number of difficulties were encountered stemming from the desire to measure these compounds in milligram amounts of vascular smooth muscle, and from the contamination of the acylcarnitine fraction by other labeled tissue components, in particular phosphatidylcholine and sphingomyelin. The method presented here overcomes these problems and allows the measurement of acylcarnitine specific activity in very small amounts of incubated tissue (see also Ref. 3).

Principle

A lipid extract of tissue labeled with [1-^{14}C]palmitate is treated with phospholipase C to cleave labeled sphingomyelin and phosphatidylcholine. ^{14}C-Labeled long-chain esters of carnitine are separated by two-dimensional thin-layer chromatography. Unlabeled sphingomyelin and phosphatidylcholine standards are added as markers. A separate TLC plate with palmitylcarnitine and the markers are run and the spots visualized with iodine vapor. On the experimental plates, the area corresponding to acylcarnitine is scraped off and eluted. An aliquot is counted and the remainder of the sample is hydrolyzed and assayed for carnitine as described below.

Tissue Preparation

Tissue [e.g., aortic rings (20–40 mg)] is incubated in Krebs–Ringer bicarbonate buffer containing [1-^{14}C]palmitate (0.1 mM, 4 μCi/μmol) bound to bovine serum albumin (BSA).[4] Following incubation the pieces of tissue are frozen in liquid nitrogen and homogenized in 2 ml of chloroform–methanol (2:1) to extract lipids.[5] To determine the efficiency of

[1] E. S. Morrison, R. F. Scott, M. Kroms, and J. Frick, *Biochem. Med.* **11**, 153 (1974).
[2] J. A. Idell-Wenger, L. W. Grotyohann, and J. R. Neely, *J. Biol. Chem.* **253**, 4310 (1978).
[3] R. Odessey and K. V. Chace, *Anal. Biochem.* **122**, 41 (1982).
[4] K. V. Chace and R. Odessey, *Circ Res.* **48**, 850 (1981).
[5] N. S. Radin, this series, Vol. 14, p. 245.

recovery, 100 or 200 pmol of palmitoyl L-carnitine dissolved in chloroform–methanol is extracted and treated in the same manner as the samples. The concentration of the standards is determined using dithionitrobenzoic acid.[6]

Thin-Layer Chromatography of Long-Chain Acylcarnitine Esters

Reagents

Phospholipase C (Sigma Chemical Co.), 7.7 U/mg. Make a solution of 1 mg/ml (7.7 U/ml) in 0.1 M Tris buffer (pH 7.2) containing 20 mM CaCl$_2$.

Palmitoyl-, stearoyl-, linoleoyl-, oleoyl-, myristoyl-, and octanoylcarnitines are purchased from PL Biochemicals.

Silica gel G plates used are Fisher brand Redi-Plates.

Procedure

The phospholipase reagent (0.3 ml) is shaken with 2.0 ml of homogenate and incubated at room temperature for 3 hr. The mixture is centrifuged at low speed for 10 min. Unlabeled sphingomyelin and phosphatidylcholine standards (100 μg) are added to the chloroform-soluble phase as markers. Standards and samples are then subjected to two-dimensional thin-layer chromatography.[7] The plates are chromatographed using the acid solvent [chloroform/methanol/acetic acid/water (50:25:8:4)] in the first dimension followed by the basic solvent [chloroform/methanol/ammonium hydroxide (14 N)/water (50:35:3:3)] in the second dimension. Since the amount of acylcarnitine present in the tissue is too small to be visualized by iodine vapor, its location on the plate is determined relative to the positions of the sphingomyelin and phosphatidylcholine markers.

After the location of the acylcarnitine area is determined, it is marked with a grid of 1- or 2-cm squares. A column made from a Pasteur pipette plugged with glass wool is filled with the powder scraped from each square and eluted with chloroform/methanol/ammonium hydroxide (14 N) (56:42:2) (1.5 ml/cm^2 of powder). Aliquots (0.1–0.2 ml) of each eluent are counted to determine the location of the labeled pamitoylcarnitine. The remainder of the eluent (containing [14C]palmitoylcarnitine) is evaporated to dryness under nitrogen at 55° and 0.2 ml of BSA (10 mg/ml) is added to redissolve the acylcarnitine. This solution is incubated at 55° for 1 hr with frequent mixing to ensure that all the acylcarnitine is dissolved.

[6] D. J. Pearson, J. F. A. Chase, and P. K. Tubbs, this series, Vol. 14, p. 612.
[7] G. Wittels and R. Bressler, *J. Lipid Res.* **6,** 313 (1965).

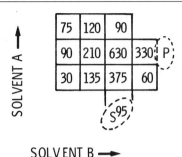

SOLVENT B ➡

FIG. 1. Segment of a chromatogram of the lipid extract of rabbit aorta incubated with [1-^{14}C]palmitate. A, the acidic solvent system; B, the basic solvent system. P and S show the location of the phosphatidylcholine and sphingomyelin spots, respectively. Following chromatography, 1-cm squares are scraped from the plate and counted. The number in each square represents the cpm above background in a 0.2-ml aliquot. The peak counts are present in the area where palmitoylcarnitine is found.

If the specific activity of acylcarnitine is to be determined, a 50-μl aliquot is counted, and a 100-μl aliquot is assayed for acylcarnitine (see below).

Remarks

Samples and standards are digested with phospholipase C (EC 3.1.4.3) to hydrolyze sphingomyelin and phosphatidylcholine[8] which migrate near the area where the acylcarnitine is found. When pieces of aorta are incubated in medium containing [1-^{14}C]palmitate, 15× more label is incorporated into sphingomyelin and 700× more label is found in phosphatidylcholine than is present in the palmitoylcarnitine spot. These large amounts interfere with the determination of acylcarnitine specific activity. To overcome this problem, the homgenates are preincubated with phospholipase C.[8] This treatment reduces the counts in sphingomyelin and phosphatidylcholine to near backround levels (~20 cpm) (Fig. 1). The peak of the label is now found in the area of palmitoylcarnitine. To further prevent contamination with surrounding areas, only the central portion of the peak is scraped for assay. Approximately 86% of the palmitoyl carnitine added can be recovered.

The chromatography procedure is modified from previous work[7] by using the acid solvent in the first dimension followed by the basic solvent in the second dimension. This modification yields greater recovery of standard palmitoyl carnitine. In both solvent systems, the R_f value of the

---

[8] M. Kates, "Laboratory Techniques in Biochemistry and Molecular Biology," p. 569. Am. Elsevier, New York, 1972.

acylcarnitine standards increases as the chain length of the fatty acid moiety increases. After chromatography in both dimensions, the R_f values for the phospholipids and acylcarnitine are variable. However, the relative R_fs of the phosphoatidylcholine, sphingomyelin, and acylcarnitine spots are more consistent. Since palmitoylcarnitine moves with an R_f equal to that of sphingomyelin in the basic solvent and with an R_f to equal to that of phosphatidylcholine in the acidic solvent unlabeled phosphatidylcholine and sphingomyelin are used as markers.

Although the different acylcarnitines moved at different rates, a mixture containing several saturated and unsaturated acylcarnitines could not be resolved. For this reason, the whole area containing acylcarnitine is divided into a grid and assayed for radioactivity and carnitine content. However all the acylcarnitine esters tested (8–18 carbons) are clearly resolved from sphingomyelin and phosphatidylcholine. Phosphatidylethanolamine, cardiolipin, di- and triglycerides, and free fatty acids all run much faster than acylcarnitine.

Assay of Long-Chain Acylcarnitines

Principle

$$\text{Acylcarnitine} \xrightarrow[70°]{\text{NaOH}} \text{carnitine + fatty acid-salt} \tag{1}$$

$$\text{L-Carnitine + acetyl-CoA} \longrightarrow \text{acetyl-L-carnitine + CoASH} \tag{2}$$

Reagents

NaOH (1 N): make solution of 40 mg/ml

Bovine serum albumin (fatty-acid free): 100 mg/ml

MOPS–HCl: add 2.62 g MOPS (morpholinopropanesulfonic acid) to 50 ml 1 N HCl (final conc. 0.25 M MOPS)

MOPS–buffer (100 mM): 2.09 g of MOPS is dissolved in 80 ml of water and the pH is adjusted to 8.0 with 1 N NaOH and the volume brought to 100 ml

[³H]Acetyl-CoA: [*methyl*-³H]acetyl-CoA (1.4 Ci/mmol, 100 μmol/ml)

Acetyl-CoA (2.5 Na/mol, 2.5 H_2O/mol): 10.64 mg is dissolved in 0.5 ml of MOPS buffer

DTNB [5,5′-dithiobis(2-nitrobenzoic acid)]: 39.6 mg of DTNB is dissolved in 10 ml MOPS buffer

Carnitine acetyltransferase (Sigma, 400 U/ml, 10 mg/ml). Since ammonium sulfate interfers with the transferase reaction, 0.5 ml of the enzyme suspension is centrifuged to remove the $(NH_4)_2SO_4$. The pellet is redissolved in 0.5 ml of MOPS buffer and dialyzed overnight against 500 ml of MOPS buffer.

Procedure

To hydrolyze the carnitine esters, 10 μl NaOH solution and 100 μl BSA solution is added to 100 μl of each sample. After incubation for 1 hr at 70°, the pH of the samples is lowered to between 6 and 8 with MOPS–HCl.

The entire sample is mixed with 0.09 ml acetyl-CoA solution, 0.11 ml [*methyl*-³H]acetyl-CoA and 0.09 ml DTNB solution. The samples are warmed to 37° and 10 μl carnitine acetyltransferase is added to start the reaction. After 10 min, 0.3 ml of the reaction mixture is transfered to a 0.7 × 2.5 cm column of AG1-X8 resin (Bio-Rad). [³H]Acetylcarnitine is eluted from the column with 2 ml of water into a counting vial. The samples are counted and corrected for quenching.

No [¹⁴C]palmitoylcarnitine or palmitoyl-CoA could be detected in the AG1-X8 effluent. However [³H]acetyl-CoA often contains ³H which eluted with the water wash (probably ³H₂O). Therefore sample values are corrected by subtracting a blank consisting of [³H]acetyl-CoA in assay mixture without enzyme.

Remarks

BSA is required (perhaps to stabilize the enzyme) for the carnitine acetyltransferase reaction to proceed to completion. The BSA is subsequently removed by chromatography through AG1-X8 resin.

To increase the sensitivity of the carnitine assay, MOPS buffer is substituted for Tris. When Tris–HCl buffer is used, the apparent amount of product formed increased linearly with time for at least 2 hr, whether or not carnitine is added to the assay mixture. This is probably due to a slow reaction of Tris with acetyl-CoA which forms a product which elutes with acetylcarnitine. Using MOPS buffer, the reaction goes to completion within 5 min. The acetylcarnitine formed is equal to acylcarnitine added. The reaction product is stable at room temperature for at least 1 hr.

In the carnitine assay DTNB is used to trap the product, CoASH, and shift the equilibrium to the right. However, DTNB also slowly inhibits the enzyme carnitine acetyltransferase,[6] so that a large amount of enzyme must be used for the assay. The K_m for acetyl-CoA is 34 μM[9] and the transferase reaction does not go to completion if the acetyl-CoA is much below 4 μM. Consequently, to improve sensitivity, the specific activity of acetyl-CoA can only be increased by using higher amounts of labeled substrate. This would also increase the blank. Thus the sensitivity of the assay could only be improved by reducing the blank by prior purification of [³H]acetyl-CoA.

[9] J. F. A. Chase, *Biochem. J.* **104**, 510 (1967).

Conclusions

This method for the determination of the specific activity of long-chain acylcarnitine has the advantage of removing other labeled compounds which interfere with the specific activity measurements. The improved sensitivity of the assay allows the determination of picomolar amounts of acylcarnitine in tissue. Since individual fatty acylcarnitines are not resolved by this method, only the average long-chain acylcarnitine specific activity can be calculated.

This method of determining the specific activity of the fatty acids in the cell could give erroneous results if intermediates of fatty acid oxidation formed carnitine esters in significant amounts. Intermediates of [1-^{14}C]palmitate oxidation with 12 and 14 carbon atoms would not be labeled but would be assayed in the acylcarnitine pool, thus leading to an artifactually low specific activity. However, carnitine esters of intermediates of fatty acid oxidation appear only in state 4 (ADP limited) conditions.[10] Total ADP levels in rabbit aorta appear to be sufficiently high[4] to ensure that this state does not exist. However free mitochondrial ADP levels remain to be determined.

[10] M. Lopes-Cardozo, N. Klazinga, and S. G. van den Bergh, *Eur. J. Biochem.* **83,** 629 (1978).

[31] Synthesis of Carnitine Precursors and Related Compounds

By Charles J. Rebouche

L-Carnitine contains a quaternary nitrogen atom which is formed biosynthetically early in the pathway. Thus several intermediates in the pathway are quaternary amines which can readily be synthesized from the commercially available parent primary amines, by the general procedure of Mazzetti and Lemmon.[1] These intermediates include ε-*N*-trimethyl-L-lysine (from α-*N*-acetyl-L-lysine), γ-*N*-trimethylaminobutyraldehyde (from 4-aminobutyraldehyde diethyl acetal), and γ-butyrobetaine (from γ-aminobutyric acid). Other compounds of interest which are synthesized by this general method include δ-*N*-trimethylaminovaleric acid (from δ-aminovaleric acid) and ε-*N*-trimethylaminocaproic acid (from ε-aminocaproic acid).

[1] F. Mazzetti and R. M. Lemmon, *J. Org. Chem.* **22,** 228 (1957).

Another carnitine precursor, β-hydroxy-ε-N-trimethyl-L-lysine, is prepared following the general procedure of Otani and Winitz,[2] by condensation of γ-N-trimethylaminobutyraldehyde with glycine in the presence of potassium carbonate and copper sulfate, yielding a diastereomeric mixture.[3] Alternatively, small amounts of the natural isomer (presumed to be *erythro*-β-hydroxy-ε-N-trimethyl-L-lysine[4]) are synthesized enzymatically with a crude preparation of ε-N-trimethyl-L-lysine hydroxylase from liver or kidney of any of several mammalian species[5] (see below).

In *Neurospora crassa* (but apparently not in mammals) ε-N-methyl-L-lysine and ε-N-dimethyl-L-lysine are direct precursors of L-carnitine.[6] The latter compound is prepared by reductive methylation of α-N-acetyl-L-lysine with formaldehyde and H_2/Pd catalyst.[7] ε-N-Methyl-L-lysine is prepared by reaction of iodomethane with α-N-benzoyl-ε-N-p-toluenesulfonyl-L-lysine, followed by hydrolysis with fuming hydrobromic acid.[7] Retention of configuration at C-2 is obtained by either procedure.

α-Keto-ε-N-trimethylaminocaproic acid, a catabolic product of ε-N-trimethyl-L-lysine, is synthesized enzymatically by reaction of ε-N-trimethyl-L-lysine with commercial L-amino acid oxidase.[8]

Carnitine is synthesized, as a racemic mixture, by several related methods. For example, DL-carnitine is prepared by reaction of epichlorhydrin with NaCN followed by condensation with trimethylamine, and subsequent hydrolysis.[9] Recently, syntheses specific for the D and L isomers of carnitine, starting with arabinose or ascorbic acid, were described.[10]

At this writing the following compounds are available commercially: L-lysine, ε-N-methyl-L-lysine, ε-N-dimethyl-L-lysine, α-N-acetyl-L-lysine, γ-aminobutyraldehyde diethyl acetal, γ-butyrobetaine (3-carboxypropyltrimethylammonium chloride), L-carnitine, DL-carnitine, δ-aminovaleric acid, and ε-aminocaproic acid. The reader is referred to CHEM SOURCES—U.S.A. (Directories Publishing Co., Inc., Ormand Beach, FL), 1984 edition, for suppliers.

In recent years radiotracers have been used extensively in studies of carnitine biosynthesis and metabolism in microorganisms, experimental

[2] T. O. Otani and M. Winitz, *Arch. Biochem. Biophys.* **102,** 464 (1963).
[3] R. A. Kaufman and H. P. Broquist, *J. Biol. Chem.* **252,** 7437 (1977).
[4] R. A. Novak, T. J. Swift, and C. L. Hoppel, *Biochem. J.* **188,** 521 (1980).
[5] J. D. Hulse, S. R. Ellis, and L. M. Henderson, *J. Biol. Chem.* **253,** 1654 (1978).
[6] C. J. Rebouche and H. P. Broquist, *J. Bacteriol.* **126,** 1207 (1976).
[7] L. Benoiton, *Can. J. Chem.* **42,** 2043 (1964).
[8] C. J. Rebouche and A. G. Engel, *J. Biol. Chem.* **255,** 8700 (1980).
[9] F. Binon, P. Bruckner, and G. Deltour, *in* "Recent Resarch on Carnitine" (G. Wolf, ed.), p. 7. MIT Press, Cambridge, Massachusetts, 1965.
[10] K. Bock, I. Lundt, and C. Pedersen, *Acta Chem. Scand., Ser. B* **B37,** 341 (1983).

animals, and humans. Currently, only L-[*methyl*-^{3}H]- and L-[*methyl*-^{14}C]carnitine are available commercially. Radiolabeled carnitine precursors must be prepared in the laboratory.

In general, compounds synthesized by exhaustive methylation with iodomethane, as described above, can be synthesized by the same method with ^{14}C or ^{3}H in the methyl moieties. The procedure only requires substitution of iodo[^{3}H]- or iodo[^{14}C]methane for unlabeled iodomethane. Similarly, *methyl*-^{14}C-labeled and *methyl*-^{3}H-labeled ε-*N*-dimethyl-L-lysine are synthesized by substituting appropriately labeled formaldehyde for unlabeled formaldehyde. ε-*N*-Trimethyl-L-lysine has been prepared with ^{3}H in the methyl groups[11] and ^{14}C in the methyl groups[12] and at C-1.[13] The racemic mixture has been synthesized with ^{3}H at C-3.[14] *methyl*-^{3}H-labeled and *methyl*-^{14}C-labeled β-hydroxy-ε-*N*-trimethyl-L-lysine are synthesized enzymatically from appropriately-labeled ε-*N*-trimethyl-L-lysine.[8] L-Carnitine has been synthesized with ^{3}H in the methyl groups[15] and ^{14}C in the methyl groups,[16–18] at C-1,[19] and uniformly labeled in the carbon chain.[19] Similarly, γ-butyrobetaine has been prepared with ^{3}H in the methyl groups[20] and at C-2–C-3,[21] and ^{14}C in the methyl groups,[19] C-1[22] and uniformly labeled in the carbon chain.[19]

Representative Synthetic Procedures

ε-N-Trimethyl-L-lysine

The general procedure for quaternarization of primary amines is illustrated by the synthesis of ε-*N*-trimethyl-L-lysine. Two grams of α-*N*-acetyl-L-lysine (10.6 mmol) is mixed with 12.88 g (40.8 mmol) of Ba(OH)$_2$·8H$_2$O and 28 ml of 80% aqueous methanol. Eight milliliters (129 mmol) of iodomethane is added and the flask is stoppered and covered with aluminum

[11] D. W. Horne and H. P. Broquist, *J. Biol. Chem.* **248**, 2170 (1973).
[12] R. A. Cox and C. L. Hoppel, *Biochem. J.* **136**, 1083 (1973).
[13] J. B. Hochalter and L. M. Henderson, *Biochem. Biophys. Res. Commun.* **70**, 364 (1976).
[14] R. Stein and S. Englard, *Anal. Biochem.* **116**, 230 (1981).
[15] O. Stokke and J. Bremer, *Biochim. Biophys. Acta* **218**, 552 (1970).
[16] R. R. Ramsay and P. K. Tubbs, *FEBS Lett.* **54**, 21 (1975).
[17] H. Schulz and E. Racker, *Biochem. Biophys. Res. Commun.* **89**, 134 (1979).
[18] S. T. Ingalls, C. L. Hoppel, and J. S. Turkaly, *J. Labelled Cmpd. Radiopharm.* **19**, 535 (1982).
[19] D. B. Goodfellow, C. L. Hoppel, and J. S. Turkaly, *J. Labelled Cmpd. Radiopharm.* **19**, 365 (1982).
[20] T. Bohmer, K. R. Norum, and J. Bremer, *Biochim. Biophys. Acta* **125**, 244 (1966).
[21] S. Englard, L. J. Horowitz, and J. T. Mills, *J. Lipid Res.* **19**, 1057 (1978).
[22] J. Bremer, *Biochim. Biophys. Acta* **57**, 327 (1962).

foil. The mixture is stirred at room temperature for 16 hr. The flask is cooled in ice and 5 ml of 18 N H_2SO_4 is added slowly with stirring. The precipitate is removed by centrifugation. The supernatant is evaporated *in vacuo* to a small volume (approximately 5 ml) and is applied to a 2.5 × 50 cm column of AG1-X8 anion exchange resin[23] (200–400 mesh, Cl^- form, equilibrated with water). The column is eluted with water, and the desired material is located as follows. An aliquot (50 μl) of each fraction (5–10 ml) is added to 50 μl of a saturated solution of Reinecke salt in 1 N HCl (prepared fresh daily). Quaternary amines produce a heavy white precipitate. Fractions containing the reaction product, α-N-acetyl-ε-N-trimethyl-L-lysine, are evaporated to an oil. The oil is taken up in 50 ml of 6 N HCl and the solution is refluxed for 16 hr to hydrolyze the ester. The HCl is removed *in vacuo* and the resulting oil is taken up in water and applied to a 2.5 × 25 cm column of AG50W-X8 cation exchange resin[23] (200–400 mesh, H^+ form, equilibrated with water). The column is washed with 300 ml of 1 N HCl and eluted with 250 ml of 4 N HCl. The desired product is located in column fractions by precipitation with Reinecke salt. Appropriate fractions are evaporated to an oil. The residue is repeatedly taken up in water, then absolute ethanol, and finally in benzene and reduced to dryness *in vacuo* to remove first, traces of HCl, and then water. The product is a very hygroscopic white solid. Typical yield is 80%, based on α-N-acetyl-L-lysine. Purity is determined by thin-layer and paper chromatography (see the table). The product is visualized with ninhydrin, Dragendorf reagent, or I_2 vapor.

ε-N-[methyl-^{14}C]Dimethyl-L-lysine

α-N-Acetyl-L-lysine (8.57 mg, 45.4 μmol) and palladium on charcoal catalyst (0.1 g) are added to 1.52 ml of water containing [^{14}C]formaldehyde (1.0 mCi, 0.68 mg). The mixture is hydrogenated in a Parr apparatus under 40 lb H_2 pressure for 6 hr. Formaldehyde (0.2 ml of a 37% solution) is added and the hydrogenation is continued for 1 hr. The reaction mixture is filtered through glass wool and the filtrate is evaporated to dryness. The residue is taken up in 10 ml of 6 N HCl and heated to reflux temperature for 16 hr. After cooling, the HCl is removed by evaporation *in vacuo*. The residue is dissolved in a small volume of water and is applied to a 0.9 × 15 cm column of Aminex A-5 cation exchange resin[23] (Na^+ form, equilibrated with 0.35 N sodium citrate, pH 6.48). The column is eluted with the same buffer (flow rate, 0.5 ml/min). One-milliliter fractions are collected. The radioactive product is located by liquid scintillation counting of 5 μl

---

[23] Obtained from Bio-Rad Laboratories, Richmond, California.

CHROMATOGRAPHIC PROPERTIES OF CARNITINE PRECURSORS AND
RELATED COMPOUNDS

Compound	Thin-layer chromatography[a] (R_f)			Paper chromatography[b] (R_f)	
	A[c]	B[d]	C[e]	1[f]	2[g]
ε-N-Trimethyl-L-lysine	12	36	35	13	64
ε-N-Dimethyl-L-lysine	62	45	31	—	—
ε-N-Methyl-L-lysine	37	60	16	—	—
β-Hydroxy-ε-N-trimethyl-L-lysine	11	—	—	15	—
γ-Butyrobetaine	40	38	47	53	—
L-Carnitine	39	50	32	42	—
α-Keto-ε-trimethyl aminocaproate	25	33	—	34	—
δ-Trimethylamino valerate	40	38	—	58	—

[a] Silica gel G-coated (250 μm) glass plates obtained from Analtech, Inc., Newark, DE, developed to a height of 15 or 17.5 cm.

[b] Whatman (Clifton, NJ) #3MM paper, descending, developed to 25 cm.

[c] Developing solvent: methanol/concentrated NH₄OH, 75:25.

[d] Developing solvent: methanol/acetone/concentrated HCl, 90:10:4.

[e] Developing solvent: phenol (liquified)/1-butanol/concentrated NH₄OH, 50:50:20.

[f] Developing solvent: 1-butanol/glacial acetic acid/water, 60:15:25.

[g] Developing solvent: 2-propanol/formic acid/water, 4:1:1.

aliquots of appropriate fractions (numbers 40–60). The radioactive peak fractions are pooled and evaporated to a small volume. Sodium citrate is removed by desalting on a column (1.2 × 15 cm) of AG50W-X8 cation exchange resin (200–400 mesh, H⁺ form, equilibrated with water). The column is washed with 100 ml of 1 N HCl and 50 ml of water and is eluted with 1 N NH₄OH. Radioactive fractions are combined and evaporated to dryness *in vacuo*. The residue is taken up in water. Radiochemical purity is determined by thin-layer chromatography (see the table). The product is located by liquid scintillation counting of 1-cm segments of the chromatogram.

β-Hydroxy-ε-N-[methyl-³H]trimethyl-L-lysine

This compound is prepared enzymatically from ε-N-trimethyl-L-lysine. Freshly excised rat liver (23.77 g) is homogenized in 71.3 ml (3 volumes, w/v) of 0.25 M sucrose containing 0.05% bovine serum albumin (fraction V, Sigma Chemical Co., St. Louis, MO) and 2 mM HEPES · NaOH, pH 6.8 (Buffer A), using a Potter-Elvejhem homogenizer with a motor-driven Teflon pestle. An additional 6 volumes (w/v) of Buffer A are added and the homogenate is centrifuged at 560 g for 15 min. The supernatant is centrifuged at 9000 g for 15 min. The pellet is resuspended in 112 ml of Buffer A and centrifuged at 9000 g for 15 min. The supernatant and fluffy layer are carefully removed and the pellet is resuspended in 56 ml of Buffer A and again centrifuged at 9000 g for 15 min. After removal of the supernatant and fluffy layer, the mitochondrial pellet is resuspended in 22 ml of Buffer A. All of the above procedures for preparation of mitochondria are performed at 0–4°.

For hydroxylation of ε-N-trimethyl-L-lysine, the following reaction mixture is prepared: 22 ml mitochondrial suspension, 1.1 ml 50 mM sodium ascorbate, 1.1 ml 60 mM α-ketoglutarate, 1.1 ml 3 mM FeSO₄, 131.6 μl ε-N-[*methyl-³H*]trimethyl-L-lysine (50 Ci/mol, 0.836 mCi/ml), and 0.9865 ml water. The mixture is incubated for 90 min at 37° with vigorous shaking. The reaction is stopped by addition of 26.4 ml of absolute ethanol. After centrifugation at 1860 g the pellet is washed with 13.2 ml of 50% aqueous ethanol. The combined supernatants are evaporated to dryness *in vacuo*. The residue is taken up in a small amount of water and filtered through glass wool. The filtrate is applied to a AG50W-X8 cation exchange resin column (2.5 × 50 cm, 200–400 mesh, H⁺ form, equilibrated with water). The column is washed with 500 ml of water and eluted with 1 liter of 2 N HCl followed by a linear gradient of 600 ml each of 2 and 5 N HCl. Ten-milliliter fractions are collected, and radioactivity in 0.1-ml aliquots of each fraction is determined by liquid scintillation counting. The product is recovered from fractions 38–62 by evaporation *in vacuo*. Unreacted ε-N-[*methyl-³H*]trimethyl-L-lysine elutes from the column in fractions 70–96. Typical yield of β-hydroxy-ε-N-[*methyl-³H*]trimethyl-L-lysine is 85%. Radiochemical purity of the product is determined by thin-layer and paper chromatography (see the table) and by column chromatography.[8]

γ-[1-¹⁴C]Butyrobetaine

γ-Amino[1-¹⁴C]butyric acid (100 μCi, 49.9 Ci/mol) in 0.1 N HCl (obtained from New England Nuclear, Boston, MA) is evaporated to dryness under N_2. $Ba(OH)_2$ (3.43 mg, 20 μmol) and iodomethane (3 μl, 24 μmol) in 1 ml of 50% aqueous methanol are added, and the mixture is stirred with a magnetic stirrer at room temperature in a glass-stoppered flask. At 3.5 and 15 hr additional 3 μl aliquots of iodomethane are added to the reaction mixture. At 18 hr Ba^{2+} is precipitated by addition of 11.1 μl of 1.8 M H_2SO_4. The precipitate is removed by centrifugation (1860 g, 15 min). The supernatant is applied to a 0.9 × 25 cm column of AG50W-X8 cation exchange resin (200–400 mesh, H^+ form, equilibrated with 1 N HCl). The column is eluted with 1 N HCl. The product elutes at 148–180 ml of column eluate. Appropriate column eluate fractions are combined and evaporated to a small volume *in vacuo* and to dryness under a gentle stream of N_2. The residue is taken up in water and filtered through a 0.22-μm (pore size) disposable filter. The yield is 70–95% based on radioactivity. Radiochemical purity is determined by thin-layer and paper chromatography (see the table).

L-[methyl-³H]Carnitine

The procedure is essentially that of Stokke and Bremer.[15] A methyl group is removed from L-carnitine propyl ester by the procedure of Jenden *et al.*[24] and is replaced with a [³H]methyl moiety from iodo[³H]methane. This sequence of reactions proceeds with retention of configuration at C-3.[15]

Two grams (10 mmol) of L-carnitine hydrochloride is mixed with 80 ml of *n*-propanol. HCl gas is bubbled through the mixture for 5 min, and then the solution is heated to reflux for 10 min. The solution is again saturated with HCl gas and heated to reflux. The solvent and HCl are evaporated *in vacuo* to a clear oil. About 100 ml of 2-butanone (distilled from molecular sieves) is added and the mixture is again evaporated *in vacuo* to an oil. This process is repeated, finally yielding a white solid. The product is suspended in 150 ml of redistilled 2-butanone, and 8 g of sodium benzenethiolate (prepared from thiophenol[24]) is added. The mixture is heated under reflux for 25 hr. The mixture is allowed to cool and the solvent is evaporated *in vacuo*. The residue is taken up in 100 ml of 0.5 N NaOH and is heated under reflux for 2 hr. The mixture is acidified with concentrated HCl and extracted twice with approximately 150 ml portions of ether. The aqueous fraction is evaporated to dryness. The oily white salt remaining is triturated in 50 ml of methanol and filtered. The filtrate is evaporated to dryness in vacuo. The oil, β-hydroxy-γ-dimethylaminobutyric acid hydrochloride, is dissolved in methanol (50–100 mg/ml).

[24] D. J. Jenden, I. Hanin, and S. I. Lamb, *Anal. Chem.* **40**, 125 (1968).

One-tenth milliliter of β-hydroxy-γ-dimethylaminobutyric acid hydrochloride (73.5 mg/ml in methanol) and 0.1 ml of 1.84 mM NaOH are added to 1.8 ml of 50% aqueous methanol. This solution is added to a vial containing 25 mCi of iodo[^{3}H]methane (21.8 μmol). During the time the vial is open, it is cooled in a dry ice–acetone bath. The vial is stoppered and the temperature of the solution is allowed to rise to room temperature. The solution is maintained at room temperature for 20 hr, with occasional mixing. The contents of the vial are transferred to a 50 ml round bottom flask. The solution is acidified with a few drops of 1 N HCl and is evaporated to dryness in vacuo. The residue is taken up in 2 ml of water and is applied to a AG50W-X8 cation exchange resin column (0.9 × 50 cm, H$^+$ form, 200–400 mesh, equilibrated with water). The column is washed with 40 ml of water and eluted with 1.5 N HCl. The product is eluted at 116–160 ml of column eluate. These fractions are evaporated to dryness and the product is repeatedly taken up in water and then absolute ethanol and evaporated in vacuo to remove final traces of HCl. The residue is taken up in 10 ml of water and filtered through a 0.22-μm (pore size) disposable filter. Specific radioactivity is determined by liquid scintillation counting and enzymatic carnitine determination.[8] Radiochemical purity is determined by thin-layer and paper chromatography (see the table).

[32] Synthesis of Radioactive (−)-Carnitine from γ-Aminobutyrate

By Shri V. Pande

$$\gamma\text{-Aminobutyric acid} + [^3\text{H}]\text{methyl iodide} \xrightarrow[\text{methanol}]{\text{K}_3\text{PO}_4} \gamma\text{-}[\textit{methyl-}^3\text{H}]\text{butyrobetaine} \quad (1)$$

$$\gamma\text{-}[\textit{methyl-}^3\text{H}]\text{Butyrobetaine} \xrightarrow[\substack{\text{ascorbate, Fe}^{2+} \\ \alpha\text{-ketoglutarate}}]{\substack{\gamma\text{-butyrobetaine} \\ \text{hydroxylase}}} [\textit{methyl-}^3\text{H}](-)\text{-carnitine} \quad (2)$$

Principle. γ-Aminobutyrate is methylated to γ-butyrobetaine using radioactive methyl iodide under conditions giving high yields with respect to both these precursors [Eq. (1)]. γ-Butyrobetaine is then quantitatively converted to (−)-carnitine using a 50–60% ammonium sulfate fraction of rat liver supernatant as the source of γ-butyrobetaine hydroxylase (EC 1.14.11.1) [Eq. (2)]. The product is purified by column chromatography on a cation exchanger and is then desalted by passing through a column of ion-retardation resin.[1]

Reagents

[³H]Methyl iodide of specific activity $\leq$ 600 Ci/mol
Methyl iodide
0.6 M γ-aminobutyric acid
1.33 M K_3PO_4
2 M KOH in 50% methanol
6% $HClO_4$
1 M potassium phosphate, pH 7.0
20 mM ferrous ammonium sulfate, freshly prepared
200 mM α-ketoglutarate freshly prepared (the monopotassium salt is dissolved in 500 mM potassium phosphate, pH 7.0)
600 mM ascorbate (the monosodium salt freshly dissolved in 20 mM potassium phosphate, pH 7.0)
20 mg/ml, (65,000 U/mg) catalase
Resins, Chelex 100 (Na^+ form, 200–400 mesh)
AG 50W-X8 (H^+ form, 200–400 mesh)
AG 11A8

γ-Butyrobetaine Hydroxylase. A 50–60% ammonium sulfate fraction of rat liver supernatant is employed and is obtained as follows. Thirty percent (w/v) homogenate of liver from adult rat is prepared in 210 mM mannitol, 70 mM sucrose, 0.1 mM EDTA, 10 mM Tris–HCl (pH 7.4), 0.1 mM dithiothreitol using a Potter-Elvehjem homogenizer and centrifuged at 17,000 g for 10 min. All operations are carried out at 0–4°. To the supernatant, solid ammonium sulfate is added gradually with magnetic stirring to bring the ammonium sulfate to 45% saturation and the separating precipitate is removed by centrifugation at 27,000 g for 5 min. To the resulting supernatant are added, with mixing, KCl to attain a 100 mM concentration and then ammonium sulfate to attain 50% saturation. The separated protein is again removed by centrifugation as above. The ammonium sulfate concentration of the superantant is now raised to 60%, and the precipitate separating this time on centrifugation is taken up in a minimum volume (about 0.1 ml for each rat liver) and dialyzed overnight against at least three changes of 500-fold volumes of 75 ml KCl, 25 mM potassium phosphate, pH 7.4, 0.1 mM dithiothreitol, and 50 μM EDTA. The γ-butyrobetaine hydroxylase activity of the resulting preparation remains stable for several weeks when stored at $-20°$. These preparations show specific activity of 2.4–3.4 mU when assayed at 0.1 mM γ-butyrobe-

---

[1] A. Daveluy, R. Parvin, and S. V. Pande, *Anal. Biochem.* **19,** 286 (1982).

taine as described.[2] Liver from one adult rat yields enough enzyme (180–260 mU) for synthesizing about 10 μmol of carnitine.

Procedure

Methylation of γ-Aminobutyrate to γ-Butyrobetaine. Radioative methyl iodide is supplied in break-seal ampules containing pieces of metallic copper to retard CH_3I oxidation. Even when protected from light and stored at $-20°$, the product decomposes slowly on storage as evidenced by the intensification of the yellowish-brown color. For high radioisotopic yields, the purchased product should be a colorless liquid when chilled and it should be used up as soon as possible. For this, about 5 mm of the lower end of the ampule containing radioactive methyl iodide is dipped in liquid nitrogen while the remainder of the tube is exposed to room temperature to bring the radioactive methyl iodide to the bottom of the tube. About an hour later, the tube is transferred to an ice-bath, 1 ml of a chilled mixture consisting of 200 μl of 0.6 M γ-aminobutyric acid, 300 μl of 1.33 M K_3PO_4, and 500 μl of methanol is placed around the sealed constriction, and the seal is carefully broken, so that enough of the mixture falls below to provide a γ-aminobutyrate to [³H]CH_3I molar ratio of ≥ 0.4, as only a slight excess of γ-aminobutyrate (the theoretical minimum ratio needed for complete quaternization is 0.33) suffices to give high radioisotopic yield with respect to [³H]CH_3I while keeping the quantity of the γ-butyrobetaine formed to a minimum; the latter helps by keeping the requirement of γ-butyrobetaine hydroxylase needed for carnitine synthesis to a minimum.

The final concentration of the radioactive methyl iodide in the above methylation mixture that contains 120 mM γ-aminobutyrate and 400 mM K_3PO_4 should be 250–300 mM to ensure a high ($\geq 90\%$) radioisotopic yield. The radioisotopic yield decreases with a lowering of the final concentration of radioactive methyl iodide in the above reaction mixture owing most likely to the mass action effect.

In practice when the seal is broken, the volume of the reaction mixture that falls below in the [³H]CH_3I compartment is not possible to control precisely; hence a rapid approximation of the volume that has fallen below is necessary to allow more reaction mixture to drop, when necessary, by immediately enlarging the seal-hole. The radioactive methyl iodide is then brought into the reaction mixture by brief vortexing and warming to room temperature. The stoppered tube, protected from light, is left overnight at room temperature.

[2] S. V. Pande and R. Parvin, *Biochim. Biophys. Acta* **617**, 363 (1980).

The next morning, the leftover mixture in the upper compartment is removed and the opening to the lower compartment is enlarged, through which are added unlabeled CH_3I and 2 M KOH in 50% methanol (for about each 100 μl of the reaction mixture are added 6 μl of methyl iodide and 20 μl of the methanolic KOH). The contents are mixed and the tube is left in the dark at room temperature for 6 hr. To the tube is then added 5 M H_3PO_4 (0.4 μl of the acid for each μl of the KOH added above), and, after mixing, the contents are dried using a vacuum centrifuge. The tube is then rinsed 20 times with 0.5 ml of water each time and the rinsings are passed through a 1-ml column of Chelex 100 in a 3-ml disposable plastic syringe (B-D Plastipak). This step frees the solution from ionic copper and provides γ-butyrobetaine that is suitable, without further purification, for the enzymatic conversion to carnitine as described below. When needed for other purposes, purified salt-free radioactive γ-butyrobetaine can be obtained by passing this solution through a retardation column exactly as described for (−)-carnitine below.

Conversion of γ-Butyrobetaine to (−)-Carnitine. The protocol outlined below is for converting 50 μmol of γ-butyrobetaine to (−)-carnitine. A portion of the eluate from the above Chelex step containing 50 μmol of γ-butyrobetaine is brought to 12.9 ml with water in a tube kept at 0°. To this are added the following with mixing and exactly in the order described: 0.5 ml of 1 M potassium phosphate (pH 7.0), 0.83 ml of catalase (16.7 mg, 1086 KU), 0.42 ml of 20 mM freshly dissolved ferrous ammonium sulfate, 0.5 ml of 200 mM α-ketoglutarate freshly prepared, 0.56 ml of 600 mM ascorbate freshly prepared, and 1 ml of γ-butyrobetaine hydroxylase (410 mg protein, 1.1 U). The contents are then left at 25° with magnetic stirring. After ≥6 hr, 1.7 ml of 70% perchloric acid is added, the contents are chilled, then centrifuged at 6000 g for 10 min at near 0°. The supernatant is removed and the pellet is washed twice by being suspended in 6 ml of 6% perchloric acid followed by centrifugation. The combined supernatants are neutralized to the phenolphthalein end point using KOH. The tubes are chilled, centrifuged, and the supernatant is removed. The precipitate is washed twice with 5 ml of chilled water each time. These combined supernatants are applied on 48 × 2 cm column of AG 50W-X8 (H$^+$ form), and about 300 ml of water is passed through the column followed by 1 M HCl at a rate of about 1 ml/min. Carnitine elutes between about 4.2 and 5.5 column volumes of HCl. These fractions are pooled and dried in a rotary evaporator at near 50° until the HCl odor is no longer discernible. The residue is taken up in ≤2 ml of water and is applied on a 51 × 2 cm column of AG 11A8. Water is then passed through the column at a flow rate of about 1 ml/min. Carnitine emerges between 0.38 and 0.56 column volumes, just prior to elution of salts. These fractions are pooled and lyophilized to obtain salt-free [³H](−)-carnitine in radioisotopic yields of up to 90% with respect to the [³H]methyl iodide.

Characterization of Radioactive (−)-Carnitine. Under the conditions described above both the conversion of γ-aminobutyrate to γ-butyrobetaine and of the latter to (−)-carnitine generally proceed to completion. Moreover, the two column chromatographic steps employed in the isolation of the synthesized (−)-carnitine are known to separate the betaines from the corresponding mono and dimethyl derivatives. These steps together with the ability of γ-butyrobetaine hydroxylase to hydroxylate γ-butyrobetaine stereospecifically ensure that the procedure yields radioactive (−)-carnitine of high purity uncontamined from the starting materials and/or the partial methylation products.

The exact (−)-carnitine content of the synthesized tritiated product is estimated using carnitine acetyltransferase, N-ethylmaleimide, and [¹⁴C]acetyl-CoA as described by Parvin and Pande,[3] except that the results are calculated using the radioactive counts accumulating in that portion of the ¹⁴C counting window to which tritium does not contribute. The total [³H]carnitine content calculated should agree with the weight of the dried final product as the ion-retardation step yields a salt-free (−)-carnitine. In this procedure of synthesis the specific radioactivity of the (−)-carnitine obtained considerably exceeds that of the radioactive methyl iodide employed. A further purification or characterization with respect to possible radioactive contaminants is generally unnecessary. Where needed this may be accomplished by incubating the product with excess acetyl-CoA, carnitine acetyltransferase, and N-ethylmaleimide under conditions giving quantitative acetylation of (−)-carnitine[3] and then chromatographing on a column of AG 50W-X8 (NH₄⁺ form, 200–400 mesh) using 0.2 M ammonium formate buffer of pH 4.2 as eluant.[4] In this system a near complete shift of the radioactive carnitine peak (at about 4 column volumes) to that of acetylcarnitine (at about 2.8 column volumes) is observed.[1] Purer radioactive (−)-carnitine may then be obtained by hydrolyzing the nonoverlapping portions of the acetyl(−)-carnitine peak collected.

Comments

The ability of γ-butyrobetaine hydroxylase to hydroxylate γ-butyrobetaine stereospecifically ensures a high stereochemical purity of the product synthesized by the present method; conversely in the procedures that

[3] R. Parvin and S. V. Pande, *Anal. Biochem.* **79**, 190 (1977).
[4] T. Bohmer and J. Bremer, *Biochim. Biophys. Acta* **152**, 559 (1968).

employ a β-hydroxy intermediate as the methyl group acceptor, the stereochemical purity of the methylated product is limited by that of the methyl group acceptor employed. Because the present method converts γ-aminobutyrate to $(-)$-carnitine in high yield, the procedure is suitable also for obtaining radioactive $(-)$-carnitine having radioactivity in the nonmethyl portion of the molecule by employing a suitably labeled γ-aminobutyrate as the precursor. For obtaining high radioisotopic yields with respect to methyl iodide, use of K_3PO_4 as the base during methylation should be preferred; we have obtained $\geq 90\%$ yields by this procedure during methylation also of 2-N-acetyl-L-lysine. A considerable self-decomposition of [^{14}C]carnitine occurs on storage owing to its marked radiation sensitivity; it is necessary therefore that, before use, such samples be repurified, e.g., by chromatography on H$^+$ exchanger.[5]

Alternative Procedures. Methyl-labeled radioactive $(-)$-carnitine has been frequently prepared by the method of Stokke and Bremer[6] which involves demethylation of commercial $(-)$-carnitine to obtain $(-)$dimethylamino-β-hydroxybutyrate which is then methylated back to $(-)$-carnitine using radioactive methyl iodide. Minor modifications of this procedure have subsequently been described.[7,8] Radioactive $(-)$-carnitine can also be prepared by resolving the commercial radioactive $(\pm)$-carnitine; in these procedures, the $(\pm)$carnitine is incubated with an acyl-CoA and an appropriate carnitine acyltransferase, the acyl$(-)$-carnitine formed is separated from the unreacted $(+)$carnitine, and the radioactive $(-)$carnitine is obtained by hydrolyzing the acyl$(-)$-carnitine.[9-11] These latter procedures are particularly useful for obtaining radioactive $(+)$-carnitine.

[5] H. Loster, H. Seim, and E. Strack, *J. Labelled Compd. Radiopharm.* **20,** 1035 (1983).
[6] O. Stokke and J. Bremer, *Biochim. Biophys. Acta* **218,** 552 (1970).
[7] J. S. Willner, S. Gimburg, and S. DiMauro, *Neurology* **28,** 721 (1978).
[8] S. T. Ingalls, C. L. Hoppel, and J. S. Turkaly, *J. Labelled Compd. Radiopharm.* **19,** 535 (1981).
[9] J. Kerner and L. L. Bieber, *Anal. Biochem.* **134,** 459 (1983).
[10] R. R. Ramsay and P. K. Tubbs, *FEBS Lett.* **54,** 21 (1975).
[11] H. Schulz and E. Racker, *Biochem. Biophys. Res. Commun.* **89,** 134 (1979).

[33] S-Adenosylmethionine : ε-N-L-Lysine Methyltransferase

By Harry P. Broquist

The biosynthesis of carnitine from lysine and methionine in *Neurospora crassa* and the rat is now known in some detail.[1,2] A critical intermediate is ε-N-trimethyllysine which is subsequently converted to γ-butyrobetaine followed by hydroxylation of the latter to give carnitine.[1,2] Trimethyllysine is formed by the methylation of the ε-amino group of lysine. An important difference between mammalian systems studied to date and *N. crassa* is that such methylation occurs posttranslationally in the former case,[3] protein bound trimethyllysine subsequently being released by lysosomal hydrolases. In *N. crassa,* however, lysine is methylated stepwise via monomethyllysine, dimethyllysine, to trimethyllysine, Reactions A, B, and C in Fig. 1, in a soluble enzyme system as described herein.[4,5]

Growth of *Neurospora crassa;* Preparation of Initial Extract

Neurospora crassa lysine auxotroph (Fungal Genetics Stock Center No. 33933) is grown in 5 gallon carboys, 15 liters of Vogel's medium[6] supplemented with 2 mM L-lysine per carboy, with aeration for 40 hr. The mycelia are harvested by filtration, pressed dry, frozen in liquid nitrogen, and pulverized to a powder. All subsequent operations are carried out at 4°, and all buffers contain 2×10^{-5} M dithiothreitol and 2×10^{-4} M EDTA. The *Neurospora* powder is extracted into 1 M potassium phosphate buffer, pH 7.4 (1 g of powder, 1.25 ml buffer) with intermittent stirring. The methyltransferase protein is extremely sensitive to endogenous proteases, hence the initial extract and each succeeding protein fraction are treated with phenylmethylsulfonyl fluoride.[5] The initial extract is then centrifuged (17,300 g, 30 min), the supernatant recentrifuged (105,000 g, 1 hr); the resulting supernatant represents crude extract (1, Table I). Operationally the volumes of protein fractions, particularly in early steps, are so large that to make them manageable, they are pro-

[1] H. P. Broquist, *Fed. Proc., Fed. Am. Soc. Exp. Biol.* **41,** 2840 (1982).
[2] J. Bremer, *Physiol. Rev.* **63,** 1420 (1983).
[3] W. K. Paik and S. Kim, *Adv. Enzymol. Relat. Areas Med. Biol.* **42,** 227 (1975).
[4] C. J. Rebouche and H. P. Broquist, *J. Bacteriol.* **126,** 1207 (1976).
[5] P. R. Borum and H. P. Broquist, *J. Biol. Chem.* **252,** 5651 (1977).
[6] R. H. Davis and F. J. de Serres, this series, Vol. 17, Part A, p. 80.

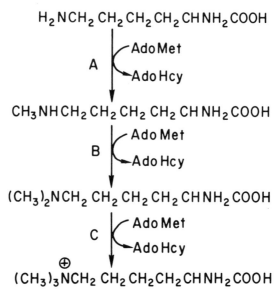

FIG. 1. ε-N-Trimethyllysine formation by S-adenoxylmethionine:ε-N-lysine methyltransferase in *Neurospora crassa*. AdoMet, S-adenosylmethionine; AdoHyc, S-adenosylhomocysteine.

cessed in successive appropriate aliquots, such fractions pooled to give representative material for analytical work, then again processed via aliquots if necessary.

Lysine Methyltransferase Assay; Substrate Sources

S-Adenosyl-L-[*methyl-*³H]methionine may be purchased from Amersham/Searle; ε-N-monomethyl-L-lysine and ε-N-dimethyl-L-lysine from Cyclo Chemical Co. Mono- and dimethyllysine may also be synthesized by the method of Benoiton.[7] ε-N-Trimethyl-L-lysine is synthesized by the general method of Mazzetti and Lemmon[8] which involves ε-N-methylation of α-N-acetyl-L-lysine followed by hydrolysis to yield trimethyllysine.

The S-adenosylmethionine:ε-N-lysine methyltransferase assay protocol is given in Table II. The methyllysine transferase assay measures the formation of radioactive methylated lysine derivatives following incubation of appropriate lysine substrates with [*methyl-*³H]AdoMet. Some di-

[7] L. Benoiton, *Can. J. Chem.* **42**, 2043 (1964).
[8] F. Mazzetti and R. M. Lemmon, *J. Org. Chem.* **22**, 228 (1957).

TABLE I

PURIFICATION OF S-ADENOSYLMETHIONINE:ε-N-L-LYSINE METHYLTRANSFERASE FROM Neurospora crassa 33933

Purification step	Volume (ml)	Protein (mg/ml)	Methyltransferase activity			Purification (fold)	Specific activity ratio	
			Lys → Lys(Me) A	Lys(Me) → Lys(Me)$_2$ B	Lys(Me)$_2$ → Lys(Me)$_3$ C		C/A	C/B
			(nmol/product/hr/mg protein)					
1. 105,000 g supernatant	8,000	15.35			0.15	1		
2. Sephadex G-25	11,330	6.93	0.35	1.11	2.87	19	8.2	2.6
3. Protamine sulfate	12,115	5.59	0.33	1.39	2.60	17	7.9	1.9
4. 0–50% ammonium sulfate	900	17.85	0.48	3.60	9.65	64	20.1	2.7
5. DEAE-Sephadex	71	21.16	0.48	3.55	13.50	89	28.1	3.8
6. CM-Sephadex	11.5	17.70	4.03	14.7	32.60	216	8.1	2.2
7. Sephadex G-100	6.3	6.30	8.39	28.8	77.20	511	9.2	2.7
8. DEAE-Sephadex	100	0.023	20.2	155.7	473.9	3,138	23.5	3.0

TABLE II

S-ADENOSYLMETHIONINE : ε-N-L-LYSINE
METHYLTRANSFERASE ASSAY PROTOCOL

10 μl 1 M carbonate/bicarbonate buffer, pH 9.7
100 nmol lysine, monomethyl or dimethyllysine
12 nmol [methyl-³H]AdoMet (Reactions B and C,
   Fig. 1); 20 nmol [methyl-³H]AdoMet (Reaction A,
   Fig. 1)
Water and enzyme to 200 μl
Incubate 30 min, 32.5°
Add 200 μl 3 N HCl containing 40 mg Norite-A
Centrifuge (2000 rpm, 10 min)
Apply 10 μl supernatant to Silica Gel G plates
Develop in methanol : ammonia, 3 : 1 system
Visualize amino acids with ninhydrin spray
Scrape "ninhydrin spots" into scintillation vials
Add water (2.3 ml) and Aquasol (7.7 ml) to vials and
   count in a scintillation counter

methyllysine in addition to monomethyllysine is formed in Reaction A, Fig. 1, and some trimethyllysine in addition to dimethyllysine is formed in Reaction B, Fig. 1. These facts are considered in the presentation of the data, Table I, and Fig. 2, and activity is expressed in terms of nmol product per hour per mg protein.

Lysine Methyltransferase Purification

Step 2, Table I. An 80-ml aliquot of the crude extract (step 1, Table I) is chromatographed on a Sephadex G-25 column (2.6 × 100 cm) in 0.1 M PO$_4$ buffer, pH 7.4. All of the enzyme activity appears in the void volume together with about 50% of the total protein. Yields are greater than 100%; an endogenous inhibitor(s) appears to be removed.

Step 3, Table I. To aliquots (e.g., 210 ml) of step 2 extract, 0.2% protamine sulfate is added dropwise with stirring over a 30-min period at 0°. The nucleic acid precipitate is removed by centrifugation (17,000 g, 30 min).

Step 4, Table I. The enzyme solution is adjusted to 5.0 mg protein/ml. To a 300-ml aliquot in an ice bath is added solid smmonium sulfate (94 g over an hour) to 50% saturation. The suspension is stirred for an additional hour then centrifuged (17,300 g, 30 min). The pellet is resuspended in a small amount of 0.1 M PO$_4$ buffer, pH 7.4, and immediately desalted on a Sephadex G-25 column (2.6 × 100 cm, equilibrated with 0.1 M PO$_4$, pH 7.4).

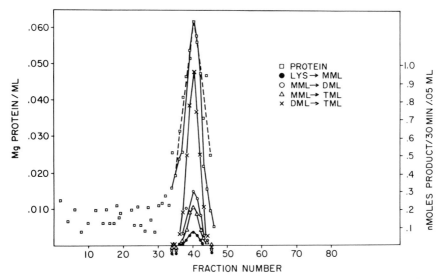

FIG. 2. Sephadex G-100 chromatography of purified *S*-adenosylmethionine: ε-*N*-lysine methyltransferase.

Step 5, Table I. Pooled aliquots (900 ml) of extract (step 4 above) are passed over a DEAE-Sephadex A-50 column (5.0 × 60 cm) equilibrated with 0.1 *M* PO₄, pH 7.4, and washed with 6 liters of 0.1 *M* PO₄ buffer, pH 7.4, containing 0.1 *M* KCl. All three lysine methyltransferase activities elute in a single peak. These eluates are concentrated 100-fold by Amicon Filtration System TC3D (Amicon PM-10 membrane) in an Amicon model 202 stirred cell and the concentrated protein desalted with Sephadex G-25 as in step 4 above.

Step 6, Table I. The protein is then chromatographed in 30-ml aliquots on a DEAE-Sephadex A-50 column (5.0 × 60 cm) and eluted with 0.1 *M* PO₄, pH 7.4, to give a protein peak in which again all three lysine methyltransferases are associated. Such fractions are pooled and concentrated (Amicon model 202 stirred cell).

Step 7, Table I. The concentrated protein (11.5 ml) is again put onto a Sephadex G-100 column (2.6 × 100 cm) and eluted with 0.01 *M* PO₄, pH 7.4, to give a single peak containing the three lysine methyltransferases. Such fractions are combined and concentrated by Amicon filtration.

Step 8, Table I. The concentrated protein (6.3 ml) is absorbed onto a DEAE-Sephadex A-50 column (1.6 × 40 cm) equilibrated with 0.01 *M* PO₄ buffer, pH 7.4. The column is eluted with a linear salt gradient (800 ml 0.01 *M* PO₄ buffer, pH 7.4, in the mixing chamber and 800 ml 0.15 *M* PO₄ buffer, pH 7.4, in the reservoir). The pooled lysine methyltransferase fractions are finally applied to a Sephadex G-100 column (1.6 × 100 cm; development: 0.01 M PO_4 buffer, pH 7.4, eluate fractions: 45 drops) to give one symmetrical peak (Fig. 2) containing all three lysine methyltransferases in an overall yield of 6%.

Lysine Methyltransferase Properties

The foregoing purification scheme yields a highly purified protein that is homogeneous following column chromatography (Fig. 2), polyacrylamide gel electrophoresis, and ultracentrifugation in which lysine methyltransferase activity for Reactions A, B, and C in Fig. 1 remains relatively constant (Table I). The rates for the reactions are A < B < C. Other examples in *Neurospora* of multiple methyltransferases thought to be catalyzed by a single enzyme include the formation of hercynine from histidine[9] and in the biosynthesis of choline.[10] S-Adenosylmethionine: ε-N-L-lysine methyltransferase as prepared herein from *N. crassa* is devoid of protein methylase III activity (W. K. Paik and S. Kim, personal communication); but *N. crassa* contains a cytochrome c-specific protein–lysine methyltransferase.[11]

The molecular weight of the lysine methyltransferase described herein is estimated to be 22,000 based on sedimentation equilibrium and molecular filtration data and 24,000 based on amino acid analysis assuming two methionine residues per mole protein. The protein appears to be devoid of subunit structure based on observations from sedimentation equilibrium analysis of the protein in 6 M guanindine hydrochloride and sodium dodecyl sulfate–polyacrylamide gel electrophoresis.

Both carnitine and trimethyllysine repress synthesis of lysine methyltransferase in growing cultures of *N. crassa*.[4]

Acknowledgment

This work described herein was supported in part by NIH Grant AM 16019.

[9] Y. Ishikawa and D. B. Melville, *J. Biol. Chem.* **245,** 5967 (1970).
[10] G. A. Scarborough and J. F. Nyc, *J. Biol. Chem.* **242,** 238 (1967).
[11] E. Durban, S. Nochumson, S. Kim, and W. K. Paik, *J. Biol. Chem.* **253,** 1427 (1978).

Section VII

Heme Porphyrins and Derivatives

[34] Reflectance Spectrophotometric and Surface Fluorometric Methods for Measuring the Redox State of Nicotinamide Nucleotides and Flavins in Intact Tissues

By Ilmo E. Hassinen

In order to test the principles involved in the regulation of cellular respiration, it is essential to determine the behavior of oxidation–reduction carriers in intact cells and tissues. Here the carriers occur at physiological concentrations and are also under the influence of physiological concentrations of enzymes, effector substances, and metabolites.

Reflectance and fluorescence methods are applicable to a number of endogenous cellular chromophores[1] and fluorochromes,[2] including coenzymes of certain oxidoreductases. Transmittance spectrophotometry of intact organs *in situ*[3] or perfused after isolation[4] has also been used successfully in a few instances.

Reflectance Spectrophotometry

The specificity of the light reflectance or transmittance measurement of a selected cellular chromophore is usually obtained by the dual-wavelength method[5] in which the measurement wavelength is near or at the absorption maximum and the reference wavelength is near or at the isosbestic point of the compound. Optimally the reference wavelength should be also an isoabsorbance wavelength of an interfering chromophore at the measurement wavelength. The dual-wavelength method minimizes the problem of changes in optical geometry, specular reflectance, or light scattering. Although the Tyndall scattering is only slightly dependent on wavelength,[6] the Rayleigh scattering is an inverse function of the fourth power of the wavelength,[6] so that compensation for scattering is obtained mainly when the measurement and reference wavelengths can be selected close to each other.

[1] D. Keilin and E. F. Hartree, *Proc. R. Soc. London, Ser. B* **127**, 167 (1939).
[2] B. Chance, P. Cohen, F. Jöbsis, and B. Schoener, *Science* **137**, 499 (1962).
[3] F. F. Jöbsis, *Science* **198**, 1264 (1977).
[4] M. Tamura, N. Oshino, B. Chance, and I. A. Silver, *Arch. Biochem. Biophys.* **191**, 8 (1978).
[5] B. Chance, *Rev. Sci. Instrum.* **22**, 634 (1951).
[6] P. Nicholls, *in* "Biochemical Research Techniques" (J. M. Wrigglesworth, ed.), p. 1. Wiley, New York, 1983.

Fluorometry

The inherent specificity of fluorescence measurement is hampered by optical quenching in intact tissues and by other endogenous chromophores such as hemoproteins (cytochromes, myoglobin, hemoglobin). The reflectance (absorbance) spectrum of the specimen can be used to evaluate the optical quenching of fluorescence and to ensure appropriate measures for its compensation. The conventional quenching correction involves subtraction from the fluorescence signal of a portion of the excitation light reflection signal.[7] This does not compensate for optical quenching at the wavelengths of the fluorescence emission. Because of the complexities introduced into the instrumentation, emission quenching correction is in fact seldom performed in spite of its advantages.

Current Instrumentation

A flexible system for spectrophotometry and fluorometry can be constructed by multiplexing illumination at several wavelengths of the measurement area, so that light absorbance spectrum (reflectance) and fluorescence measurements can be performed simultaneously. An appropriate reflectance signal can be used for quenching correction. A number of instruments have been described which allow one of these principles to be implemented. Chance et al.[8,9] designed a rotating disc apparatus in which the illumination wavelengths are alternated by optical filters inserted into a motor or air turbine-driven rotating disc. The same rotating disc also permits an alternation of fluorescence detection wavelengths by means of filters. An apparatus has also been described for simultaneous flavin absorbance and fluorescence measurements in biological samples. This consists of a dual-wavelength spectrophotometer which is combined with a independent filter fluorometer.[10] Cross-talk between the fluorometer and spectrophotometer sections is prevented by suitable guard filters in front of the detectors.[10] This limits the usable wavelength range, however, and restricts the instrument to one specific purpose.

Developments in Fluorometry

Fluorescence has also recently been measured in intact tissues with laser excitation light sources combined with micro light guides consisting

[7] A. Mayevsky and B. Chance, *Science* **217**, 537 (1982).
[8] B. Chance, D. Mayer, and L. Rossini, *IEEE Trans. Biomed. Eng.* **BME-17**, 118 (1970).
[9] B. Chance, V. Legallais, J. Sorge, and N. Graham, *Anal. Biochem.* **66**, 498 (1975).
[10] B. Chance, D. Mayer, and V. Legallais, *Anal. Biochem.* **42**, 494 (1971).

of a single optical fiber.[11,12] Problems of fluorescence quenching correction have been discussed by Kobayashi et al.[13,14] and Renault et al.,[11] who note that it would be advantageous to use excitation and emission wavelengths which are isosbestic for the interfering chromophores undergoing metabolic spectral changes.[11] It has also been demonstrated[15] that quenching correction can be achieved using a reflectance monitoring wavelength (isoabsorbance wavelength) at which the interfering chromophore has the same absorptivity as at the excitation wavelength. This makes the compensation independent of the absorbance changes in the fluorescent compound itself (the absorbance and fluorescence excitation bands of a given compound coincide).

By employing compensation by reflectance at 586 nm when fluorescence is excited at 337 nm [for NAD(P)H fluorescence] and measured at 480 nm in an isolated blood-perfused rat heart, Renault et al.[11] obtained the empirical relationship $F_0/F_b = (I_0/I_b)^{1.09}$, where F is fluorescence, I is reflectance, and the subscripts 0 and b refer to the absence and presence of blood in the perfusion medium. The practically direct proportionality between quenched fluorescence and reflectivity at 480 nm (isosbestic for hemoglobin) suggests that an acceptable degree of compensation is achieved simply by appropriate ratio recording. It has also been reported that measurement of the difference between fluorescence and reflectance works well in many instances, but this is true only for small percentage changes in fluorescence.[15] It would be advantageous, however, to use a single common photomultiplier for both fluorescence and reflectance measurement, so that the reference (i.e., reflectance) signal can be kept constant by means of the gain control devise which converts the difference recording to ratio recording. This type of measurement would be more appropriate based on the Renault relationship[11] and has been employed successfully.[14] The table summarizes the excitation and emission wavelengths and methods of quenching correction in use.[16–21]

[11] G. Renault, E. Raynal, M. Sinet, M. Muffat-Joly, J.-P. Berthier, J. Cornillault, B. Godard, and J.-J.- Pocidalo, Am. J. Physiol. **246,** H491 (1984).

[12] M. J. Sepaniak, B. J. Tromberg, and J. F. Eastham, Clin. Chem (Winston-Salem, N.C.) **29,** 1678 (1983).

[13] S. Kobayashi, K. Nishiki, K. Kaede, and E. Ogata, J. Appl. Physiol. **31,** 93 (1971).

[14] S. Kobayashi, K. Kaede, K. Nishiki, and E. Ogata, J. Appl. Physiol. **31,** 693 (1971).

[15] R. A. Kauppinen and I. E. Hassinen, Am. J. Physiol **247,** H508 (1984).

[16] B. Chance, C. Barlow, Y. Nakase, T. Takada, A. Mayewski, R. Fischetti, N. Graham, and J. Sorge, Am. J. Physiol. **235,** H809 (1978).

[17] E. M. Nuutinen, Basic Res. Cardiol. **79,** 49 (1984).

[18] J. B. Chapman, J. Gen. Physiol. **59,** 135 (1972).

[19] F. F. Jöbsis and W. N. Stainsby, Respir. Physiol. **4,** 292 (1968).

[20] R. S. Kramer and R. D. Pearlstein, Science **205,** 693 (1979).

[21] J. K. Hiltunen, V. P. Jauhonen, M. J. Savolainen, and I. E. Hassinen, Biochem. J. **170,** 235 (1978).

SELECTION OF WAVELENGTH AND METHOD OF QUENCHING CORRECTION IN
FLUOROMETRY OF NICOTINAMIDE NUCLEOTIDES AND FLAVINS IN INTACT TISSUES

| Fluorochrome | Wavelength (nm) | | Compensation method | | |
	Excitation	Emission	Reflectance wavelength (nm)	Calculation	Reference
NADH	337 N$_2$[a]	480[b]	586	Digital computer[c]	11
	356/363Ar[d]	450 ± 30	350/363	Difference	16
	360	420<	—		17
	366 Hg	460 ± 20	720	Analog computer[e]	14
	366 Hg[f]	420<	366	Difference[g]	7
	366 Hg	465 ± 40	366	Difference	18
	366 Hg	465 ± 40	366	Difference	19
	366 Hg	448	549	Ratio[h]	20
Flavin	436 Hg	570	436	Difference	8
	442 He-Cd[i]	550 ± 30	442	Difference	8
	457.9	550 ± 30	457.9	Difference	16
	465	520	—		21

[a] Nitrogen laser.
[b] Dye laser.
[c] $F_0 = F(I/68.88)^{-1.09}$, on-line digital computer.
[d] Argon laser.
[e] $I_0/I = 1 + k \ln(F_0/F)$, on-line analog computer.
[f] Mercury arc lamp, wavelength selection limited by the line spectrum.
[g] 1 : 1 subtraction of reflectance from fluorescence.
[h] Fluorescence-to-reflectance ratio recording.
[i] Helium-cadmium laser.

Optical Coupling to the Biological Specimen

Fiber optics form the most frequently used method for reproducible optical coupling between the tissue and the measuring instrument. Front-face illumination equipment for the microscope (Ultropak, Leitz) has also been employed, but this is evidently prone to artifacts due to changes in the optical geometry in measurements *in vivo* or in moving organs such as the heart. The use of Y-formed fiber optic light guides raises problems of the acceptance angle, especially if the transmitting and receiving fibers are not evenly distributed at the common end. This problem can also be solved by extension of the common branch of the light guide by means of a glass or quartz rod, which is then gently pressed against the surface to be observed.[22] Fiber optics allow measurements of very small tissue vol-

[22] I. E. Hassinen, J. K. Hiltunen, and T. E. S. Takala, *Cardiovasc. Res.* **15,** 86 (1981).

umes using single fibers of 80–200 μm diameter.[23,24] Parallax problems due to the limited acceptance angle of the fibers can be avoided by using a single fiber[11] instead of separate fibers for the transmitting and receiving light paths. This necessitates a beam splitter at the fluorometer end of the fiber to separate the illumination and emission light. The acceptance half-angle of the fiber–water interface is approximately 15° for a glass fiber with a low-refractive-index material coating for optical isolation.[25]

Micro light guides consisting of a single fiber require a very high light intensity at the fiber end which can be achieved by laser illumination.[11,12] A nitrogen laser gives a wavelength of 337 nm suitable for NAD(P)H excitation and can also be used to pump a dye laser[11] which allows variation of the wavelength by dye selection. This method has been successfully applied to the blood-perfused isolated rat heart.[11]

Two-fiber micro light guides have been employed in two- or three-dimensional mapping of the cellular redox state.[26] The fluorescence responses of the flavins and nicontinamide nucleotides to reduction can be determined. The oxidized species of flavins are fluorescent and the reduced forms are not. On the other hand the reduced forms of the nicotinamide nucleotides are fluorescent, but the oxidized forms are not. By taking the ratio of flavin to NAD(P)H fluorescence, a partial correction for optical quenching is obtained. A three-dimensional redox map of quick-frozen tissue is obtained by scanning a two-dimensional pixel matrix and gaining a third dimension by exposing sequential surfaces for scanning by cutting or milling in the frozen state.[26] A scanning fluorometer using a 0.1-mm-diameter flying spot for excitation from a laser light source has also been constructed for studying perfused organs or tissues *in vivo*.[16]

New Developments in the Spectrophotometry of Intact Tissues

Multiple-wavelength spectrophotometry is conventionally performed with multiple monochromator devices and optical multiplexing. Diode array detectors have become available in recent years. The method implies white light illumination of the sample and detection of the transmitted light by a diode array at the exit plane of the monochromator.[27] An image sensor has also been used successfully in organ spectrophotometry

[23] S. Ji, J. J. Lemasters, and R. G. Thurman, *FEBS Lett.* **113**, 37 (1980).

[24] S. Ji, B. Chance, K. Nishiki, T. Smith, and T. Rich, *Am. J. Physiol.* **236**, C144 (1979).

[25] Information Faseroptik No. 7003. Schott, Mainz.

[26] B. Quistorff, B. Chance, and H. Takada, in "Frontiers in Biological Energetics: Electrons to Tissues" (P. L. Dutton, J. S. Leigh, and A. Scarpa, eds.), Vol. 2, p. 1487. Academic Press, New York, 1978.

[27] K. V. S. Reddy and R. W. Hendler, *J. Biol. Chem.* **258**, 8568 (1983).

instead of the diode array.[28] These methods require a complicated system for data acquisition, storage, and processing, but offer in return the possibility of a practically continuous monitoring of a wide span of the spectrum. This technology, however, is less advantageous when reflectance and fluorescence are to be measured simultaneously and has not been used until now. The same is true of highly turbid biological samples.[6]

The sensitivity of reflectance spectroscopy is poor at short wavelengths and is not useful for measuring NADH concentration changes in intact tissues, as exemplified by the use of 366 nm reflectance as a nonspecific compensation signal for fluorometry. Transmittance spectrophotometry of NADH nevertheless functions satisfactorily in the isolated perfused liver, for example.[29]

Reflectance Spectrophotometer–Fluorometer of Hassinen and Jämsä[30]

Experience with the use of a laboratory-built reflectance photometer–fluorometer for studying the energy metabolism of the intact isolated perfused heart and liver and of isolated mitochondria has accumulated over a number of years. The device can be constructed using commercially available units. Some useful details and modifications to the previously published design[30] are described in the following.

Optical Design[30]

The spectroscopic section of the apparatus employs two grating monochromators with an inverse dispersion of 7 nm/mm and a relative aperture of $f/10$. High-power monochromators would be more advantageous, however, and the number of mirrors would be minimized by the use of monochromators with their entrance and exit beams perpendicular such as those manufactured by the Farrand Optical Co. (Valhalla, NY). The filament image of an automotive tungsten-halogen lamp (type 12258 of any European manufacturer, 12 V 55 W) is focused on the entrance slit of the monochromator with a quartz condenser lens (focal length 63 mm). Similar lenses are used to focus the images of the exit slits halfway between two light modulator discs and, thereafter, on the fiber optic light conductor (Fig. 1).

[28] N. Sato, T. Matsumura, M. Schichiri, T. Kamada, H. Abe, and B. Hagihara, *Biochim. Biophys. Acta* **634**, 1 (1981).

[29] T. Bücher, B. Brauser, A. Conza, F. Klein, O. Langguth, and H. Sies, *Eur. J. Biochem.* **27**, 301 (1972).

[30] I. Hassinen and T. Jämsä, *Anal. Biochem.* **120**, 365 (1982).

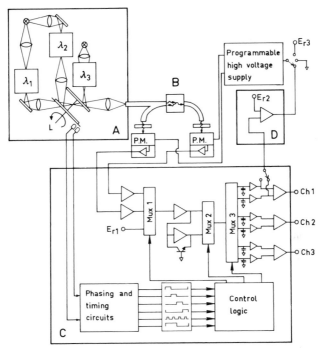

FIG. 1. A triple-wavelength reflectance spectrophotometer–fluorometer for intact tissues. (A) Optical unit; (L) dual-disc light modulator; (λ_1 and λ_2) grating monochromators; (λ_3) interference filter; (B) biological specimen connected to the optical unit and detectors by means of fiber optics; (C) electronic unit; (Mux 1, Mux 2, and Mux 3) multiplexers to gate the difference computing amplifiers and the mode of operation (linear or logarithmic) in a phase-locked fashion; (D) photomultiplier dynode voltage feedback circuit; (E_{r1}, E_{r2}, and E_{r3}) reference voltages. Adapted from Ref. 30 with permission of the publisher.

A similar type 12258 automotive lamp is used as the light source for the fluorometer section. For high throughput, optical filters are used for the isolation of the excitation wavelength. A pyrex double condenser is used to form a filament image on the fiber optics by mediation of the mirrors of the light modulator. Since it is advantageous to use an optical reference pulse to suppress the sensitivity of the apparatus to leaks of ambient light for single-ended fluorescence measurement, the apparatus incorporates a reference light pathway for monitoring the light source of the fluorometer (not shown in Fig. 1).

The light modulator incorporates two circular aluminium discs (diameter 11 cm) and has been slightly modified from that described previously by doubling the modulation frequency. Each disc is equipped with a pair of front surface mirrors (25 × 15 mm) and perforations giving two pulses of each light beam per revolution of the discs, which are mounted on the same shaft driven by an asynchronous ac motor at 2700 rpm. The doubling of the modulation–demodulation frequency reduces the interference caused by the electromagnetic field of the ac line. A synchronous motor would do the same, provided it were free of jitter.

The optical coupling between the measurement optics and the biological specimen has been tested previously.[22] The addition of a second photodetector allows simultaneous recording of surface fluorescence and transmittance at a visible wavelength range *through* the specimen.[31]

Electronic Design

Phase-sensitive detection is accomplished by a digital circuit controlled by a rotation encoder attached to one of the light modulator discs and assembled from a graphic film disc on which sector rulings of 128 divisions per revolution and one division per revolution are copied by photography and from two inexpensive IR-emissive diode–phototransistor pairs. The signals are encoded by conventional CMOS digital circuitry for gating semiconductor switches to obtain differences of three selected photomultiplier signals. Logarithmic conversion of the signals at half or full cycle of operation is provided to allow a linear absorbance scale for the photometry. Since it has been found necessary in certain cases to allow the photomultiplier gain to be regulated by the dynode voltage,[15] the photomultiplier H.V. supply has been modified for dynode voltage feedback controlled by one of the reflectance signals by means of a simple comparator amplifier. This in effect changes the difference output to a ratio output. This ratio analysis technique was found necessary when working with a fluorescent membrane potential probe (safranine) in intact tissues where variable quenching by myoglobin oxygenation changes interferes with the fluorescence.[15] The specificity of NADH fluorometry in intact tissues has been under debate until recent years. It is evident that many investigators find the simple 1 : 1 difference recording (fluorescence minus reflectance) insufficient for correcting hemodynamic or oxy-/deoxyhemoglobin artifacts.[11,20,32] The use of 366 nm reflection with an excitation fluorescence of 366 nm is not satisfactory, at least when recording the NADH fluorescence of intact cortex during experimental anoxia.[20] Kramer and Pearlstein[20] have demonstrated that 549 nm would be a suitable isosbestic wavelength and that the NADH fluorescence correction for hemoglobin quenching should be performed by monitoring the ratio of

[31] V. P. Jauhonen, E. Baraona, C. S. Lieber, and I. E. Hassinen, *Alcohol* **2**, 163 (1985).
[32] F. F. Jöbsis, J. H. Keizer, J. C. LaManna, and M. Rosenthal, *J. Appl. Physiol.* **43**, 858 (1977).

NADH fluorescence to 549 nm reflectance. They use analog-to-digital conversion and a laboratory computer, but the same can be done with higher sensitivity and less effort by simply employing a single photomultiplier tube under dynode voltage control to keep the reflectance signal constant.[15,30]

The need for correction of the fluorescence reading for scattering or specular reflectance is obviously dependent on the quality of the monochromator devices used for the isolation of the fluorescence excitation and emission bands and on the concentration of the quenching agent. As shown in some recent studies, NADH and flavin fluorescence in isolated hemoglobin-free perfused hearts can also be recorded without any correction for reflectance.[17,21]

Application of Surface Fluorescence Measurements to Intact Organs

The apparatus described above has been employed for measuring NAD(P)H redox changes in isolated perfused hearts, where the origin of the nicotinamide nucleotide fluorescence is mitochondrial.[17,18] This is in contrast to the isolated perfused liver, where both mitochondrial and cytosolic compartments contribute to the fluorescence.[32]

It has been shown previously[33] that the flavin fluorescence of isolated mitochondria is mainly due to lipoamide dehydrogenase, one of the few enzymes for which enhancement of flavin fluorescence occurs upon its binding to the apoprotein.[34] Moreover, the lipoamide dehydrogenase flavin is in near equilibrium with the mitochondrial NADH/NAD. This relationship has been employed to demonstrate an equilibrium between myocardial glutamate dehydrogenase and mitochondrial free NADH/NAD.[35] Because of the paucity of redox indicator metabolites in the myocardium, such a test has proved elusive by other methods. The only other intramitochondrial flavoprotein showing appreciable fluorescence is the electron-transferring flavoprotein (ETF),[36] but obviously this does not markedly interfere with the use lipoamide dehydrogenase fluorescence as an indicator of the redox state of the mitochondrial free $NAD^+/NADH$ couple.

Application of the surface fluorescence method to the isolated perfused liver has demonstrated that the nicotinamide fluorescence origi-

[33] I. Hassinen and B. Chance, *Biochem. Biophys. Res. Commun.* **6**, 895 (1968).
[34] G. Palmer and V. Massey, *in* "Biological Oxidations" (T. E. Singer, ed.), p. 263. Wiley (Interscience), New York, 1968.
[35] E. M. Nuutinen, J. K. Hiltunen, and I. E. Hassinen, *FEBS Lett.* **128**, 356 (1981).
[36] C. L. Hall and H. Kamin, *J. Biol. Chem.* **250**, 3476 (1975).

nates from bound, not free NADH[29] and that the midpotential of the NADH/NAD$^+$ couple is much higher in its bound moiety than in the free form.[17] The behavior of the NADH fluorescence in the heart muscle can be explained by its mitochondrial origin[17,18] without postulating any shift in redox properties.

It is also important to know that the fluorescence of NADH can in most cases be interpreted only qualitatively, because of difficulties in its calibration on an absolute redox scale. This is caused by the fact that it is mostly impossible to arrange conditions of full oxidation of the nucleotide, even though a fully reduced state can easily be obtained. When an appropriate indicator metabolite couple is available, calibration is possible but it requires a redox titration by the metabolite and a method for linearization of the data, as demonstrated by Bücher et al.[29]

Optical methods are indispensable when observing rapid phenomena at intact tissue level, where chemical rapid sampling methods are often impracticable because of the difficulties in timing and the uncertainties imposed by biological variation.

[35] Spectrophotometric Determination of Iron in Heme Proteins

By Serge N. Vinogradov

The spectrophotometric determination of iron in aqueous solutions relies on the formation of stable iron(II) complexes possessing a high absorptivity in the visible region of the spectrum. A substantial number of sensitive iron ligands have been developed over the last 40 years.[1] The majority have been compounds possessing the ferroin group, i.e., the group —N=C—C=N— which acts as a bidentate ligand toward iron (II), including 1,10-phenanthroline derivatives such as bathophenanthrolinesulfonic acid[2] and triazine derivatives such as 2,4-bis(5,6-diphenyl-1,2,4-triazin-3-yl)pyridine tetrasulfonic acid.[3]

The table summarizes the absorptivities of the ferrous complexes of several of the more sensitive and useful iron ligands and lists the cations

---

[1] "The Iron Reagents." G. F. Smith Chemical Co., Columbus, Ohio, 1980.
[2] D. Blair and H. Diehl, *Talanta* **7**, 163 (1961).
[3] G. L. Traister and A. A. Schilt, *Anal. Chem.* **48**, 1216 (1976).

SOME PROPERTIES OF SENSITIVE IRON REAGENTS

Name	Molar absorptivity	Interfering cations	Reference
4,7-Diphenyl-1,10-phenanthroline-disulfonic acid (bathophenan-throlinedisulfonic acid)[c]	22,140 (535 nm)[a] 12,250[b] (483 nm)[b]	Cu(I), Cu(II), Cr(III), Al(III), Sn(IV)	2
2,4,6-Tripyridyl-s-striazine (TPTZ)[c]	22,600 (593 nm)[a]	Cu(II), Co(II), Ni(II), Cr(III), Ag(I), Hg(II), Bi(III)	4
3-(2-Pyridyl)-5,6-bis(4-phenyl-sulfonic acid)-1,2,4-triazine (PDTS) (ferrozine)[d]	27,900 (562 nm)[a]	Cu(I), Co(II)	5
2,4-Bis(5,6-diphenyl-1,2,4-triazin-3-yl)pyridinetetrasulfonic acid (2,4-BDTPS)[c]	32,200 (565 nm)[a]	Cu(I), Co(II), Ni(II), Cr(III)	3
3-(4-Phenyl-2-pyridyl)-5,6-diphe-nyl-1,2,4-triazinetrisulfonic acid (PPDTS)[c]	30,700 (563 nm)[a]	Cu(I), Co(II), Ni(II), Cr(III)	3
3-(2-Pyridyl)-5,6-bis[2-(5-furyl-sulfonic acid)]-1,2,4-triazine (ferene S)[c]	35,500 (600 nm)[a]	Cu(I)	6
4-(2-Pyridylazo)resorcinol (PAR)[f]	56,000 (510 nm)[a]	Cu(I), Co(II), Ni(II), Cr(III), and others	7

[a] Position of the absorption peak of the Fe(II) complex.
[b] Cu(I) complex.
[c] Available from G. F. Smith Chemical Co., Columbus, OH 43223.
[d] Available from Hack Chemical Co., Ames, IA 50010.
[e] Available from Polysciences Inc., Warrington, PA 18976.
[f] Available from Aldrich Chemical Co., Milwaukee, WI 53201.

that interfere.[4-7] It is evident that most of the more recent iron reagents possess a much greater molar absorptivity in the visible region than the 11,000 at 509 nm of 1,10-phenanthroline which has been used for the determination of iron in heme proteins.[8,9] Since most of the sensitive iron reagents form complexes with other transition metal cations, albeit having molar absorptivities severalfold smaller over the same wavelength range,

[4] F. Collins, H. Diehl, and G. F. Smith, *Anal. Chem.* **31,** 1862 (1959).
[5] L. L. Stookey, *Anal. Chem.* **42,** 779 (1970).
[6] J. D. Artiss, S. N. Vinogradov, and B. Zak, *Clin. Biochem.* **14,** 311 (1981).
[7] T. Yotsuyanagi, R. Yamashita, and K. Aomura, *Anal. Chem.* **44,** 1091 (1972).
[8] D. L. Drabkin, *J. Biol. Chem.* **140,** 387 (1941).
[9] A. D. Adler and P. George, *Anal. Biochem.* **11,** 159 (1965).

it is always desirable to check for the relative content of metals other than iron in the protein sample under consideration, for example, by X-ray fluorescence.

Digestion of Protein

Cameron[10] has developed a reliable method for the digestion of heme protein samples in order to liberate the iron completely from the heme groups. The procedure consists in placing a heme protein sample in solid form or solution not exceeding 0.1 ml and containing 10–50 μg of iron, into a 10-ml volumetric flask. After addition of 0.1 ml of perchloric acid (70%) and 0.1 ml of 30% hydrogen peroxide solution the sample is digested at 100° for 30 min. At this point there must not be any precipitate and the solution should have a pale straw yellow color. If necessary, additional 0.1 ml quantities of the perchloric acid and the hydrogen peroxide solution can be added and the time of digestion extended. The sample is then ready for the complexometric determination of iron.

If the volume of the sample exceeds 0.2 ml, it should be reduced by evaporation prior to digestion, in a boiling water bath at 100° taking care to avoid bumping. This process may be accelerated by the addition of several portions of 75% ethyl acetate in ethanol with removal of the 70.2° boiling ternary azeotrope, followed by addition of portions of absolute ethanol to remove final traces of ethyl acetate as the 71.8° boiling binary ethyl acetate–ethanol azeotrope; the final ethanol addition must bring the sample essentially to dryness.[11]

It is also known that iron is released rapidly and quantitatively from hemoglobin in the presence of aqueous sodium hypochlorite.[12] A procedure for hemoglobin iron analysis has been developed in which treatment with 2.5% sodium hypochlorite[12a] was used to liberate the iron.[13]

Iron Determination Using 1,10-Phenanthroline

The procedure developed by Cameron, using 1,10-phenanthroline, provides a very reliable method for the determination of iron in a variety of hemeproteins and other iron-containing materials.[11] To the digested sample in which no trace of insoluble solids remain, 0.1 ml of 10% aqueous hydroxylammonium chloride solution is added in order to reduce the iron.

[10] B. F. Cameron, *Anal. Biochem.* **11**, 164 (1965).
[11] B. F. Cameron, *Anal. Biochem.* **35**, 545 (1970).
[12] H. V. Connorty and A. R. Briggs, *Clin. Chem. (Winston-Salem, N.C.)* **8**, 151 (1962).
[12a] Instead of reagent grade sodium hypochlorite, household chlorine bleach can be used.
[13] B. Klein, B. K. Weber, L. Lucas, J. A. Foreman, and R. L. Searcy, *Clin. Chim. Acta* **26**, 77 (1969).

After standing for 15 min, 1.0 ml of a 0.5% solution of 1,10-phenanthroline in 50% aqueous ethanol is added first, followed immediately by the addition of 1.0 ml of pyridine. The contents of the flask are mixed, made up to volume with distilled, iron-free water, and the absorbance of the solution read at 510 nm. A reagent blank, omitting the sample, should be treated identically. The corrected absorbance at 510 nm can then be used to calculate the iron content using the molar absorptivity of 11,000.

Iron Determination Using 4-(2-Pyridylazo)resorcinol

A simple procedure has been developed using Cameron's method of digesting the heme protein sample and 4-(2-pyridylazo)resorcinol (PAR) as the complexometric reagent at alkaline pH.[14] The sodium salt of the ligand is used as a 1% solution in water; it is stable for long periods of time provided it is kept in the dark.

Because of the sensitivity of PAR to iron, it is better to use 15-ml polypropylene vials instead of the glass volumetric flasks. After digestion, the iron is reduced by the addition of 0.1 ml of 10% hydroxylamine solution. After standing for 15 min, 0.1 ml of the PAR solution and 1.0 ml of 3 M ammonium hydroxide are added. The solutions are mixed, allowed to stand for 3 min, and the volume adjusted to 10 ml with distilled, iron-free water by weighing. The absorbance of the solution at 500 nm is measured using a 1-cm cell, relative to a reagent blank carried through an exactly identical procedure with omission of the sample.

If other trace metals such as copper and zinc are present in the protein sample, their interference can be masked by the addition of EDTA.[15]

With the foregoing procedure it is possible to determine as little as 1 to 2 μg of iron. An additional increase in sensitivity can be achieved by decreasing the total volume of the solution. PAR also forms a complex with iron(III) at alkaline pH with a molar absorptivity of 60,500 at 496 nm.[16] The above procedure could perhaps be simplified by omitting the reduction of the iron. It should be noted that among the sensitive iron reagents, PAR is easily the most inexpensive.

Standard iron solutions can be prepared from electrolytic grade iron wire dissolved in concentrated hydrochloric acid or from ferrous salts such as Mohr's salt, $Fe(NH_4)_2(SO_4)_2 \cdot 6H_2O$ and Oesper's salt, $Fe(CH_2NH_3)_2(SO_4)_2 \cdot 4H_2O$ (G. F. Smith Chemical Co.).

[14] S. N. Vinogradov, T. F. Kosinski, and B. Zak, *Anal. Biochem.* **120**, 111 (1982).
[15] C. S. Feldkamp, R. Watkins, E. S. Baginski, and B. Zak, *Microchem. J.* **22**, 335 (1977).
[16] D. Nonova and B. Evtimova, *J. Inorg. Nucl. Chem.* **35**, 3581 (1973).

[36] Affinity Chromatography of Heme-Binding Proteins: Synthesis and Characterization of Hematin- and Hematoporphyrin-Agarose

By Kenneth W. Olsen

Principle

In recent years, the technique of affinity chromatography has revolutionized the purification of proteins. Several general affinity columns using immobilized coenzymes, such as nucleotides[1] or NAD,[2] have been developed. These resins are able to bind many different proteins and, therefore, have been used in a great number of purification schemes. The heme prosthetic group, however, has largely been ignored as a general ligand. A heme affinity column may prove to be successful for purifying a variety of proteins, including hemopexin,[3] ligandin,[4] and tryptophan 1,2-dioxygenase (pyrrolase),[5] cytochromes, enzymes involved in heme metabolism, and heme-regulated factors in protein synthesis.[6]

This chapter discusses a method for making a hematin-agarose resin that has been used to successfully purify hemopexin.[7,8] The method can be used to synthesize any porphyrin-agarose column packing. The reactions are summarized in Fig. 1. First, aminohexyl-agarose is made using the cyanogen bromide method,[9] and then the porphyrin is attached with a carbodiimide reaction. For the hematin-agarose, it is critical that the second reaction be done in dimethylformamide so that the hematin is soluble.

Synthetic Procedures

The synthesis of porphyrin-agarose resin is a two-step procedure. In the first phase the agarose is activated by cyanogen bromide,[9] and then

[1] R. Barker, K. W. Olsen, J. H. Shaper, and R. L. Hill, *J. Biol. Chem.* **247**, 7135 (1972).
[2] C. R. Lowe, M. J. Harvey, D. B. Craven, and P. D. G. Dean, *Biochem. J.* **133**, 499 (1973).
[3] U. Müller-Eberhard and W. T. Morgan, *Ann. N.Y. Acad. Sci.* **244**, 624 (1975).
[4] E. Tipping, B. Ketterer, and P. Koskelo, *Biochem. J.* **169**, 509 (1978).
[5] H. Marver, D. P. Tschudy, M. G. Perbroth, and A. Collins, *Science* **153**, 501 (1966).
[6] S. Ochoa and C. de Haro, *Annu. Rev. Biochem.* **48**, 549 (1979).
[7] K. W. Olsen, *Anal. Biochem.* **109**, 250 (1980).
[8] R. Majuri, *Biochim. Biophys. Acta* **719**, 53 (1982).
[9] P. Axen, J. Poráth, and S. Ernback, *Nature (London)* **214**, 1302 (1967).

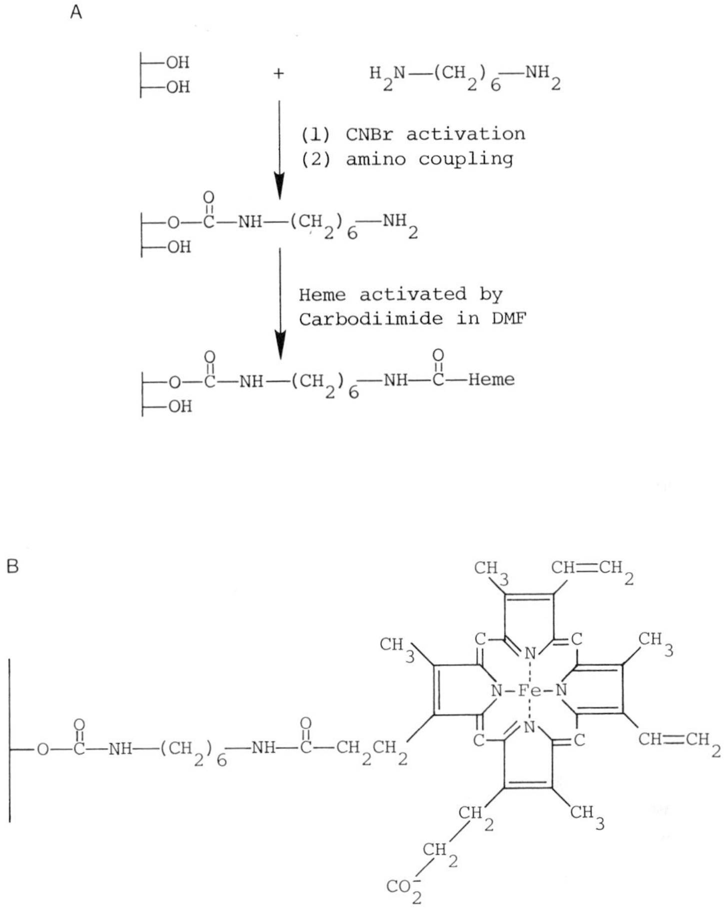

A

B

FIG. 1. (A) Synthesis of hematin-agarose. (B) Structure of hematin-agarose.

coupled to 1,6-diaminohexane.[10] In a typical synthesis, well-washed Sepharose 4B (150 ml) was activated by cyanogen bromide (37.5 g) at pH 10.5 and 25°. After thoroughly washing with ice water (1500 ml) the activated resin was reacted with an equal volume of 1,6-diaminohexane solution (34.8 g) at pH 10 and 4° overnight.[11] The aminohexyl-agarose was washed with water (2 liters) to remove excess 1,6-diaminohexane.

In the second phase of the synthesis, the porphyrin is linked to the free amino group of the aminohexyl "arm" by a carbodiimide reaction. The

[10] P. Cuatrecasas, *J. Biol. Chem.* **245**, 3059 (1970).
[11] M. Wilcheck, *FEBS Lett.* **33**, 70 (1973).

aminohexyl-agarose (100 ml) was washed four times with dimethylform-amide (DMF). The hematoporphyrin or hematin (0.5 g) was dissolved in DMF (150 ml) by stirring overnight at room temperature. In this and all subsequent steps, the porphyrins were protected from light as much as possible. The hematin solution had to be filtered on a medium-sintered glass funnel to remove insoluble particles. Each of the porphyrin solutions were mixed with 100 ml of the aminohexyl-agarose. A solution of 1-ethyl-3-(3-dimethylaminopropyl)carbodiimide (7.5 g) in 50% DMF (20 ml) was added dropwise to the mixture, while the pH was adjusted to 4.7 with 1 N HCl. The pH was readjusted to 4.7 every 30 min. After 5 hr, a second addition of carbodiimide (7.5 g) was made. Several more pH adjustments were made at 30 min intervals, and the reaction was allowed to proceed overnight at room temperature. In the morning the pH was adjusted to 4.7 several more times. After 18 hr at pH 4.7, the reaction was stopped by adding 1 N sodium hydroxide to bring the pH to 7.5.

The hematoporphyrin-agarose was washed with 4 liters of distilled water to remove the DMF and unreacted porphyrin. The resulting affinity resin was pink, and further washing did not remove any additional color.

The hematin-agarose resin required more extensive washing to remove the unreacted hematin. After adjusting the pH to 7.5, the resin was washed twice on a sintered glass funnel with DMF (200 ml). The hematin-agarose was suspended in DMF and shaken in a waterbath at 30°. The solvent was changed four times daily for the next 4 days until two consecutive washes had no color. The resin was then washed extensively with water to remove the DMF. Unreacted primary amino groups of the gel can be blocked by an additional carbodiimide reaction in acetate buffer.[8]

Both affinity resins were stored in dark bottles at 4° under toluene when not in use. Resin stored in this manner appears to be stable for at least 6 months.

Analytical Procedures

The amount of porphyrin attached to the resin was determined spectrophotometrically by modifying the method of Golovina et al.[12] A 50% (w/v) solution of polyethylene glycol 20,000 was used to suspend the affinity resins. The spectra were obtained with a Cary 219 double-team spectrophotometer using an identical amount of suspended agarose as the reference. Standard curves were constructed by adding known amounts of hematin or hematoporphyrin to the same volume of suspended agarose.

[12] T. O. Golovina, T. V. Cherednikova, A. T. Mevkh, and N. K. Nagradova, Anal. Biochem. 83, 778 (1977).

The concentration of the hematin in the stock solution was determined using an E_{403}^{mM} of 170 in 80% dimethyl sulfoxide.[13] The concentration of hematoporphyrin was calculated from the amount used to make the stock solution.

The spectrophotometric analyses of the new affinity resins showed that 0.13 μmol of hematin and 0.16 μmol of hematoporphyrin were attached per milliliter of agarose by the synthetic procedures described here. An alternative method of determining the amount of hematin bound involves hydrolysis in base followed by measuring the released hematin as a pyridine hemichrome.[14]

Applications

The usefulness of these affinity resins was tested by chromatographing several apohemoproteins or heme-binding proteins. Undenatured globin (Fig. 2A) and bovine serum albumin (Fig. 2B) can bind to hematin-agarose. Neither of these proteins was eluted by sodium phosphate buffer, but both proteins could be completely removed by any one of several deforming buffers. Similar results were obtained for the chromatography of albumin on hematoporphyrin-agarose.

The hematin-agarose resin can be used to further purify hemopexin from a crude preparation, as shown in Fig. 3. This starting material was prepared from serum by the perchloric acid precipitation method[15]; however, if the rivanol precipitation method is used, the albumin can be removed at this step.[8] The material that came straight through the column contained no hemopexin, as judged by the spectral assay procedure of Drabkin.[16] The protein eluted by the acidic buffer consisted almost exclusively of albumin and hemopexin, as measured by cellulose acetate electrophoresis at pH 8.6 by the method of Kohn.[17] A third protein, which is present in very small amounts in the acidic eluate, is probably the histidine-rich glycoprotein.[18] The spectrum of the eluate showed that no hematin was bound to these proteins. Elution by 8 M urea did not elute any additional protein from this column.

To demonstrate that the columns are specific for heme-binding proteins, lysozyme and hemoglobin were applied to hematin-agarose. Both of

[13] W. T. Morgan, H. H. Liem, R. P. Sutor, and U. Müller-Eberhard, *Biochim. Biophys. Acta* **444,** 435 (1976).

[14] K. Tsutsui and G. C. Mueller, *Anal. Biochem.* **121,** 244 (1982).

[15] H. E. Schultze, K. Heide, and H. Haupt, *Clin. Chim. Acta* **7,** 854 (1962).

[16] D. L. Drabkin, *Proc. Natl. Acad. Sci. U.S.A.* **68,** 609 (1971).

[17] J. Kohn, *Clin. Chim. Acta* **3,** 450 (1958).

[18] W. T. Morgan, *Biochim. Biophys. Acta* **533,** 319 (1978).

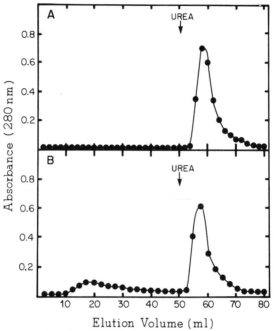

Fig. 2. Chromatography of globin and albumin on hematin-agarose. (A) Undenatured globin (10 mg) was applied to a column of hematin-agarose (10 ml) in 0.05 M sodium phosphate buffer, pH 7.6. The column was washed with this buffer (50 ml), and then the globin was eluted with 6 M urea at the point indicated. (B) Bovine Cohn fraction V (10 mg) was applied to a column of hematin-agarose (10 ml) in 0.05 sodium phosphate buffer, pH 7.6. After the contaminating proteins were washed out of the column with the same phosphate buffer, the albumin was eluted with 6 M urea.

these proteins failed to interact with the resin. Elution of these columns with 8 M urea gave no evidence that these proteins had been absorbed, although a very small amount of protein which lacked heme was eluted in the case of the hemoglobin.

The capacity of the affinity resins has been measured by frontal analysis by saturating them with bovine serum albumin. The results depend on the history of the resin being used. Porphyrin-agarose that has never been used absorbs more albumin than resin which has already been saturated with protein in previous experiments. However, the extra absorbed protein cannot be removed from the affinity column, even by 6 M guanidine hydrochloride. The unused hematin-agarose absorbed 22 mg/ml, while the used resin absorbed 16 mg/ml. From hematoporphyrin-agarose, the unused resin bound 19 mg/ml, but the used resin, only 13 mg/ml. After the first use, the binding capacity of the resin remained essentially constant.

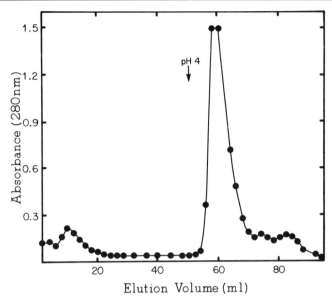

Elution Volume (ml)

FIG. 3. Chromatography of bovine hemopexin on hematin-agarose. Partially purified hemopexin (1 ml) was applied to a hematin-agarose column (10 ml) in 0.05 M sodium phosphate buffer, pH 7.6. Proteins which were not absorbed by the affinity resin were washed out of the column with the same buffer. The heme-binding proteins were then eluted with 0.1 M sodium citrate buffer, pH 4.0. The progress of this buffer through the column could be followed easily due to a reversible color change of the affinity resin from dark green to red-brown.

Although most of the interest in hematin-agarose affinity chromatography has concerned the purification of hemopexin,[7,8,14,19–21] this matrix has also been used to isolate the RNA-containing fraction of the enzyme system that converts glutamic acid into δ-aminolevulinic acid in plants.[22,23]

Alternative Methods

Several attempts have been made to purify hemoproteins by affinity chromatography. This is an obvious approach, but it has been plagued

[19] J. Suttnar, Z. Hrkal, and Z. Vodrazka, J. Chromatogr. **131**, 453 (1977).
[20] J. Suttnar, Z. Hrkal, Z. Vodrazka, and J. Rejnkova, J. Chromatogr. **169**, 500 (1979).
[21] P. Strop, J. Borvak, V. Kasicka, Z. Prusik, and L. Moravek, J. Chromatogr. **214**, 317 (1981).
[22] D.-D. Huang, W.-Y. Wang, S. P. Gough, and C. G. Kannangara, Science **225**, 1482 (1984).
[23] W.-Y. Wang, D.-D. Huang, D. Stachon, S. P. Gough, and C. G. Kannangara, Plant Physiol. **74**, 569 (1984).

with technical problems. As early as 1966, Heide[24] mentioned the possibility of isolating hemopexin with heme bound to Sephadex G-100, but no experimental details were given. More recently, Conway and Müller-Eberhard[25] coupled hemin to aminoethyl-Sepharose, but the resulting resin had a relatively low capacity for both hemopexin and albumin. A better resin was made by using 2,4-disulfonic acid deuteroporphyrin instead of hemin.[25] Rabbit hemopexin and this water-soluble porphyrin form an equimolar complex with an apparent dissociation constant of 1.8 $\times$ 10^{-6} M.[26] Although this resin requires a considerably more complex synthesis, the affinity column did bind hemopexin effectively. Apohemopexin could be eluted with 2.5 M guanidine–HCl.[8] Unfortunately, the experimental details of this work have not yet been published.

Suttnar et al.[19] have coupled hemin to BioGel P-200 by a p-nitrobenzoylazide spacer arm. This synthesis required seven reactions, and the resulting affinity column had low capacity and released some hemin when the hemopexin was eluted with glycine–HCl buffer (pH 2.4). This synthetic method can also be used to make a heme-agarose matrix, which allowed a much higher recovery of hemopexin from human serum than did the hemin-BioGel P-200 matrix.[20]

Two other heme-agarose gels have been developed that can be used to isolate hemopexin. Strop et al.[21] have attached heme to agarose via an 8-amino-2-hydroxy-4-thiooctyl linkage. This method requires six synthetic steps but the resulting gel appears to have properties similar to the matrix described here. Tsutsui and Mueller[14] coupled hemin to aminoethyl-agarose with 1,1-carbonyldiimidazole. The details of this method are given in this volume.[27] This synthetic procedure is as easy to accomplish as the one reported here. Since the hemin-agarose linkage is the same chemical type as in the present method, the lack of serum albumin binding by the Tsutsui and Mueller matrix must be due to the shorter spacer employed.[14]

Comments

The synthesis of the porphyrin affinity resins described here is a two-step procedure in which aminohexylagarose is first synthesized by the cyanogen bromide technique[9] and then hematin is coupled to the "arm" by the carbodiimide reaction. The porphyrin in the affinity resin can be

[24] K. Heide, Protides Biol. Fluids 14, 593 (1966).
[25] T. P. Conway and U. Müller-Eberhard, Fed. Proc., Fed. Am. Soc. Exp. Biol. 32, 469 (1973).
[26] T. P. Conway and U. Müller-Eberhard, Arch. Biochem. Biophys. 172, 558 (1976).
[27] K. Tsutsui, this volume [37].

attached by either or both of its propionic acid groups. The novel and essential aspect of this reaction is the use of pure DMF as a solvent for hematin. Aqueous DMF in concentrations up to 50% (v/v) has been used in similar reactions involving hydrophobic ligands, such as estradiol.[10] Although pure dioxane has been used as a solvent for the carbodiimide reaction,[11] higher concentrations of DMF have not been used, apparently for fear of damaging the agarose. The results presented here demonstrate that this is not a problem. For the synthesis of hematin-agarose the use of the pure solvent is a necessity. If it is replaced by 50% (v/v) DMF, the resulting resin is not an effective affinity column due to the low level of substitution and to the presence of considerable amounts of noncovalently attached heme in the resin. Both of these problems are caused by the low solubility of hematin in even partially aqueous solvents under acidic conditions. Thus, the use of DMF as the solvent in the coupling reaction is essential to the success of this synthesis.

[37] Affinity Chromatography of Heme-Binding Proteins: Synthesis of Hemin-Agarose

By KEN TSUTSUI

Heme (ferroprotoporphyrin IX) plays an important role as a prosthetic group of many proteins such as oxygen-carrier proteins, proteins of the electron transport system, mixed function oxidases, and peroxidases. In addition to being directly involved in biological reactions as a cofactor, heme also appears to serve as a regulatory molecule in such processes as the initiation of protein synthesis,[1] ATP/ubiquitin-dependent protein degradation,[2] inhibition of DNA polymerase,[3] transcriptional regulation of a cytochrome *c* gene,[4] and enhancement of cell differentiation.[5] Specific binding of heme to some of the proteins involved in these functions has been demonstrated. Hemopexin,[6] serum albumin,[6] histidine-rich glycoprotein,[7] and HBP.93[8] are serum proteins that have been shown to bind

[1] S. Ochoa and C. de Haro, *Annu. Rev. Biochem.* **48**, 549 (1979).
[2] A. L. Haas and I. A. Rose, *Proc. Natl. Acad. Sci. U.S.A.* **78**, 6845 (1981).
[3] J. J. Byrnes, K. M. Downey, L. Esserman, and A. G. So, *Biochemistry* **14**, 796 (1975).
[4] L. Guarente and T. Mason, *Cell* **32**, 1279 (1983).
[5] J.-J. Chen and I. M. London, *Cell* **26**, 117 (1981).
[6] U. Müller-Eberhard and W. T. Morgan, *Ann. N.Y. Acad. Sci.* **244**, 624 (1975).
[7] W. T. Morgan, *Biochim. Biophys. Acta* **535**, 319 (1978).
[8] K. Tsutsui and G. C. Mueller, *J. Biol. Chem.* **257**, 3925 (1982).

heme. Among these proteins, hemopexin is thought to function as a carrier of free heme in serum, but the physiological significance of the other proteins is still unclear.

In studying hemoproteins, affinity chromatography utilizing a matrix with immobilized heme is an obvious approach to identify and purify apo forms of these proteins. A variety of coenzymes have been successfully conjugated as a general ligand to various matrices to purify the corresponding enzymes.[9] However, immobilized hemin derivatives of practical value have not been available until recently although several attempts have been made. Some physicochemical properties of hemin seem to be the reason for this delay. The conjugated double bond system of the porphyrin ring and the two carboxyl groups of the propionic acid side chains are the only possible functional groups of hemin which can be directly utilized for covalent immobilization. In early attempts, a diazo-coupling reaction between matrix-bound diazo groups and the porphyrin ring of hemin was employed.[10] The affinity resins prepared by diazo coupling were successfully applied to the purification of hemopexin, but this procedure has not been widely used, probably because of the laborious synthetic steps involved and the relatively low capacity of the resulting resin. Although the side-chain carboxyl groups are readily available for coupling, they were not utilized in early studies because of the possibility of decreasing or even losing the affinity of immobilized ligand for specific proteins by the modification of these charged groups. Besides problems with coupling, hemin has a limited solubility in neutral aqueous media and is prone to aggregate.[11] The insolubility of hemin is the main obstacle to the synthesis of a highly substituted matrix when coupling agents which require an aqueous environment are to be used. This problem is not easily solved by exploiting a water–organic solvent mixture in coupling reactions. For example, the immobilization of hemin was rather low ($\sim$0.1 μmol of hemin per ml of agarose) when 1-ethyl-3-(3-dimethylaminopropyl)carbodiimide was used in a 1:1 aqueous mixture of dimethylformamide (DMF) and aminohexyl-agarose.[12] However, the resulting resin was shown to effectively bind globin and heme-binding proteins of serum despite the fact that the side-chain carboxyl groups were modified through immobilization.

Some features of the present method for preparing hemin-agarose may be described as follows. (1) To avoid the solubility problem, the coupling reaction is performed in a pure nonaqueous solvent (100% DMF) in which

[9] M. Wilchek and W. B. Jakoby, this series, Vol. 34, p. 3.
[10] J. Suttnar, Z. Hrkal, Z. Vodrážka, and J. Rejnková, *J. Chromatogr.* **169**, 500 (1979).
[11] S. B. Brown, T. C. Dean, and P. Jones, *Biochem. J.* **117**, 733 (1970).
[12] K. W. Olsen, *Anal. Biochem.* **109**, 250 (1980).

hemin is highly soluble. (2) The immobilization is accomplished through the propionic acid group of hemin. (3) To activate the carboxyl group, 1,1'-carbonyldiimidazole (CDI) is employed instead of more commonly used agents such as carbodiimides. This agent (CDI) is more convenient to use because there is no need to adjust the pH during the reaction. (4) A cross-linked agarose (Sepharose CL-6B) which is quite resistant to organic solvents and mechanical agitation is used as the matrix.[13] (5) The time required to remove the unreacted hemin which is tightly adsorbed on the hemin-agarose is significantly reduced by introducing a wash with 25% pyridine. (6) Compared to other methods, a significantly higher coupling efficiency (1–2.5 μmol hemin/ml agarose) is obtained.

Preparation of Hemin-Agarose

Sepharose CL-6B (Pharmacia) is activated with BrCN by the method of March *et al.*,[14] and then allowed to react with diaminoethane to prepare aminoethyl-Sepharose. One hundred milliliters of Sepharose CL-6B is washed with distilled water on a sintered glass funnel and suspended in 100 ml of distilled water. To the Sepharose suspension placed in a 1-liter Erlenmeyer flask, 200 ml of 2 M Na_2CO_3 is added and stirred gently until equilibration, after which 10 ml of a BrCN solution (25 g BrCN dissolved in 12.5 ml of acetonitrile) is added and the suspension agitated for 2 min at room temperature. The reaction is terminated by quickly washing the beads with 2 liters of ice cold 0.1 M $NaHCO_3$ (pH 9.5) on a sintered glass funnel. The activated beads are immediately transferred to a beaker containing 100 ml of 2 M diaminoethane in 0.1 M $NaHCO_3$ (pH 10.0). The slurry is gently stirred for 16 hr at 4°. The uncoupled diaminoethane is removed by washing the beads with 2 liters of 0.1 M $NaHCO_3$ (pH 9.5) and 2 liters of distilled water.

Hemin (0.5 g) is dissolved in 37.5 ml of DMF (spectrophotometric grade) and filtered through a membrane filter resistant to organic solvents (Millex-SR, Millipore Co.) to remove insoluble material. A powder of CDI (0.5 g) is added to the solution, and the mixture is heated at 80° for 15 min,[15] and then allowed to cool at room temperature for 30 min. The

[13] In our original study [K. Tsutsui and G. C. Mueller, *Anal. Biochem.* **121**, 244 (1982)], we used an uncross-linked agarose beads which tended to aggregate after coupling, probably because of minor damage to the beads caused by organic solvents used in the synthesis. No such problems have been encountered with Sepharose CL-6B.

[14] S. C. March, I. Parikh, and P. Cuatrecasas, *Anal. Biochem.* **60**, 149 (1974).

[15] Progression of the reaction is revealed by the change in the color of the hemin solution from a brown to a red tinge. The color change is probably caused by the coordination of the heme iron with imidazole released from CDI.

resulting hemin acylimidazolide in DMF is added to 100 ml[16] of aminoethyl-Sepharose (placed in a 300-ml Erlenmeyer flask), which has been washed successively with 33% DMF (400 ml), 66% DMF (400 ml), and finally 100% DMF (800 ml).[17] The coupling reaction is allowed to proceed at room temperature for 18 hr while the flask is shaken gently on a rotary type shaker. The unbound hemin is removed by washing the colored beads with decreasing concentrations of DMF: 100% DMF (800 ml), 66% DMF (400 ml), and 33% DMF (400 ml), followed by washing with distilled water (2 liters). The residual hemin adsorbed on the resin is removed by rewashing it with 25% pyridine until the wash has no color (about 2 liters required), and finally with distilled water. The hemin-Sepharose should be stored at 4° in the dark with a few drops of toluene as a preservative.

The coupling efficiency is routinely monitored by measuring the amount of hemin immobilized on Sepharose. The conjugated hemin is released from Sepharose by heating the beads (50 μl) in 0.1 M NaOH (3 ml) at 75° for 30 min.[18] After cooling to room temperature, 1 ml of pyridine (spectrophotometric grade) is added and the Sepharose beads (now colorless) are removed by centrifugation. The hemin derivative released from the matrix is converted to pyridine hemochrome by reducing the central iron with sodium dithionite, and the absorbance is recorded immediately at 420 nm. For standards, aliquots of 1 mM hemin in pyridine (freshly prepared) are mixed with aminoethyl-Sepharose suspended in 0.1 M NaOH, and the mixture is treated in the same way as the hemin-Sepharose. The preparation procedure described here should yield 1–2 μmol of immobilized hemin per milliliter of Sepharose.

General Properties of Hemin-Agarose

Some properties of the hemin-agarose resin prepared by the present synthetic procedure but using an uncross-linked agarose matrix have been

---

[16] A higher ratio of hemin to aminoethyl group was used originally in order to increase the probability of one free carboxyl group being left on each immobilized hemin molecule. However, it was found that the relative amount of hemin to aminoethyl-Sepharose could be reduced to $\frac{1}{4}$ without affecting the binding capacity of the resultant hemin-Sepharose when it was used for purification of serum heme-binding proteins.

[17] DMF and pyridine used in the washing steps are of reagent grade.

[18] A thick suspension of hemin-Sepharose is taken up into a 100-μl capillary micropipette. One end of the capillary is sealed with a piece of paraffin and the beads are packed by centrifugation. The capillary is cut at the 50 μl volumetric marker line and the lower portion containing the packed beads is then cut into several pieces and transferred to a test tube. The beads inside the capillary are easily suspended in 0.1 M NaOH by briefly vortexing the tube.

described.[19] Intactness of the ligand during the coupling reaction is demonstrated by the identical pyridine hemochrome absorption spectra of hemin and the hemin derivative released from the resin by alkali treatment. Although the propionic acid side chains are modified by the spacer (diaminoethane), the resonance structure of the porphyrin ring, including the coordination with the central iron, remains intact.

An unexpected character of hemin-agarose is that it binds significant amounts of hemin in a high-ionic-strength buffer containing 0.5 M NaCl. The binding very likely occurs through the interaction of free hemin with matrix-bound hemin since aminoethyl-agarose has no affinity to hemin. About 1–2 mol of hemin are bound per mol of immobilized hemin at saturation. The bound hemin is readily released by washing the resin with 8 M urea, DMF, or pyridine. Because of this extensive binding of hemin to the affinity resin, it is not possible to specifically elute proteins bound on the hemin-agarose using hemin solution as a competing ligand. However, this phenomenon could be exploited to concentrate hemin and possibly its related compounds from a large volume of samples such as deproteinized body fluids.

For affinity chromatography on the hemin-agarose, a high-ionic-strength buffer (0.5 M NaCl and 10 mM sodium phosphate, pH 7.5) has been routinely used in the protein adsorption step to minimize nonspecific ionic interaction between protein and the affinity resin. Under this condition, specific retention of apohemoprotein to a column of hemin-agarose can be demonstrated using heme-depleted hemoglobin. Passage of globin solution through the column results in a highly effective binding of globin (4–5 mg protein/ml packed gel) whereas essentially no adsorption occurs with hemoglobin. Aminoethyl-agarose does not bind globin at all. A variety of conditions have been tested for the elution of bound globin. Nearly complete elution is attainable with 8 M urea or 1% SDS, but 20–25% of the globin remains bound on the resin when 0.1 M acetic acid is used. However, as described below, the low pH elution protocol has been successfully used in purifying serum heme-binding proteins. A highly chaotropic salt such as 3 M NaSCN fails to elute globin, suggesting that the interaction between globin and matrix-bound hemin is mainly hydrophobic in nature.

Serum albumin, although being one of the major heme-binding proteins in serum, does not bind to the hemin-agarose prepared by the present procedure.[19] In contrast to this, a similar hemin-agarose adsorbent prepared by Olsen using a carbodiimide coupling procedure has been shown to bind serum albumin.[12] In this case, a matrix with a longer spacer

[19] K. Tsutsui and G. C. Mueller, *Anal. Biochem.* **121**, 244 (1982).

(diaminohexane) was used to conjugate the hemin, and a low-ionic-strength buffer was employed in the binding study. These factors may be responsible for the observed differences in the performance of the two resins.

Affinity Chromatography of Serum Heme-Binding Proteins on Hemin-Agarose

In the schemes previously applied to the purification of hemopexin, precipitation of serum with rivanol to remove serum albumin was a mandatory step since the hemin-conjugated affinity resins used in these studies bound serum albumin. The preparation of hemin-agarose does not bind albumin so that serum samples can be applied directly onto the affinity column without any pretreatment. After extensive washing, hemopexin and HBP.93 are eluted together with 0.1 M acetic acid, and the two proteins are separated by subsequent chromatography on SP-Sephadex. By this improved two-step procedure employing hemin-agarose affinity chromatography, hemopexin and HBP.93 are easily purified from whole serum in high yields.

As an example of the application of the present hemin-Sepharose resin, our previous procedure for the simultaneous purification of hemopexin and HBP.93 from rabbit serum[8] is slightly modified and described here in detail. A water suspension of the hemin-Sepharose CL-6B is degassed under reduced pressure and poured into a column of 1.5 × 30 cm to make a settled bed volume of 50 ml. The following steps are performed at 4°. The column is conditioned with 500 ml of buffer A containing 0.5 M NaCl and 20 mM sodium phosphate (adjusted to pH 7.5 with NaOH). One hundred milliliters of pooled rabbit serum[20] is dialyzed against buffer A and then diluted 5 times with the same buffer. The diluted serum is cleared by centrifugation at 5000 rpm for 10 min and loaded on the hemin-Sepharose column at a flow rate of 25 ml/hr. The column is then washed with 1 liter of buffer A at an increased flow rate (50 ml/hr). To elute proteins adsorbed on the resin, a continuous flow of 0.1 M acetic acid is applied while six 20-ml fractions are collected. Absorbance at 280 nm of each fraction is measured, and fractions containing eluted proteins (usually fraction No. 1 through 4) are pooled and dialyzed immediately against 2 liters of buffer B (0.1 M NaCl and 20 mM sodium acetate adjusted to pH 5.2 with acetic acid). The dialysis buffer is changed two more times. Hemopexin and HBP.93 comprise more than 95% of the protein eluted from the affinity resin (Fig. 1b). The hemin-Sepharose column can be reused without any problem simply by reequilibrating it with buffer A. After several operations, however, the flow rate of the column may de-

---

[20] Nonhemolyzed rabbit serum obtained frozen from Pel-Freez Biologicals, Inc.

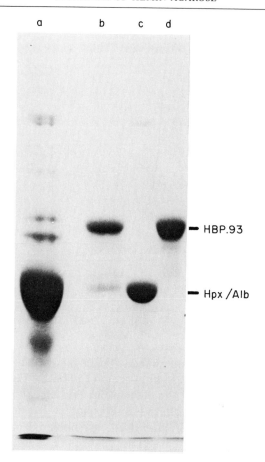

Fig. 1. Simultaneous purification of hemopexin and HBP.93 on hemin-agarose and subsequent separation on SP-Sephadex. Protein samples were subjected to electrophoresis in 7.5% polyacrylamide gel containing 0.1% SDS. (a) Rabbit whole serum, (b) proteins eluted from hemin-Sepharose, (c) first peak of SP-Sephadex, (d) second peak of SP-Sephadex. Stained bands for HBP.93, hemopexin (Hpx), and serum albumin (Alb) are labeled. Hemopexin and albumin migrate to the same position in this gel system. The unusually low amount of hemopexin retained on hemin-Sepharose (lane b) reflects the fact that a batch of moderately hemolyzed serum was used in this experiment.[21]

crease. A brief washing with 1% SDS at room temperature helps to alleviate this problem.

SP-Sephadex C-50 (Pharmacia) is hydrated in buffer B, packed in a column (1.5 × 30 cm, 50 ml bed), and equilibrated with 1 liter of buffer B. After clarification by centrifugation at 5000 rpm for 10 min, the dialyzed protein sample eluted from the hemin-Sepharose is applied to the column at a flow rate of 30 ml/hr. The column is washed with 100 ml of buffer B,

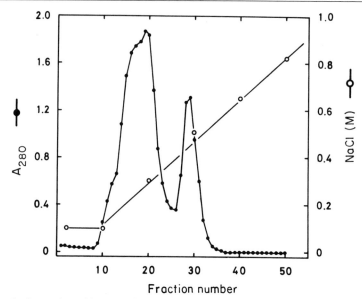

Fɪɢ. 2. Separation of hemopexin and HBP.93 by SP-Sephadex chromatography. First peak (fraction No. 10–24) and second peak (No. 27–35) correspond to hemopexin and HBP.93, respectively.

and the adsorbed proteins are eluted with 300 ml of a NaCl linear gradient (0.1–1.0 M NaCl in 20 mM sodium acetate, pH 5.2). Five-milliliter fractions are collected at a flow rate of 30 ml/hr. Protein concentrations of the fractions are determined from the absorbance at 280 nm (Fig. 2). Fractions containing hemopexin and HBP.93 are pooled separately, dialyzed extensively against distilled water (3 changes, 2 liters each time), and finally concentrated using a Diaflo membrane PM10 (Amicon). The concentrated protein solutions are lyophilized in small vials. The high purities of these preparations are confirmed by SDS–polyacrylamide gel electrophoresis (Fig. 1c and d). Yields of these proteins from 100 ml of rabbit serum are usually 50–60 mg for hemopexin[21] and 25–30 mg for HBP.93. Heme-binding properties of the purified proteins are demonstrated to be intact by spectrophotometric titrations with hemin.

Acknowledgment

The author wishes to express his gratitude to Dr. Gerald C. Mueller, in whose laboratory at the University of Wisconsin the essential part of this work was initiated and performed.

[21] When hemolyzed serum is used, the yield of hemopexin is significantly lower. This reduction seems to be caused by a partial saturation of hemopexin due to the transfer of heme from hemoglobin to hemopexin during the preparation or serum storage.

[38] Porphobilinogen Synthase: A Specific and Sensitive Coupled-Enzyme Assay

By DAVID F. BISHOP and ROBERT J. DESNICK

Porphobilinogen synthase (δ-aminolevulinic acid dehydratase)[1] is the second enzyme in the mammalian heme biosynthetic pathway. The enzyme catalyzes the condensation of two molecules of δ-aminolevulinic acid (ALA)[2] to produce the monopyrrole, porphobilinogen (PBG). ALA-dehydratase has been purified to homogeneity from several mammalian sources including human erythrocytes[3] and bovine liver.[4-6] The human and bovine enzymes have been shown to be homooctomers with subunit molecular weights of about 31,000 and 35,000, respectively. Both purified enzymes have eight zinc atoms per octomer and contain pyridoxal phosphate. Lead and other heavy metals markedly inhibit ALA-dehydratase activity by replacing the zinc atoms, which are essential for optimal activity.[3,5,6] Inhibition of the human erythrocyte enzyme has been used as a sensitive indicator of lead poisoning.[7] In addition, the inherited deficiency of ALA-dehydratase results in a hepatic porphyria characterized by excessive urinary ALA excretion and chronic neurologic manifestations.[8]

Principle

As shown in Fig. 1, ALA is converted to the monopyrrole, PBG, by ALA-dehydratase. PBG is then converted to the linear tetrapyrrole, hydroxymethylbilane, by PBG-deaminase. It has been shown that PBG-

[1] Enzymes: porphobilinogen synthase (ALA-dehydratase, 5-aminolevulinate hydro-lyase, EC 4.2.1.24); porphobilinogen deaminase (PBG-deaminase, porphobilinogen ammonia-lyase, EC 4.3.1.8).

[2] Abbreviations: ALA, δ-aminolevulinic acid; PBG, porphobilinogen; URO, uroporphyrin; TCA, trichloroacetic acid; DTT, dithiothreitol; PMSF, phenylmethylsulfonyl fluoride; BSA, bovine serum albumin.

[3] P. M. Anderson and R. J. Desnick, *J. Biol. Chem.* **254**, 6924 (1979).

[4] E. L. Wilson, P. E. Burger, and E. B. Dowdle, *Eur. J. Biochem.* **29**, 563 (1972).

[5] W. H. Wu, D. Shemin, K. E. Richards, and R. C. Williams, *Proc. Natl. Acad. Sci. U.S.A.* **71**, 1767 (1974).

[6] I. Tsukamoto, T. Yoshinaga, and S. Sano, *Biochim. Biophys. Acta* **570**, 167 (1979).

[7] R. A. Mitchell, J. E. Drake, L. A. Wittlin, and T. A. Rejent, *Clin. Chem. (Winston-Salem, N.C.)* **23**, 105 (1977).

[8] M. Doss, R. von Tiepermann, J. Schneider, and H. Schmid, *Klin. Wochenschr.* **57**, 1123 (1979).

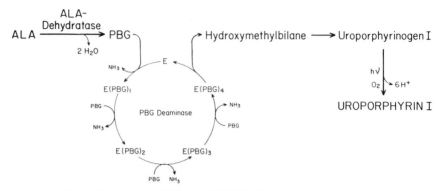

FIG. 1. Conversion of ALA to URO I in the coupled-enzyme assay.

deaminase catalyzes the sequential formation of stable enzyme–substrate intermediates (i.e., mono-, di-, tri-, and tetrapyrroles) in the stepwise synthesis of the linear tetrapyrrole.[9,10] Thus, to ensure linearity with time in the conversion of PBG to hydroxymethylbilane, PBG-deaminase first must be saturated with these intermediates. Hydroxymethylbilane nonenzymatically cyclizes to uroporphyrinogen I, which then can be oxidized quantitatively to the highly fluorescent compound, uroporphyrin I (URO I) by exposure to ultraviolet light at low pH.[11] The amount of URO I produced is measured fluorometrically with excitation and emission wavelengths of 405 and 600 nm, respectively.

Reagents for the Coupled-Enzyme Assay

PBG-deaminase, 50–150 units/ml
Dithiothreitol (DTT), 20 mM
ALA, 5 mM, pH 3 (do not neutralize)
Sodium phosphate, 0.4 M, pH 6.7
Trichloroacetic acid (TCA), 50% w/v
HCl, 1.0 N
URO I, 0.5 μg/ml in 1 N HCl (Porphyrin Products, Logan, UT)

Rapid Purification of PBG-Deaminase

For assay of PBG-deaminase activity, the reaction mixture contains 50 μl of enzyme source, 0.5 ml of 0.1 M Tris–HCl buffer, pH 8.1, and 0.2 ml of 0.5 mM PBG. After incubation in the dark at 37° for 30 min, 250 μl of

[9] A. R. Battersby, C. J. R. Fookes, G. W. J. Matcham, and E. McDonald, *Nature (London)* **285,** 17 (1980).
[10] P. M. Anderson and R. J. Desnick, *J. Biol. Chem.* **255,** 1993 (1980).
[11] P. F. Giampietro and R. J. Desnick, *Anal. Biochem.* **131,** 83 (1983).

50% TCA is added. The mixture is centrifuged (1000 g, 10 min) and the uroporphyrinogen I in the supernatant is photooxidized and quantitated as described below for the coupled-assay procedure. One unit of PBG-deaminase is that amount of enzyme required to form 1 nmol of URO I per hr at 37°.

PBG-deaminase can be purified from human erythrocytes in 4–5 days by the following procedure. All steps are at 4°. Outdated human erythrocytes obtained from the blood bank (500 ml) are washed by mixing with 2 volumes of 0.9% NaCl followed by centrifugation for 15 min at 600 g. The buffy coat is removed and discarded, and the erythrocytes are washed again and frozen overnight. The thawed cells are osmotically lysed by stirring for 1 hr with 3 volumes of distilled water containing 3 mM DTT and 10 μM phenylmethylsulfonyl fluoride (PMSF). The lysed cells are centrifuged at 12,000 g for 1 hr, and the supernatant is dialyzed first against water and then against 7 mM potassium phosphate buffer, pH 6.8, containing 1 mM DTT and 10 μM PMSF. The dialyzed lysate is applied (5 ml/min) onto a 5 × 30 cm column of DEAE-cellulose (Whatman DE-52) which had been equilibrated with dialysis buffer to pH 6.80. Hemoglobin remaining in the column is removed with 1 liter of the above dialysis buffer. PBG-deaminase is then eluted with 1 liter of dialysis buffer containing 0.08 M NaCl, pH 6.8. The fractions containing PBG-deaminase activity are pooled and concentrated to 50 ml by ultrafiltration in an Amicon stirred cell using a PM 10 membrane (Amicon, Danvers, MA). The concentrated enzyme is applied at a flow rate of 0.5 ml/min to tandem octyl- and phenyl-Sepharose CL-4B columns (each 0.9 × 15 cm, Pharmacia, Piscataway, NJ), equilibrated with 20 mM potassium phosphate, pH 8.0, 1 mM DTT, and 10 μM PMSF (buffer A), containing 0.25 M NaCl. The tandem columns are washed with 50 ml each of buffer A, pH 8.0, containing 0.25 M KCl followed by buffer A containing 0.12 M KCl and finally by buffer A alone. The octyl-Sepharose column is removed and PBG-deaminase is eluted from the phenyl-Sepharose column with a four-chambered gradient containing 25 ml each of 0, 40, 60, and 80% ethylene glycol in buffer A each adjusted to pH 8.0. Fractions containing PBG-deaminase activity are pooled and stored in 1.0-ml aliquots at −20°. A typical purification profile is shown in Table I.

URO I Standard Curve

The URO I concentration in the stock solution (5 μg/ml) is determined by its molar absorptivity of 5.05 × 10^5 M^{-1} cm^{-1} at 406 nm.[12] For the

---

[12] National Research Council. Committee on Specifications and Criteria for Biochemical Compounds (1972). "Specifications and Criteria for Biochemical Compounds." Natl. Acad. Sci., Washington, D.C.

TABLE I
RAPID PURIFICATION OF PBG-DEAMINASE FROM HUMAN ERYTHROCYTES

Step	Volume (ml)	Total activity (units)	Total protein (mg)	Specific activity (units/mg)	Purification (fold)	Yield (%)
Lysate	1500	9020	91,800	0.098	1	100
DEAE-cellulose	42	8530	836	10.2	104	95
Octyl/phenyl-Sepharose	60	4380	17.1	256	2610	49

standard curve, stock URO I is diluted with 1.0 N HCl to 0.9–90 pmol/150 μl. Each 10 × 75-mm disposable glass test tube contains 150 μl of URO I (diluted to the specified concentration), 250 μl of 0.4 M sodium phosphate buffer, pH 6.7, 300 μl of 5 mM ALA, 125 μl of 20 mM DTT, 100 μl of 50% TCA, and 75 μl of glass-distilled water. Exposure to light should be minimized by covering the stock URO I container and the individual test tubes with aluminum foil. The standard curve should be linear to about 100 pmol URO I/ml after which self-quenching occurs.[13,14]

URO I is quantitated in a fluorometer equipped with a standard photomultiplier tube (e.g., Turner Model 111) using a Corning 7-51/Wratten 2C primary filter (405 nm) and a Wratten 25 secondary filter (590+ nm). Disposable test tubes (10 × 75 mm) can be used as cuvettes in a Turner fluorometer with an adapting sleeve for 1.0 ml samples. For increased sensitivity, a variable wavelength instrument such as the Perkin-Elmer Model 204-A may be used.

Sample Preparation

Fresh heparinized whole blood is centrifuged at 800 g for 10 min, the plasma and leukocytes removed, and the erythrocytes washed three times with 0.9% NaCl. The erythrocytes are lysed by the addition of 2 volumes of buffer containing 1.0 mM potassium phosphate, pH 7.6, 1 mM DTT, 1 mM MgCl$_2$, 10 μM PMSF, and 0.5% Triton X-100. The lysate is centrifuged at 35,000 g for 20 min, and the supernatant is removed and either assayed immediately or stored at −20° until used. Liver tissue is diced and homogenized in 3 volumes of 0.1 M sodium phosphate buffer, pH 6.8,

[13] S. Schwartz, M. H. Berg, I. Bossenmaier, and H. Dinsmore, *Methods Biochem. Anal.* **8**, 221 (1960).
[14] D. F. Bishop, K. R. Devey, L. McBride, and R. J. Desnick, *Anal. Biochem.* **113**, 68 (1981).

centrifuged at 17,000 g for 30 min, and the supernatant is used as enzyme source. Cultured fibroblasts are harvested by trypsinization and centrifugation at 800 g for 10 min and then washed twice with 0.9% NaCl. Cells are lysed by the addition of an equal volume of 1.0 mM potassium phosphate buffer, pH 6.8, containing 1 mM DTT, followed by three cycles of freezing (dry ice/acetone) and thawing (37°). Lysates are centrifuged at 35,000 g for 20 min and the supernatants are assayed immediately or stored at −20°.

ALA-Dehydratase Coupled-Enzyme Assay

The enzyme source (50 μl of an appropriate dilution) is preincubated at 37° for 15 min with 140 μl of 20 mM DTT, 310 μl of 0.4 M sodium phosphate buffer, pH 6.7, and 50 μl of distilled water in a 12 × 75-mm test tube. The tubes are placed in an ice bath and 3.8 units of purified PBG-deaminase, 0.3 ml of 5 mM ALA, and water are added to a final volume of 0.9 ml. The reaction is initiated by transferring the tubes to a 37° water bath. After 90 min, the reaction is terminated by the addition of 0.1 ml of 50% TCA followed by centrifugation at 800 g for 10 min. The supernatants are transferred to 10 × 75-mm test tubes, and the uroporphyrinogen I is photooxidized to URO I by exposure to long-wave UV lights (e.g., Model XX15, Ultraviolet Products, Inc., San Gabriel, CA) on each side of the test tube rack. Fluorometric quantitation is as described above. The picomoles of URO I produced are multiplied by 4 to obtain the picomoles of PBG formed and then divided by 0.47 to correct for the incomplete (47%) conversion of PBG to URO I at equilibrium by the enzyme couple (see Comments). One unit of ALA-dehydratase activity is defined as the amount of enzyme required to form 1 nmol of PBG per hr at 37°.

Comments

Fresh erythrocytes may be used instead of outdated cells for the purification of PBG-deaminase. Since the specific activity is higher in fresh erythrocytes, the yield will be slightly improved. However, there are no differences in the properties of the enzyme isolated from fresh or outdated cells.[10] Freezing the washed erythrocytes has been found to facilitate the subsequent pelleting of the lysed cell membranes. The pH of the DEAE-cellulose step is critical and must be between 6.75 and 6.85 to prevent binding of hemoglobin.

To optimize the sensitivity of the coupled-enzyme assay, the amount of PBG-deaminase (3.8 units) required for maximal conversion of PBG to uroporphyrinogen I was determined. As shown in Fig. 2, increasing the

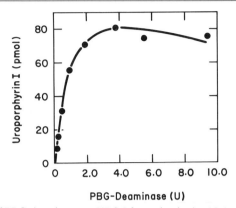

FIG. 2. Effect of PBG-deaminase on URO I formation in the ALA-dehydratase coupled-enzyme assay. The conditions for the assay were as described in the text. The amount of ALA-dehydratase was 0.68 units in each assay while the PBG-deaminase varied from 0.12 to 9.4 units.

amount of PBG-deaminase to more than 4 units per assay resulted in slightly less URO I formed, presumably due to the formation of additional stable PBG-deaminase–substrate intermediates.[9,10] Thus, the presence of additional coupling enzyme requires more of the PBG generated by ALA-dehydratase to saturate the PBG-deaminase with substrate intermediates.

In order to determine the percentage conversion of PBG to uroporphyrinogen I, known amounts of PBG (ranging from 50 to 630 pmol) were added to 3.8 units of PBG-deaminase, and the reaction mixture was incubated for 60 min at 37°. Approximately 11.8 pmol of URO I was detected for each 100 pmol of PBG over the range of 100–630 pmol of added PBG. This corresponded to a 47% conversion of PBG to uroporphyrinogen I, assuming a mole for mole photooxidative conversion of uroporphyrinogen I to URO I. Therefore, to determine the amount of PBG synthesized by ALA-dehydratase, the amount of URO I detected in the coupled assay was multiplied by 4 (4 mol of PBG per mol of URO I) and divided by 0.47.

Using purified ALA-dehydratase, the coupled-enzyme assay was linear with protein concentration over the range of 10–100 pmol of URO I formed. These values correspond to 0.085–0.85 units of ALA-dehydratase per assay. The addition of 0.25–2.5 mg of bovine serum albumin (BSA) to the reaction mixture had little, if any, effect on URO I formation. However, when 5.0 or 10.0 mg of BSA was added, the URO I formation was inhibited about 10 and 20%, respectively.

After incubation for 30, 60, and 90 min, the ALA-dehydratase activities determined by the coupled assay were about 60, 90, and 100%, re-

TABLE II
LEVELS OF ALA-DEHYDRATASE ACTIVITY IN
VARIOUS SOURCES

Enzyme source	Enzyme activity (units/mg protein)
Human erythrocytes	
Mean ± 1 SD	4.80 ± 1.71
Range (n = 12)	3.13–9.56
Rat liver	
Mean ± 1 SD	7.40 ± 2.09
Range (n = 3)	4.56–9.53
Friend erythroleukemia cells (line 586)	
Uninduced	19.3
DMSO-induced	92.2
Mouse RAG cells	11.5
Cultured human skin fibroblasts	
Mean ± 1 SD	1.10 ± 0.62
Range (n = 6)	0.63–1.97
Cultured human amniotic cells	
Mean ± 1 SD	0.488 ± 0.132
Range (n = 3)	0.340–0.661

spectively, of the activity detected by the standard colorimetric assay.[9] This nonlinearity is due to the initial time required for saturation of the PBG-deaminase with substrate intermediates.[11] Therefore, a 90-min incubation period was used for the coupled assay. The levels of endogenous PBG-deaminase in normal cells and tissues do not alter the reliability of the coupled-enzyme assay.

The chief advantage of this coupled-enzyme assay is its increased sensitivity. As shown in Table II, the coupled-enzyme assay was able to detect the low activity levels in cultured fibroblasts and amniocytes; these activities were undetectable by the standard colorimetric methods.[15] Thus, the specificity, sensitivity, and rapidity of this fluorometric coupled-enzyme assay make possible the reliable determination of low levels of ALA-dehydratase activity in small amounts of crude tissue homogenates or in cultured cells.

Acknowledgment

This investigation was supported in part by a grant (1 R01 AM26824) from the National Institutes of Health and a grant (1-535) from the March of Dimes Birth Defects Foundation.

[15] S. Sassa, *Enzyme* **28,** 133 (1982).

[39] High-Performance Liquid Chromatographic Analysis of Porphyrins and Their Isomers with Radial Compression Columns

By Andreas Seubert and Sigrid Seubert

Porphyrins form the backbone of pigments such as heme and chlorophylls. They also are present in a wide variety of biocatalysts such as vitamin B_{12} and cytochromes. Therefore, porphyrins are of interest to many scientists for the study of the mechanisms of these metabolic pathways. Of special interest are disturbances of heme biosynthesis, characterized by an excess of porphyrins in different tissues, blood, and excreta. These disturbances of biosynthesis, called porphyrias, are in most cases a metabolic abnormality, due to an enzyme deficiency and detectable in many tissues. Analysis of excreted or accumulated porphyrins is useful in the diagnosis of porphyrias.[1] Besides uroporphyrin, coproporphyrin, and protoporphyrin, there exist intermediate porphyrins with three, five, six, and seven carboxylic groups. The naturally occurring porphyrins are of the I and III isomer types. Although the determination of isomers for diagnostic purposes is of little importance, it nonetheless is essential for scientific reasons.

Many investigators are engaged in developing methods for measuring porphyrins in different tissues and fluids.[2] In addition to thin-layer chromatography,[3] high-performance liquid chromatographic (HPLC) methods for the separation of free acids and porphyrin methyl esters are presently employed.[4,5] Columns filled with silica or reverse-phase material are utilized.

In this chapter we will describe two methods for the separation of porphyrin methyl esters and their isomers. One method makes possible the separation of porphyrins with two to eight carboxylic groups in a short time. The other method allows, in addition to this differentiation, a separation of the isomers of uroporphyrin and heptacarboxy porphyrin. This method requires significantly more time. Both methods are used routinely and are well suited for separation of porphyrins in biological materials.

[1] L. Eales, in "The Porphyrins" (D. Dolphin, ed.), Vol. 6, Part A, p. 663. Academic Press, New York, 1979.
[2] G. H. Elder, Clin. Haematol. 9, 371 (1980).
[3] M. Doss, Z. Klin. Chem. Klin. Biochem. 8, 197 (1970).
[4] M. Chiba and S. Sassa, Anal. Biochem. 124, 279 (1982).
[5] A. Seubert and S. Seubert, Anal. Biochem. 124, 303 (1982).

Materials

The methyl esters of the following porphyrins were obtained from Porphyrin Products (Logan, UT): uroporphyrin I, heptacarboxylic porphyrin I, hexacarboxylic porphyrin I, pentacarboxylic porphyrin I, coproporphyrin I and III, mesoporphyrin IX, and protoporphyrin IX. Uroporphyrin III octamethyl ester from Turaco feathers was obtained from Sigma Chemical Company (St. Louis, MO). Talc, glacial acetic acid, 2 M sodium hydroxide, and concentrated sulfuric acid were of analytical grade; solvents were LiChrosolv (Merck, Darmstadt, West Germany).

High-Pressure Liquid Chromatography Equipment

In the author's laboratory a model 6000 A pump, model U6K injector, and model RCM-100 module (Waters Associates, Milford, MA) were used. Detection was accomplished either with a model 440 absorbance detector, filter 405 nm (Waters Associates) or a model 650 fluorescence spectrophotometer with micro flow cell unit for LC and photomultiplier R 928 (Perkin-Elmer Corporation, Norwalk, CT).

In our laboratory Radial-PAK silica cartridges, 100 × 8 or 5 mm internal diameter, 10 or 5 μm particle size, combined with RCSS Guard-PAK, silica (Waters Associates), were used.

For the separation two systems were employed as follows:

System 1: n-heptane/ethyl acetate/dichloromethane/methanol
(60 : 25 : 12.5 : 2.5)
10 μm Radial-PAK silica cartridge

System 2: n-heptane/ethyl acetate/dichloromethane/
methanol (60 : 25 : 15 : 0.2)
5 μm Radial-PAK silica cartridge

Routinely a column was developed with a flow rate of 2 ml/min. Fluorescence was monitored with an excitation wavelength at 403 nm and an emission at 630 nm; both slits were 10 nm.

Preparation of Samples

Owing to the variable sample workup of porphyrins no universal preparation for chromatographic analysis is possible. The methods described proved to be the best in routine work.

Urine. About 2 g talc and 0.2 ml acetic acid were put into an Erlenmeyer flask. Then, depending on the porphyrin content, 1–10 ml of urine were added and shaken for about 1 min. The talc with the adsorbed porphyrins was evacuated through a filter funnel (porosity 4) and dried directly on this filter funnel at 95° for 1 hr. The filter funnel with the talc was placed in a wide-necked glass. Ten milliliters methanol/sulfuric acid (95 : 5, v/v) was added to the hot talc to esterify the adsorbed porphyrins. After the methanol–sulfuric acid mixture had passed through the filter, the filtrate was warmed together with the filter funnel to 60° for 20 min. Then about 10 ml of dichloromethane was added to the filter to wash possibly adherent porphyrin methyl esters from the talc. The filtrates were washed twice with water and neutralized with 3% sodium bicarbonate solution. The organic phase was dried with sodium sulfate and evaporated at 20° in vacuum.

Feces. Methanol/sulfuric acid (20 ml) was added to a sample of wet feces and mixed well. The mixture was warmed to 60° for 20 min. The residue was filtered off and washed with dichloromethane. The washing solution and the filtrate were combined and worked up as was described for urine.

Other Biological Materials (e.g., Liver, Kidney, Blood, Plasma). Liver and kidney were freeze-dried and pulverized with a glass rod. In

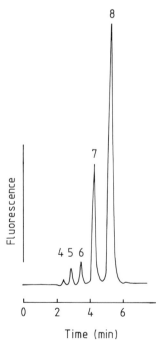

Fig. 1. Separation of porphyrin methyl esters in an hemofiltrate with System 1. Peaks: 8, uroporphyrin octamethyl ester; 7, heptacarboxylic porphyrin heptamethyl ester; 6, hexacarboxylic porphyrin hexamethyl ester; 5, pentacarboxylic porphyrin pentamethyl ester; 4, coproporphyrin tetramethyl ester.

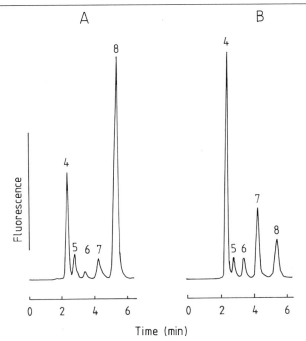

FIG. 2. Separation of an uroporphyrinogen decarboxylase assay with System 1. (A) Substrate, uroporphyrinogen I; (B) substrate, uroporphyrinogen III. For key, see Fig. 1.

this way one can achieve total homogeneity. An aliquot of the homogenate was suspended in saline and treated with a needle probe of an ultrasound generator. This homogenate was extracted three times with ethyl acetate/acetic acid (4 : 1, v/v) as with erythrocytes and plasma. The extracts were shaken with 1.5 N hydrochloric acid until no fluorescence could be seen in the organic layer. The hydrochloric acid, which contained the porphyrins, was neutralized with 2 N sodium hydroxide. Then 0.2 ml acetic acid and approximately 1 g of talc were added. The mixture was shaken for 1 min. The preparatory procedure was the same as for urine.

Examples of Application

Separation of Porphyrins in Hemofiltrates

In hemodialysis patients, blisters often appear which clinically resemble porphyria cutanea tarda (PCT).[6] Because these patients often are

---

[6] L. C. Harber and D. R. Bickers, *J. Invest. Dermatol.* **82**, 207 (1984).

unable to produce any urine, a diagnostic evaluation is necessary to differentiate between PCT and pseudoporphyria. The separation of porphyrins in the hemofiltrate is a very helpful diagnostic tool. Figure 1 shows the baseline separation of the porphyrins with 4, 5, 6, 7, and 8 carboxylic groups in an hemofiltrate. System 1 and fluorescence detection were used. Mainly uroporphyrin and heptacarboxyporphyrin can be seen. This pattern, combined with strongly elevated plasma porphyrins, indicates PCT in this patient. Separation of the plasma porphyrins shows a similar pattern of the porphyrins. The recoveries in routine assays were more than 90%.

Separation for Uroporphyrinogen Decarboxylase Assay

The separation of porphyrins in the investigation of the enzymes of heme biosynthesis is helpful. Determination of the uroporphyrinogen de-

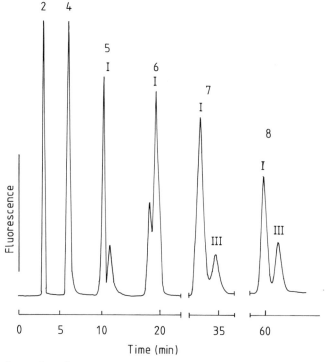

FIG. 3. Separation of a standard mixture of porphyrin methyl esters with System 2. Peaks: 2, mesoporphyrin IX dimethyl ester; I, III: isomers of type I and III; for key, see Fig. 1.

carboxylase activity in human erythrocytes shows considerable variations in the distribution of the decarboxylation products by using uroporphyrinogen I and III. As is shown in Fig. 2, uroporphyrinogen I is
more slowly decarboxylated than uroporphyrinogen III. The enzymatic
decarboxylation of uroporphyrinogen III yields about 51% coproporr-
phyrin, in addition to 23% heptacarboxyporphyrin and 11% penta- and
hexacarboxyporphyrin. From uroporphyrinogen I only 21% coproporr-
phyrin and 15% hepta-, hexa-, and pentacarboxyporphyrin are produced.

Separation of Isomers

Determination of the isomer porphyrins is of interest. The separation
of the I and III isomers of uroporphyrin and heptacarboxyporphyrin is
performed with System 2. The 5 μm silica columns and the reduced
amount of methanol in the eluent additionally make possible the separation of these isomers in the same run, as well as the differentiation of the
porphyrins with 2–6 carboxylic groups. Isocoproporphyrin can clearly
be distinguished from coproporphyrin. The coproporphyrin isomers,
however, cannot be separated. Figure 3 shows the pattern of the porphy-

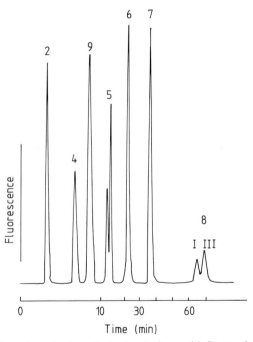

FIG. 4. Separation of porphyrin methyl esters in feces with System 2. Peak 9: isocoproporphyrin; for key, see Fig. 1 and 3.

rins of a standard mixture. The hexa- and pentacarboxyporphyrins are remarkable. There are two peaks for each substance. They represent the different isomers of series I, as was identified by comparison with chemically synthesized samples.

Note. The use of new columns with 5 μm material results in better separations and longer retention times. To get the same retention times as in our previous paper[5] one had to add 0.5 ml methanol to the eluent.

Separation of Fecal Porphyrins

In Fig. 4 the distribution pattern of porphyrins in feces of an hexachlorobenzene-poisoned rat is shown. In addition to uroporphyrin I and III, one finds heptacarboxyporphyrin I, hexacarboxyporphyrin I, and two peaks of pentacarboxyporphyrin. A large amount of isocoproporphyrin is also found.

Acknowledgment

This work was supported by P. G. Unna-Foundation.

[40] High-Performance Liquid Chromatography of Porphyrin Methyl Esters

By James G. Straka

Porphyrins and hexahydroporphyrins (or porphyrinogens) are the tetrapyrrolic intermediates of chlorophyll, heme, and corrin (vitamin B_{12}).[1,2] The separation and identification of the intermediates have elucidated the pathway for biosynthesis of these compounds. Hereditary deficiencies of enzymes in the heme biosynthetic pathway in man[3] and animals[4,5] give rise to a family of diseases, the porphyrias. Each of the porphyrias is characterized by the accumulation of certain precursors in the blood,

[1] J. Lascelles, "Tetrapyrrole Biosynthesis and Its Regulation." Benjamin, New York, 1964.

[2] B. F. Burnham, *in* "Metabolic Pathways" (D. M. Greenberg, ed.), Vol. 3, p. 404. Academic Press, New York, 1969.

[3] S. Sassa and A. Kappas, *Adv. Hum. Genet.* **11**, 121 (1981).

[4] G. R. Ruth, S. Schwartz, and B. Stephenson, *Science* **198**, 199 (1977).

[5] J. G. Straka, G. R. Ruth, S. Schwartz, L. A. Anderson, B. Stephenson, and J. R. Bloomer, *Clin. Res.* **31**, 770A (1983).

urine, feces, or tissues of individuals with the diseases. Characteristic relationships among the porphyrin precursors exist for each of the diseases, which form the basis for the diagnosis and classification of the porphyrias.[3,6,7] In addition, the separation and quantification of these intermediates provide the analytical basis for assays of the enzymes involved in heme biosynthesis.[8-11]

Biologically significant porphyrins and porphyrinogens have from two to eight carboxylate substituents at the periphery of the tetrapyrrole ring. Separation and quantification of the corresponding porphyrin methyl esters by high-performance liquid chromatography are rapid and reproducible. The chromatography utilizes normal phase columns and isocratic elution systems. The systems are simple and may be easily modified for a particular analysis. Furthermore, the use of methyl esters permits a separation of the porphyrins from other components in some samples (e.g., from feces or urine) which may either obscure the analysis or shorten column life. The disadvantages of using methyl esters are that sample preparation (esterification and extraction) can be time consuming, and that certain separations, such as the separation of isomer I from isomer III type porphyrins are not yet possible with the esters. Isomer separation has been achieved using reverse-phase HPLC.[12,13]

This chapter is divided into two major sections. The first deals with sample preparation from a variety of sources; the second deals with chromatographic systems and their development.

Sample Preparation and Esterification

Reagents

5% (w/v) H_2SO_4 in methanol. Slowly add 2.7 ml concentrated H_2SO_4 to about 80 ml absolute methanol. After allowing the solution to cool, bring the volume to 100 ml with methanol.

10–12% BF_3 in methanol (Sigma Chemical Co., St. Louis, MO). *Note:* BF_3 in the presence of water generates HF, a volatile and

[6] M. Doss, W. Meinhof, D. Look, H. Henning, P. Nawrocki, W. Dölle, G. Strohmeyer, and L. Filippini, *S. Afr. Med. J.* **45**, Spec. Issue, 50 (1971).
[7] C. H. Gray, C. K. Lim, and D. C. Nicholson, *Clin. Chim. Acta* **77**, 167 (1977).
[8] J. G. Straka, J. P. Kushner, and M. A. Pryor, *Enzyme* **28**, 170 (1982).
[9] B. Grandchamps and Y. Nordmann, *Enzyme* **28**, 196 (1982).
[10] J. G. Straka, J. P. Kushner, and B. F. Burnham, *Anal. Biochem.* **111**, 269 (1981).
[11] G. Elder, J. O. Evans, and S. A. Matlin, *Clin. Sci. Mol. Med.* **51**, 71 (1976).
[12] C. K. Lim and T. J. Peters, *Clin. Chim. Acta* **139**, 55 (1984).
[13] C. K. Lim and T. J. Peters, this volume [44].

toxic gas. Appropriate caution should be exercised when using this reagent mixture.

CH_2Cl_2 or $CHCl_3$, reagent grade. If $CHCl_3$ is used, the ethanol used for stabilization must be extracted[10] by washing once with an equal volume of 10 mM EDTA and twice with water ($CHCl_3$ is occasionally contaminated with copper salts[14] which are removed by the EDTA wash). The washed $CHCl_3$ is passed through Whatman 1 P/S (phase-separator) paper and dried over anhydrous $CaCl_2$ or Na_2SO_4. The $CHCl_3$ may then be stabilized by adding 2% (v/v) 1-hexene. This reagent is stable for about 2 weeks if stored at room temperature in a dark bottle. CH_2Cl_2 may be used directly.

Talc (USP grade). Metal ions which may be a problem can be removed by suspending the talc in 10 mM Na_2EDTA. It is then collected on a Büchner funnel and washed at the pump successively with two portions of water followed by two portions of methanol. It is dried by drawing air through at the pump.

Trimethyl orthoformate (Eastman, Rochester, New York).

9% (w/v) $HClO_4$ in methanol. Slowly add 8.0 ml 70% $HClO_4$ to 80 ml absolute methanol. Bring to 100 ml with methanol after cooling.

Procedures

Three types of biological samples which contain porphyrins are generally encountered: dilute aqueous, concentrated aqueous, and semisolid (e.g., tissues, cells, or feces). Each of these samples is prepared for chromatography according to the unique problems associated with the type and source of the sample. Quantitative esterification requires that the water present in the samples be eliminated. Water may be conveniently removed from an esterification mixture by reaction with trimethyl orthoformate [$HC(OCH_3)_3$]. In the presence of an acid catalyst, 1 mol of H_2O (1 volume) is consumed as it hydrolyzes 1 mol of trimethyl orthoformate (~7 volumes), producing 2 mol of methanol and 1 mol of methyl formate. Due to the large volume ratio of orthoester : H_2O (7 : 1), the amount of water in the sample should be estimated and reduced if possible prior to reaction.

The methods described below use 10% BF_3 in methanol as the esterifying reagent, which we have found to be the most effective. In every case, however, 5% H_2SO_4 in methanol may be used with nearly equal effectiveness. The esterification procedure for all the prepared samples, unless otherwise stated, is performed as described in the Esterification section.

---

[14] D. T. Zelt, J. A. Owen, and G. S. Marks, *J. Chromatogr.* **189**, 209 (1980).

Sample Preparation

Dilute Aqueous Porphyrin Solutions. Dilute solutions of porphyrins (e.g., urine, tissue culture supernatants, tissue or cell extracts, serum) must be concentrated prior to esterification. A convenient and inexpensive method utilizes the adsorption of porphyrins from aqueous solutions onto talc. Concentration and recovery of porphyrins are equivalent for all porphyrins having 2–8 carboxylate substituents. Routinely, 60–90% recovery is obtained. To the stirred aqueous porphyrin solution is added dry talc. (The amount of talc finally added is empirically determined, but 5 g should be sufficient for 100 ml of normal urine.) The solution is titrated to pH 3.5 ($\pm$0.3) with glacial acetic acid or 2 M sodium acetate, as appropriate; the additional use of mineral acid or strong base may be required if the original solution was strongly buffered. After thorough mixing, the talc is collected by vacuum on a medium-porosity sintered glass Büchner funnel (centrifugation may also be used). To the filtrate is added a second portion of talc and the pH of the suspension is adjusted if necessary. The talc is collected on a second funnel or by centrifugation, and the talc is examined for fluorescence under a long-wavelength ultraviolet lamp (e.g., a Woods lamp). This is repeated until no pink fluorescence is observed adsorbed to the talc. The portions of talc containing porphyrins are pooled with the initial portion (using a 0.1 M sodium acetate buffer, pH 3.5 if necessary). Dry air is drawn through the resulting cake of talc for 10–30 min until no bubbling is observed on the underside of the filter. The funnel is transferred to a clean dry side-arm flask or tube. Porphyrins are eluted from the talc by successive portions of 10% BF_3 in methanol. Elution is continued until the talc (or the eluate) shows no fluorescence when observed under UV light. Since the talc retains some water prior to elution, about 1 ml of trimethyl orthoformate should be added per gram of talc used. Esterification is performed as described.

If the talc was collected by centrifugation rather than by filtration, the pellet should be drained thoroughly. The talc is suspended in 5–10 volumes of BF_3–methanol containing 10–20% trimethyl orthoformate (v/v), and the suspension esterified directly.

When the sample volume is modest (5–20 ml, e.g., tissue culture supernatants), the sample may be dried by lyophilization.[14] The esterifying agent (BF_3–methanol) is added directly to the dry residue.

Concentrated Aqueous Porphyrin Solution. If the porphyrin solution is sufficiently concentrated ($\gtrsim$0.5 nmol/ml), the porphyrins can be esterified directly as described for the assay mixture for uroporphyrinogen decarboxylase.[15] The method precipitates proteins with 9% (w/v) $HClO_4$

[15] J. G. Straka and J. P. Kushner, *Biochemistry* **22,** 4664 (1983).

in methanol; for other applications, simply acidify the original mixture with 50–100 μl concentrated HCl, H_2SO_4, or $HClO_4$. To 0.5 ml of porphyrin solution add 0.5 ml 9% $HClO_4$ in methanol. Slowly add 3.5–4.0 ml trimethyl orthoformate, followed by 1 ml 10% BF_3 in methanol.

Semisolid Sample Containing Porphyrin. The sample (0.5 g) is triturated or homogenized in 5 ml of 5% H_2SO_4 in methanol. To the resulting suspension is added 3.5 ml of trimethyl orthoformate, and the entire suspension esterified. Note that H_2SO_4–methanol is preferred here. The water content of feces, tissue, or cells would react with BF_3 releasing HF. The latter is toxic and etches glass and would damage a glass homogenizer if this were used to disperse the sample.

Esterification

Samples prepared as described above are transferred to a glass culture tube with a Teflon-lined screw cap (e.g., Corning Pyrex brand, serial number 9826-series). After being tightly sealed, the tube is placed in a boiling water bath for 30 min. The tubes are removed and allowed to cool to room temperature. The porphyrin esters are extracted as described below. (Esterification is complete after 2–3 hr at 65° or overnight at room temperature.)

Extraction of Porphyrin Esters

Suspended solid material (e.g., proteinaceous precipitate, tissue or fecal residue, talc, etc.) is removed by centrifugation following esterification. After decanting the supernatant fluid, the pellet is triturated with fresh absolute methanol (or acid–methanol) and recentrifuged. The methanolic supernatants are pooled and transferred to a separatory funnel or a glass conical centrifuge tube (45 ml capacity) with a ground-glass stopper.

The porphyrin esters are transferred to CH_2Cl_2 (or $CHCl_3$, prepared as described in the Reagents section) by adding 5–10 ml CH_2Cl_2 to the methanolic solutions and mixing thoroughly. Sufficient water (usually 2–4 volumes) is added to effect a phase separation. After shaking vigorously, the phases are allowed to separate. Low-speed centrifugation at room temperature may be employed to effect a more complete separation of phases. The upper (aqueous) phase is drawn off and examined under UV light for fluorescence. The aqueous phase is reextracted with CH_2Cl_2 as necessary, and the organic phases are pooled.

The organic phase (lower) is washed free of acid with one 2-volume portion of 1 M sodium acetate (or 2 M NH_4OH), followed by two 2-volume portions of water. All aqueous phases are discarded. The CH_2Cl_2

phase is filtered by gravity through Whatman 1 P/S paper which has been saturated with CH_2Cl_2. The organic phase is collected, and the paper washed with 1–2 ml of fresh CH_2Cl_2; water accumulated in the funnel is discarded. The porphyrin ester solution is blown to dryness under a gentle stream of dry N_2, Ar, or air. To prevent water from distilling into the sample during the drying process, the sample container should be placed in a room-temperature water bath.

Preparation of Samples for Chromatography

Many samples esterified, extracted and dried as described above are ready for chromatography directly. Certain samples, e.g., fecal, whole blood, and concentrated urinary porphyrin esters, contain material which bind irreversibly to silica columns and should be prechromatographed prior to HPLC. This is conveniently done by dissolving the residue in a minimal volume of CH_2Cl_2 or ethanol-free $CHCl_3$ (see Reagents section) and applying it to a silica Sep-Pak cartridge (Waters Associates, Milford, MA). The cartridge is washed with CH_2Cl_2 until no blue or blue-green fluorescence is seen in the eluate. The porphyrin esters are eluted with several portions of ethyl acetate; elution with ethyl acetate is continued until no additional pink fluorescence is seen in the eluate. The pooled ethyl acetate fractions are blown to dryness under N_2, Ar, or air, as described above. The resulting dry residue is free of material (including unesterified porphyrins) which would bind tightly to the HPLC column, thereby decreasing its chromatographic efficiency.

Chromatography

Separation of porphyrin esters by HPLC is performed on silica columns using mixtures of organic solvents. The porphyrins elute from the column with increasing polarity. Chromatographic systems are therefore designed and can be modified predictably for highly specific applications.

Samples of dried porphyrin esters are dissolved completely in a known volume of $CHCl_3$ (see Reagents section). A measured portion is injected onto the column and chromatographed. The volume of $CHCl_3$ used for dissolving the sample must be empirically determined and depends on porphyrin concentrations, the type of injector used, the column dimensions and resolution capabilities, and the type of detector used.

Columns

Columns used are packed with microparticulate porous silica (usually 5–10 μm particle size). Pellicular packings (50–75 μm), although less

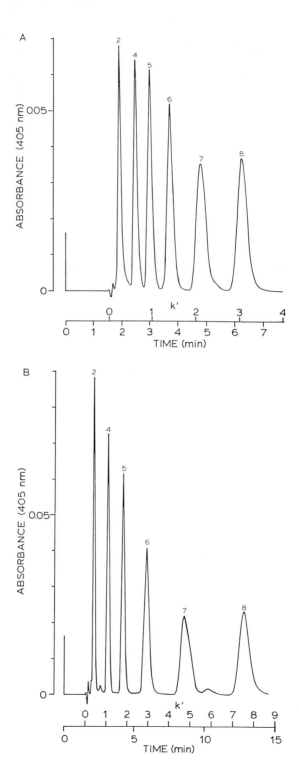

fragile, do not give satisfactory resolution for most routine analytical work. Prepacked columns are available commercially (e.g., μPorasil, 30 cm $\times$ 4 mm, i.d.; Waters Associates, Milford, MA) which work very well for these applications. They do not suffer appreciably from lot to lot variation, as is often observed with "home-made" columns.

Solvent Systems

The ideal chromatography solvent should (1) permit high resolution in a short period of time, (2) be composed of components of similar volatility so that the composition of the mixture does not change significantly with time, (3) be easily and predictably manipulated by the chromatographer, (4) not interfere with (or if possible, enhance) detection of components, (5) not alter column performance with time, and (6) be relatively nontoxic.

A number of isocratic systems containing two[7,10,11,16–20] or three [10,14,21] components and gradient systems comprising two[16] to four[22] components have been described. Two-component isocratic systems are, I feel, best suited for separation of porphyrin esters, since they are simple and can be designed to meet most of the criteria outlined above.

Two solvent systems are illustrated in Figs. 1 and 2. The chromatograms in Fig. 1 were run in *n*-heptane : ethyl acetate at two different volume ratios (40 : 60 and 50 : 50). Note that by increasing the amount of

[16] N. Evans, D. E. Games, A. H. Jackson, and S. A. Matlin, *J. Chromatogr.* **115**, 325 (1975).
[17] N. Evans, A. H. Jackson, S. A. Matlin, and R. Towill, *J. Chromatogr.* **125**, 345 (1976).
[18] A. H. Jackson, K. R. N. Rao, and D. M. Supphayen, *J. Chem. Soc., Chem. Commun.* p. 696 (1977).
[19] A. H. Jackson, *Semin. Hematol.* **14**, 193 (1977).
[20] A. R. Battersby, D. G. Buckley, G. L. Hodgson, R. E. Markwell, and E. McDonald, *in* "High Pressure Liquid Chromatography in Clinical Chemistry" (P. F. Dixon, C. H. Gray, C. K. Lim, and M. S. Stoll, eds.), p. 63. Academic Press, New York, 1976.
[21] R. E. Carlson and D. Dolphin, *in* "High Pressure Liquid Chromatography in Clinical Chemistry" (P. F. Dixon, C. H. Gray, C. K. Lim, and M. S. Stoll, eds.), p. 87. Academic Press, New York, 1976.
[22] H. de Verneuil, S. Sassa, and A. Kappas, *J. Biol. Chem.* **258**, 2454 (1983).

FIG. 1. HPLC of an equimolar mixture of porphyrins having 2-, 4-, 5-, 6-, 7-, and 8-carboxymethyl substituents using *n*-heptane/ethyl acetate solvent. Each peak represents 160 pmol of porphyrin and is labeled with the number of substituents. Chromatography was performed at a flow rate of 2.0 ml/min using a 6000 A pump, a U6K injector, a M440 detector equipped with a 405 nm interference filter, and a μPorasil column (30 cm $\times$ 3.9 mm, i.d.), all obtained from Waters Associates, Milford, MA. $k' = (V_e - V_0)/V_0$, where V_0 is the void volume of the column and V_e is the elution volume for a component of the mixture. (A) *n*-Heptane/ethyl acetate = 40 : 60 (v/v). (B) *n*-Heptane/ethyl acetate = 50 : 50 (v/v).

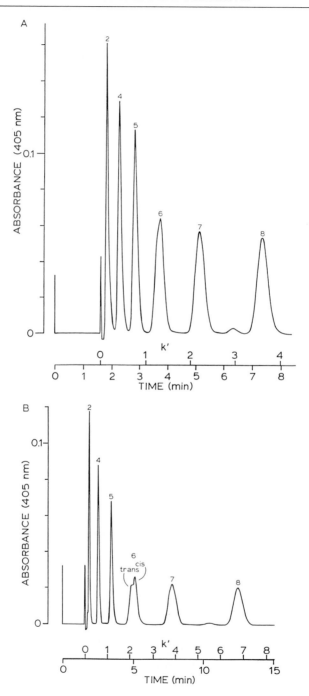

hydrocarbon solvent in the mixture, resolution was improved with an increase in overall run time. Varying the volume ratio of components of a system comprising cyclohexane : ethyl acetate (v/v) allows the resolution of the closely related dicarboxylate meso- and protoporphyrin esters (85 : 15), tricarboxylate hardero- and isoharderoporphyrin esters (80 : 20), and tetracarboxylate copro- and isocoproporphyrin esters (60 : 40);[17] similar separations have been observed using the n-heptane : ethyl acetate system as well.[23] Note that with these systems, the first components of interest elute well after the solvent front, thereby separating them from components in the sample which are not retarded by the column. Furthermore, column life is extended with these solvents. Using care in sample preparation and n-heptane : ethyl acetate (55 : 45), between 600 and 1000 analyses were routinely obtained with a single μPorasil column without a change in its chromatographic behavior. Thus, these hydrocarbon : ethyl acetate solvents are very well suited to the separation of porphyrin esters, and satisfy the criteria outlined above.

The use of a second solvent system is illustrated in Fig. 2. The solvent (toluene : acetonitrile) is less well suited for routine use for several reasons: (1) absorbance of the porphyrins is somewhat attenuated in this solvent, (2) resolution and retention are very sensitive to small changes in solvent ratios, and (3) both components are toxic and may pose a greater health risk than the solvents described above. However, this system does have the capability of providing very high resolution, with baseline separation (Fig. 2B). Note that the two components of the hexacarboxylate porphyrin I methyl esters are partially resolved. These represent the isomers in which the two methyl substituents are located on adjacent ("cis") and on opposite ("trans") pyrrole rings.

Use of solvents which contain free amines (e.g., triethylamine) or aqueous components should be avoided, since these interact strongly with silica, changing its chromatographic properties, and may shorten column life. Additionally, the use of methanol as a modifier in three-component solvents (see Refs. 10 and 14) appears to give rise to rounded chromatographic peaks which are not baseline resolved. In thin-layer

[23] J. G. Straka, unpublished data (1979).

FIG. 2. HPLC of an equimolar mixture of porphyrins having between two and eight carboxymethyl substituents in toluene/acetonitrile solvent. Each peak represents approximately 200 pmol of porphyrin. Chromatographic conditions and figure representations are as described in the legend to Fig. 1. (A) Toluene/acetonitrile = 15 : 85 (v/v). (B) Toluene/acetonitrile = 18 : 82 (v/v).

chromatographic systems the methanol probably forms an inherent gradient as the chromatogram develops. As a consequence of this, the resultant spots on thin-layer chromatograms developed with these solvents are sharp and well resolved.[6] Since it cannot form an inherent gradient in HPLC, methanol does not contribute to resolution, may affect it adversely, and should therefore be avoided.

Detection and Quantification

Porphyrins and porphyrin esters are extremely photoactive compounds and can be detected either by measuring absorbance or fluorescence. The methyl esters of naturally occurring porphyrins in organic solvents have a characteristic strong absorption band (the Soret band) near 405 ± 5 nm, with extinction coefficients of 1.6–2.2 × 10^5 M^{-1} cm^{-1}.[24] Therefore a fixed-wavelength detector with a 400–405 nm interference filter or a variable-wavelength detector set to 405 nm is used to detect as little as 10 pmol of a given porphyrin ester in the chromatogram. Furthermore, porphyrins fluoresce at 600–700 nm when excited near the Soret band. A fluorescence detector ($\lambda_{ex} \cong 400$ nm; $\lambda_{em} > 500$ nm) affords additional sensitivity so that as little as 0.1 pmol of porphyrin ester may be detected in a single chromatographic band.

The amount of porphyrin ester represented by a chromatographic peak is directly proportional to both the area under the peak and the peak height over a broad range, using either fluorescence or absorbance detection. A standard curve for porphyrin esters is constructed by chromatographing known amounts of the pure esters or a standardized equimolar mixture of esters (all available from Porphyrin Products, Logan, UT) and plotting peak height or area against the mass of each compound injected. Although the standard curve should be checked periodically, it does not vary significantly as long as chromatographic conditions are kept constant. Chromatographic peaks in samples are then quantified by comparison against the standards. The addition of a known quantity of a marker porphyrin to the sample prior to esterification may be used as an internal standard. This also serves to quantify recovery through the esterification–extraction procedure. The choice of standard depends on the nature of the sample. For example, mesoporphyrin can be used in urine specimens (since dicarboxylate porphyrins are not found in urine) and uroporphyrin in fecal specimens (since this compound is rarely excreted in the stool).

---

[24] K. M. Smith, ed., "Porphyrins and Metalloporphyrins," p. 871. Elsevier, Amsterdam, 1975.

General Considerations

Reliable separation and quantification of porphyrin esters require rigorous control of chromatographic conditions. For this reason, the following should always be observed. (1) A standard mixture of porphyrin esters (Porphyrin Products, Logan, UT) should be chromatographed at least once a day to monitor instrument and column performance and to adjust chromatographic conditions if necessary. (2) All solvents should be HPLC grade or better. Solvent mixtures should be thoroughly mixed and then filtered and degassed by vacuum prior to use. (3) All samples should be rendered free of particulate material which could block the column by filtration through 0.45-μm filters. Regenerated cellulose or PTFE filters (Schleicher and Schuell, Keene, NH) are most effective. After filtration, the samples should be blown dry, then redissolved in a known volume of filtered $CHCl_3$ (stabilized with 1-hexene) prior to chromatography. (4) System backpressure should be routinely monitored. An increase in pressure or sudden changes in pressure may indicate column inlet filter blockage, accumulation of material on the column bed, or chromatographic system malfunction, and measures should be taken to identify and correct the problem.

Acknowledgments

This work was supported in part by National Institutes of Health Grants 5-T32-AM07115, AM20503, and AM26466. The author thanks Dr. James P. Kushner of the University of Utah College of Medicine and Dr. Joseph R. Bloomer of the University of Minnesota for their help and encouragement.

[41] Synthesis of [¹⁴C]Coproporphyrin III by Yeast Cell-Free Extracts

By PIERRE LABBE

Recently, Grandchamp and Nordmann[1] and De Verneuil et al.[2] described modifications of the classical method pioneered by Dresel and Falk[3] for preparing ¹⁴C-labeled methyl esters of coproporphyrin III, pro-

[1] B. Grandchamp and Y. Nordmann, Biochem. Biophys. Res. Commun. 74, 1089 (1977).
[2] H. De Verneuil, B. Grandchamp, and Y. Nordmann, Biochim. Biophys. Acta 611, 176 (1980).
[3] E. I. B. Dresel and J. E. Falk, Biochem. J. 56, 156 (1954).

toporphyrin IX, as well as different isomer types of tetra- to octacarboxyl-porphyrins. By using purified outdated human blood hemolysates instead of crude hemolysates to catalyze total biosynthesis of the porphyrin ring, these authors were able to prepare with good yields highly labeled porphyrins from δ-amino[14C]levulinic acid which is the only physiological precursor commercially available.

In our previous studies concerning the yeast heme pathway, crude cell-free extracts were shown to produce in a very reproducible way relatively large amounts of porphyrins when incubated with δ-aminolevulinic acid (ALA). When incubated in the absence of oxygen with ALA, cell-free extracts from *Saccharomyces cerevisiae* synthesize large quantities of coproporphyrinogen III, yielding coproporphyrin III after photooxidation. Maximal activity is attained with low concentrations of ALA (0.25 mM) and the percentage of ALA converted to coproporphyrin is as high as 60%. When incubated in the presence of oxygen, the cell-free extracts catalyze protoporphyrin synthesis with the same high efficiency. Under both conditions, only small amounts of uroporphyrin and porphyrins with seven, six, or five carboxyl groups are produced. Moreover, cell-free extracts from wild strains are very stable on prolonged low-temperature storage (≥ 1 year, −30°) and during the incubation in the presence of ALA (24 hr, 30°, pH 7.0). Taking into account these properties of yeast cell-free extracts, we developed a simple procedure for obtaining rapidly and inexpensively pure [14C]coproporphyrin III[4] which was used for extensive coproporphyrinogen III oxidase activity measurements by the sensitive radioassay of Grandchamp and Nordmann.[1]

The method of preparation of the yeast cell-free extract and the complete procedure for preparing pure coproporphyrin III are described in this chapter.

Yeast Culture Conditions

Yeast Strain. Saccharomyces cerevisiae diploid wild strain PS194, kept at 4° on nutrient sloped agar.

Growth Medium. For 1 liter of deionized water: Difco yeast extract, 10 g; $(NH_4)_2SO_4$, 1.2 g; KH_2PO_4, 1 g; glucose, 50 g (sterilized separately, 121°, for 30 min).

Growth Conditions. Three liters of complete growth medium in 6-liter glass spherical flasks is inoculated with 1–2 ml from a small preculture (20 ml growth medium in a 100-ml Erlenmeyer flask inoculated from the

---

[4] H. Chambon and P. Labbe, *Anal. Biochem.* **126**, 81 (1982).

sloped agar with a platinum wire). The cultures are grown at 28–30° with magnetic stirring and continuous injection of sterile air (approximately 1 liter/liter/min) preferably through a sintered glass.

Harvesting. At the end of the log phase, cells are collected and washed once with deionized water by centrifugation (5 min, 3000 *g*). Wet weight yeast cells (20–25 g) per liter of growth medium are obtained.

Preparation of Cell-Free Extract

All conventional methods of yeast cells disruption can be used. When especially designed grinding machines are not available, the inexpensive handshaking grinding technique described by Lang *et al.*[5] can be successfully utilized. Whatever the breaking technique, it is essential to use potassium phosphate buffer instead of sodium phosphate buffer to ensure maximal porphobilinogen synthetase activity. Here the breaking method is described using a widely used cell grinder, the Braun MSK Cell Homogenizer.

Reagents. Potassium phosphate buffer (0.1 *M*) pH 7.6 (ice cold).

Glass Beads. Size 0.45–0.50 mm (B. Braun, Melsungen, West Germany). The beads are used as delivered without further purification.

Braun Homogenizer MSK. Glass flasks of 75 ml capacity. Homogenizer cooled with liquid CO_2.

Centrifuge. Sorvall model RC-2B or RC-5, rotor SS-34, 50 ml polycarbonate or polyethylene tubes.

Procedure (0°). Wet weight yeast cells (20 g) in suspension in 20 ml potassium phosphate buffer are introduced in a precooled (0°) 75-ml flask containing 60 g (32 ml) glass beads. Breaking is carried out for 60 sec at high speed (4000 oscillations/min). The supernatant is recovered and glass beads are washed 3–4 times with small volumes of the buffer, 5 ml each time. Supernatant and washings are centrifuged at 1500–2000 *g* for 10 min. The supernatant is saved and the pellet, suspended in a small volume of buffer, is centrifuged again (10 min, 2000 *g*). The combined supernatants represent the cell-free extract. The pH of the cell-free extract is adjusted near 7.6 with 3 *M* KOH. The cell-free extract is used immediately or frozen (−20°) as 40-ml aliquots in well closed plastic tubes. The cell-free extract fraction usually contains 20–30 mg/ml of protein; 10 g (wet weight) of yeast cells yield 500–600 mg of proteins.

Comments. When kept frozen (−20°), the cell-free extract is very stable (no loss in activity after 2 years). It is important to keep in mind

[5] B. Lang, G. Burger, I. Doxiadis, D. Y. Thomas, W. Bandlow, and F. Kaudewitz, *Anal. Biochem.* **77**, 110 (1977).

that yeast cell-free extracts contain some free ALA, which is quantified as described previously.[6] Endogenous ALA is approximately 50 μM.

[^{14}C]Coproporphyrin III Production

To avoid conversion of coproporphyrinogen III to protoporphyrin IX by the action of coproporphyrinogen oxidase and protoporphyrinogen oxidase, the cell-free extract is incubated with [^{14}C]ALA in anaerobiosis. Anaerobic incubations are maintained for 24 hr by using syringes filled with the incubation mixture to avoid contact with air and by taking advantage that the cell-free extract itself contains large amounts of endogenous reduced substrates that promote anaerobiosis in the incubation mixture within a few seconds.

Reagents

Molar potassium phosphate buffer, pH 7.6
0.1, 3, and 6 M HCl
0.05 M KOH
Ethyl acetate–acetic acid, 3 : 1 (v/v)
$CH_3OH–H_2SO_4$, 95 : 5 (v/v)

Chromatographic Solvent Mixtures

Ethyl acetate–cyclohexane, 40 : 60 (v/v)
Benzene–methyl ethyl ketone, 40 : 3 (v/v)

Chemicals

[^{14}C]ALA, 50 mCi/mmol (Ref. CMM11): CEA, Département des Molécules marquées BP8 91190 Gif/Yvette France
ALA (HCl) and pure porphyrin methyl esters: Porphyrin Products, Logan, UT
Aquasol liquid scintillation mixture: New England Nuclear
Silica gel: Merck (Silica gel 60; 70–230 mesh)
Silica gel plates: Schleicher and Schull (TLCF 1500, Ready plastic sheets)

Miscellaneous

60-ml capacity plastic disposable syringes
Temperature-controlled water bath (30°)
Glass chromatography column (2 cm internal diameter × 40 cm height)

---

[6] R. Labbe-Bois and C. Volland, *Arch. Biochem. Biophys.* **179,** 565 (1977).

Glass chromatography tank
Rotary evaporator
Conventional recording double beam spectrophotometer
Conventional liquid scintillation counting equipment

Procedure. The procedure described below will result in 50 ml (final volume) of the incubation mixture made from 40 ml of cell-free extract.

Incubation. After thawing, the cell-free extract is made 0.25 M with respect to potassium phosphate, pH 7.6, by adding 8.5 ml of M potassium phosphate buffer, pH 7.6, and 1.5 ml H$_2$O, and then 0.25 mM with respect to ALA. Endogenous ALA already present ($\sim$0.05 mM) has to be taken into account in calculations. If final coproporphyrin III specific radio activity ranging from 40 to 20 μCi/μmol is required dilute 1 μmol of [^{14}C]ALA (50 μCi/μmol) 10–20 times with 0.1 M cold ALA.

Anaerobic incubation is carried out in a 60-ml plastic syringe filled with the incubation mixture and closed by a needle inserted in a rubber stopper. The syringe is immersed in a water bath (30°) and incubated for 24 hr in the dark. Under these conditions, colorless porphyrinogens are produced. After incubation, the contents of the syringe are transferred to a beaker kept in crushed ice. Quantitation of synthesized porphyrinogens and chemical extraction are *immediately* undertaken.

Comments. The high molarity (0.25 M) of the buffer pH 7.6, is necessary to prevent strong acidification during the incubation. Incubations longer than 24 hr do not enhance porphyrinogens formation.

Quantitation. The content of porphyrinogens, expressed as coproporphyrinogen, is estimated by spectrophotometry of an acidic extract of the incubation mixture (0.1 in 5.9 ml 3 M HCl) rapidly spun down to eliminate proteins and put under dim daylight to ensure coproporphyrinogen photooxidation into coproporphyrin without porphyrin destruction. From time to time, the spectrum of the acid extract is recorded from 370 to 440 nm. When the absorbance does not increase anymore, that means that all coproporphyrinogen has been converted into coproporphyrin. Coproporphyrin III concentration is roughly calculated using $\varepsilon_{mM} = 489$ for the Soret band near 405 nm.

Extraction, Esterification. To the incubation mixture kept in a beaker, 30 ml of ethyl acetate–acetic acid mixture (3 : 1) is rapidly added. The mixture is continuously and slowly stirred for 30 min under daylight until the mixture becomes red. After 5 min centrifugation at 10,000 g, the upper organic phase is saved and the lower phase is reextracted by 30 ml of the solvent mixture in the same way. The combined organic phases are evaporated to dryness under reduced pressure. To accelerate the process, heating (no more than 40°) is used. The oily residue is submitted to overnight esterification at 4° after careful grinding of the residue with a glass rod in 25 ml of the esterification mixture (CH$_3$OH/H$_2$SO$_4$, 95 : 5).

Porphyrin methyl esters are extracted by 25 ml CHCl$_3$ after addition of 50 ml of water. The CHCl$_3$ lower phase is repeatedly washed with water until clear. If not, a few grains of anhydrous Na$_2$SO$_4$ are added. After standing, the filtered liquid is evaporated to dryness under reduced pressure.

Purification of Coproporphyrin III Methyl Ester. A glass chromatographic column is prepared by pouring onto glass wool plus glass beads a slurry of silica gel in ethyl acetate/cyclohexane (40 : 60). A gel height of 30 cm is sufficient for efficient separation.

Porphyrin esters dissolved in the minimal amount of the above-described mixture are applied to the top of the column. Elution is performed with the same solvent mixture. The first very faint band eluted corresponds to pure protoporphyrin ester, then coproporphyrin ester is eluted followed by the ester of the porphyrin with five carboxyl groups. The fraction corresponding to coproporphyrin methyl ester is collected, evaporated under vacuum, and kept at −20° in the dark.

Chemical and Radiochemical Purity. The purity of the coproporphyrin methyl ester is inexpensively checked by TLC. An aliquot of the purified labeled coproporphyrin ester in CHCl$_3$ is applied to the coated plastic film which is developed for 1 hr in the dark using benzene methyl ethyl–ketone mixture (40 : 3). A mixture of pure cold protoporphyrin and coproporphyrin esters is run simultaneously. After development and drying, the spots are detected under ultraviolet light ($\lambda = 365$ nm). The area of the plate corresponding to the separation of the putative radioactive porphyrins is cut into regular 1-cm-wide pieces which are put into 10-ml Aquasol and counted.

Note. If HPLC equipment is available, the chemical purity of porphyrin methyl esters before and after silica gel chromatography is rapidly checked as shown in Fig. 1 on a column packed with porous microparticulate silica and a solvent mixture made of ethyl acetate–cyclohexane.

Preparation of Coproporphyrin-Free Acid. The dried ester is hydrolyzed overnight in the dark at room temperature with the minimum amount of 6 M HCl. The acid is eliminated under vacuum and the dried residue is dissolved in 0.05 M KOH.

Quantitation of Coproporphyrin-Free Acid Concentration and Measurement of Specific Radioactivity. Coproporphyrin III-free acid concentration is calculated from an aliquot of the alkaline solution diluted in 0.1 M HCl. Identical aliquots were counted in 10 ml Aquasol. Because coproporphyrin III concentration in Aquasol is very low (<1 nmol/ml), no color quenching correction is needed.

Once specific radioactivity is measured, the 0.05 M KOH coproporphyrin III solution is kept frozen in the dark.

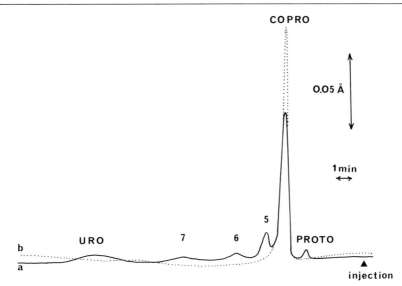

FIG. 1. HPLC separation of porphyrin methyl esters before (a) and after (b) silica gel column separation. Conditions of porphyrins production are described in the table. COPRO, PROTO, URO, coproporphyrin III tetramethyl ester, protoporphyrin IX dimethyl ester, and uroporphyrin octamethyl ester. 5, 6, 7, the methyl esters of the porphyrins with five, six, or seven carboxyl groups. Waters HPLC equipment (6000 A delivery system, U6K injector) and a Waters μPorasil column (0.39 × 30 cm; 10 μm particle size) were used. Solvent mixture was ethyl acetate–cyclohexane, 55:45 (v/v). Flow rate, 1 ml/min; λ = 405 nm.

The table gives some quantitative data concerning a typical preparation of [^{14}C]coproporphyrin III tetramethyl ester.

Concluding Remarks

Samples of coproporphyrin III (1 μmol each) with specific radioactivities ranging from 20 to 50 μCi/μmol are easily and reproducibly prepared within 3 days. One batch is sufficient for 1000 radiochemical coproporphyrinogen oxidase assays as described previously.[7]

This procedure is easily adaptable for [^{14}C]protoporphyrin IX preparation since the presence of oxygen is the only major difference from the above-described procedure. In practice, good aerobiosis is achieved by shaking the 50-ml incubation mixture for 24 hr at 30° *in the dark* in a 500-ml Erlenmeyer flask. A reciprocal shaker working at a frequency avoiding foaming ensures good aerobiosis while addition of chloramphenicol (0.02 mg/ml) prevents possible bacterial proliferation. All subsequent steps of

[7] B. Grandchamp and Y. Nordmann, *Enzyme* **28**, 196 (1982).

[^{14}C]ALA CONVERSION TO [^{14}C]COPROPORPHYRIN III BY A CRUDE
CELL-FREE EXTRACT OF *Saccharomyces cerevisiae*[a]

ALA specific radioactivity (μCi/μmol)	Labeled porphyrins recovered after 24 hr anaerobic incubation, expressed as coproporphyrin (nmol)	ALA converted to porphyrins (%)	Pure labeled coproporphyrin III recovered[b] (nmol)	Labeled ALA recovered as coproporphyrin III (%)	Coproporphyrin III specific activity (μCi/μmol)
5	970	62	770	49	38.9

[a] Initial ALA concentration was 0.25 mM for a 50 ml (final volume) of incubation mixture incubated anaerobically for 24 hr at 30° in the dark in 0.25 M potassium phosphate buffer, pH 7.6.

[b] Only isomer III was found when pure coproporphyrin free acid was checked by reverse HPLC with a Waters μBondapak C$_{18}$ column (0.39 × 30 cm) and a solvent mixture made of 15% acetonitrile/85% 10 mM sodium phosphate buffer, pH 6.85, containing 0.5 mM Na$_4$EDTA (flow rate 1.2 ml/min, λ = 405 nm).

purification are unchanged except that pure protoporphyrin IX dimethyl ester is used for chromatographic controls. ε_{mM} at λ = 409 nm (=262) of protoporphyrin (in CHCl$_3$ or in 2.7 M HCl) is used for calculations. Under these conditions, the final yield in protoporphyrin IX is similar to that of coproporphyrin III.

It should be noted that the ^{14}C-labeled methyl esters of uroporphyrin I and pentacarboxylic porphyrin I can be also prepared from cell-free extracts warmed 5 min at 60° before the 24 hr anaerobic incubation at 30°. However a 2-fold decrease in yield of total porphyrin esters should be expected.

[42] Isotachophoretic Analysis of Porphyrin Precursors: 5-Aminolevulinic Acid Derivatives

By YUZO SHIOI and TSUTOMU SASA

5-Aminolevulinic acid (ALA) and porphobilinogen (PBG) are important precursors in the biosynthesis of tetrapyrroles such as hemes, corrins, bilins, and chlorophylls and are found to be present widely in living tissues. The isolation and quantitative determination of ALA derivatives in urine and other tissues have always been complex and time consuming. This is mainly due to the necessity of prior separation of individual compounds before colorimetric determination because of the interference of other precursors and impurities.[1,2]

Isotachophoresis has recently been applied in the analysis of ALA and its derivatives.[3] Generally, the main advantage of the isotachophoretic method is that it is less affected by the presence of extraneous substances. Furthermore, components of a mixture can be separated simultaneously and detected directly using ionic gradient without elaborate procedures for detection. This method is, therefore, more suitable for the detection of ALA and its derivatives in samples of biological origin, for instance, tissue homogenate or urine.

We describe here the technique of isotachophoresis in separation and quantification of ALA derivatives.

Isotachophoresis

Principle

Isotachophoresis is an electrophoretic technique that can be used for the qualitative analysis of charged molecules in a discontinuous electrolyte system.[4,5] Thus, the method can readily be applied for the simultaneous analysis of chemically analogous compounds.

Standards

1–20 mM ALA
1–20 mM PBG
1–20 mM 4,5-dioxovaleric acid

Equipment and Analytical Procedure

Isotachophoresis was carried out with a Model IP-2A (Shimadzu, Kyoto) equipped with a potential gradient detector. Analytical separation was run in capillary tubes (first stage: Teflon 40 × 1.0 mm i.d.; second stage propylene 100 × 0.5 mm i.d.) at 20°. The electrolyte system is described in the table. To obtain clear separation of ALA derivatives, the pH of the leading electrolyte was critical. The pH of electrolyte was adjusted to each value with 0.1 M Ammediol, $Ba(OH)_2$, or CH_3COOH according to requirements of the table. Before starting, the Model IP-2A

[1] D. Mauzerall and S. Granick, *J. Biol. Chem.* **219**, 435 (1956).
[2] G. Urata and S. Granick, *J. Biol. Chem.* **238**, 811 (1963).
[3] R. Fukae, Y. Shioi, and T. Sasa, *Anal. Biochem.* **133**, 190 (1983).
[4] G. Eriksson, *Anal. Biochem.* **109**, 239 (1980).
[5] J. P. M. Wielders and J. L. M. Muller, *Anal. Biochem.* **103**, 386 (1980).

OPERATIONAL SYSTEM FOR SEPARATION OF ALA DERIVATIVES

Analysis[a]	Leading electrolytes	Terminal electrolytes	Operational current (μA)[b]		Chart speed (cm/min)
			First-stage	Second-stage	
Anion	0.01 M HCl	0.01 M β-alanine			
	Ammediol	Ba(OH)$_2$	200 (7 min)	75	2
	pH 7.2	pH 10.7			
Cation	0.01 M CH$_3$COONa	0.01 M carnitine–Cl			
	CH$_3$COOH		200 (6 min)	100	2
	pH 6.0				

was carefully rinsed and prerun with distilled water. After the system had been filled with electrolyte and stabilized, samples from 1 to 30 μl were injected. Throughout the handling, care was taken to avoid the formation of air bubbles in the capillary tubes. The separations were started at 200 μA and automatically reduced to 75 μA for anion and 100 μA for cation analysis. In anion analysis, the voltage rose from 2 to 10 kV during detection at 75 μA. The chart speed was 2 cm/min. The peaks were monitored with a two-pen recorder, Model R-112 (Shimadzu) programmed to hold a baseline and to differentiate potential gradient of peaks.

Treatment of Data

The potential unit (PU) value, which indicates the identification parameter, was calculated by the following equation: $PU = (P_X - P_L)/(P_T - P_L)$, where P_X, P_T, and P_L are potential gradient values of sample X, terminal ion, and leading ion, respectively. In practice, the potential gradient value was estimated from the height of each zone on the chart paper.

The quantitative parameter, the zone length of the sample, was measured as the distance between the differential peaks of potential gradient in each sample on the chart; the zone length is proportional to the amount of a compound ion in equilibrium with the leading ion.

Separation

The isotachopherograms of ALA derivatives both in anion and cation analyses are shown in Fig. 1. In anion analysis, chemically analogous ALA derivatives could be simultaneously separated with a single run

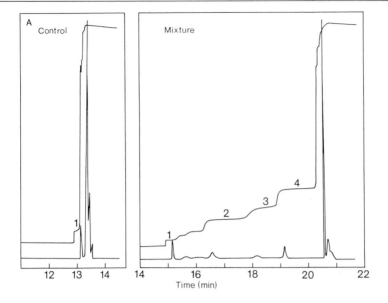

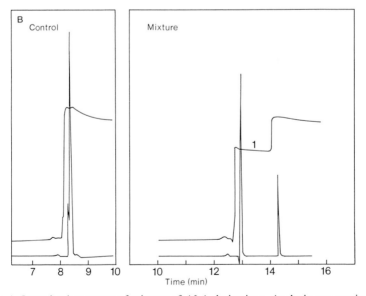

FIG. 1. Isotachopherograms of mixture of ALA derivatives. Analysis was carried out under the conditions described in the table. (A) Anion analysis: peak 1, HCO$_3^-$; peak 2, 4,5-dioxovaleric acid; peak 3, PBG; peak 4, ALA. (B) Cation analysis: peak 1, ALA.

within 25 min (Fig. 1A). The PU values by anion analysis were 0.249 for ALA, 0.178 for PBG, and 0.130 for 4,5-dioxovaleric acid. Whereas in cation analysis, only ALA acts as cation, not 4,5-dioxovaleric acid nor PBG. Consequently only ALA could be detected (Fig. 1B).

Separations of biological samples by isotachophoresis are usually more time consuming than that for a synthetic mixture due to the presence of contaminated ion species. If biological samples contain high concentrations of added ions, e.g., buffer, analysis takes much more time. Thus this method is not suitable for the routine assay of ALA derivatives in the reaction mixture. Biological samples should be prepared without adding extraneous ions. Exceptionally, only ALA was effectively separated by cation analysis without being affected by the anionic constituents because ALA acts as amphoion (cf. Fig. 1B). In the analysis of biological samples, the use of internal standard is recommended to examine the reproducibility and to readily identify the zone.

Quantification and Reproducibility

The calibration curves for ALA derivatives were linear over a wide range of concentrations up to 240 nmol for ALA, 400 nmol for PBG, and 80 nmol for 4,5-dioxovaleric acid. The quantitative analysis required a minimum of 12 nmol for ALA, 35 nmol for PBG, and 4 nmol for 4,5-dioxovaleric acid. The linearity is as good as 0.999 in each correlation coefficient.

The deviations in six successive determinations of the PU value and zone length of these compounds were within 4 and 5%, respectively. In addition, experimental errors of the PU values as determined from more than 20 different concentrations for these individual compounds were within 5%. Injection volume ranging from 0.2 to 35 μl did not affect the reproducibility of PU value and zone length.

Comments

With respect to the separation of levulinic acid, a competitive inhibitor of PBG synthase (formerly ALA dehydratase, EC 4.2.1.24), complete resolution of ALA and levulinic acid (PU value = 0.109) was obtained in anion analysis, but separation of 4,5-dioxovaleric acid and levulinic acid was unsuccessful.

[43] Preparation of Stereospecifically Labeled Porphobilinogens

By M. Akhtar and C. Jones

The stereospecifically labeled samples of porphobilinogen (PBG) have proved to be invaluable for studying the stereochemical aspects of porphyrin (heme and chlorophyll) and corrin (vitamin B_{12}) biosynthesis in cell-free systems.[1-4] In whole-cell systems, the PBG precursors 5-amino-levulinic acid (ALA) and succinic acid have proved more useful, as the cell membrane is not normally permeable to PBG. These syntheses of PBG have been developed in two laboratories, in Southampton and Cambridge, and specific stereochemical problems have required a range of suitably labeled substrates. The resulting syntheses are multistep and frequently complex. In practice, due to problems in chemical synthesis and the relatively ready availability of the enzyme porphobilinogen synthase (ALA dehydratase), most syntheses of labeled PBG have involved building up the molecule enzymatically[5] from ALA. This leads to complex labeling patterns as label is introduced into both halves of the PBG molecule (as it is formed from the dimerisation of ALA) (see Fig. 1).

All three side chains, the acetic acid, the propionic acid, and the aminomethyl, and all four prochiral centers, C-6, C-8, C-9, and C-11, have been stereospecifically labeled. The acetic and propionic acid side chains have generally been labeled together, whereas labeling at the C-11 has also led to labeling at C-2 in the pyrrole ring. These experiments have generally included a ^{14}C reference label to measure rates of incorporation or to assay for loss of tritium by changes in the tritium/carbon ratio. This label will not be mentioned in the text to avoid unnecessary complexity. In most of the work performed in our laboratory, ALA itself has been built up enzymatically from succinyl-CoA and glycine (Fig. 2).

This chapter will deal first with the use of labeled succinic acid derivatives, which introduce label into the acetate and propionate side chains of PBG, and then with the use of labeled glycine, which is incorporated into

[1] M. M. Abboud, P. M. Jordan, and M. Akhtar, *Nouv. J. Chim.* **2,** 419 (1978).
[2] A. R. Battersby, A. L. Gutman, C. J. R. Fookes, H. Gunther, and H. Simon, *J. Chem. Soc., Chem. Commun.* p. 645 (1981).
[3] C. Jones, P. M. Jordan, and M. Akhtar, *J. Chem. Soc., Perkin Trans. 1* p. 2625 (1984).
[4] J. S. Seehra, P. M. Jordan, and M. Akhtar, *Biochem. J.* **209,** 709 (1983).
[5] H. A. Sancovich, A. M. Ferramola, A. M. del C. Battle, and M. Grinstein, this series, Vol. 17, Part A, p. 220.

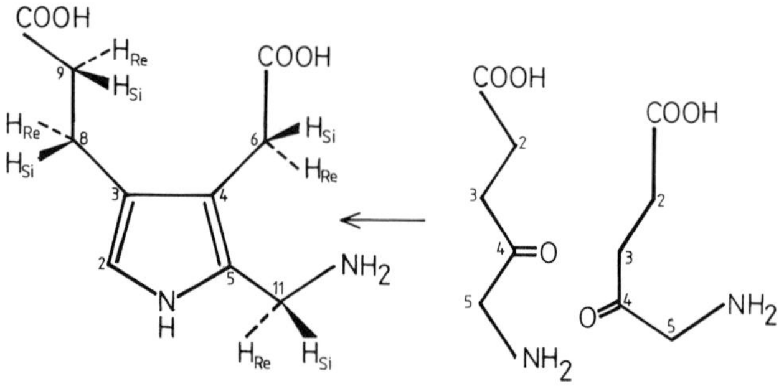

FIG. 1. ALA dehydratase-catalyzed formation of porphobilinogen (PBG, **2**).

the aminomethyl side chain of PBG. For convenience, prototype methods will be described in detail and possible variations considered briefly.

Labeling the Acid Side Chains of PBG Using Enzymatic Approaches

Labeling of the acid side chains was achieved biosynthetically with the stereospecific label originating from succinic acid. The labeled succinic acid was prepared in turn by decarboxylation of a stereospecifically labeled 2-oxoglutarate, prepared by essentially literature methods.[6] As these syntheses pass through a succinic acid intermediate, symmetrical apart from the isotopic labeling, the final labeling is distributed equally between C-2 and C-3 of the ALA, and hence between C-6, C-8, and C-9 of the PBG (Fig. 3).

Preparation of [6R,8R,9R-^{3}H$_3$]PBG (2a)[1,4]

Reagents

Disodium 2-oxoglutarate
Tritiated water
Triethanolamine–HCl, pH 7.4 buffer, 0.5 M
Isocitrate dehydrogenase (from pig heart, Sigma)
NADPH, 6 mM in distilled water (pH to 7 if necessary)
MgCl$_2$·6H$_2$O, 6 mM in distilled water
Hydrogen peroxide, 5%
Dowex chloride 1-X2 (mesh size 50–100, Sigma)
Dowex chloride 1-X8 (mesh size 20–50, BDH)
HCl, 0.1 M in water

[6] G. E. Lienhard and I. A. Rose, *Biochemistry* **3**, 185 (1964).

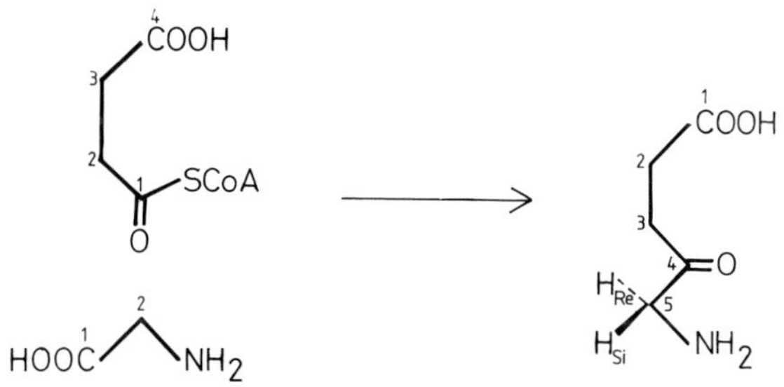

Fig. 2. ALA synthetase-catalyzed formation of ALA.

Disodium 2-oxoglutarate (100 mg = 526 μmol) was dissolved in tritiated water (5 Ci/ml: 1 ml) and sealed *in vacuo*. The tube was autoclaved at 110° for 10 min, cooled, opened, and the tritiated disodium 2-oxoglutarate was isolated by repeated lyophilization. The specific activity of the 2-oxoglutarate was typically 20–25 μCi/μmol.

Disodium 2-Oxo[3R-³H₁]glutarate and [2R-³H₁]Succinate

Disodium 2-oxo[3RS-³H₂]glutarate (60 μmol; 1.42 mCi) was dissolved in a solution prepared from 0.5 M triethanolamine–HCl buffer, pH 7.4 (2 ml), NADPH (1 ml), magnesium chloride (1 ml), and isocitrate dehydrogenase (25 units) made to 10 ml with water. Samples (0.1 ml) were removed at timed intervals and applied onto a column of Dowex chloride (1-X2, 100 mesh; 0.5 × 4 cm) that was washed with water (10 ml), and the radioactivity in the eluent was measured to determine tritium released in the medium. After 50% of the tritium in the 2-oxo[3-³H₂]glutarate had been released into the medium, the reaction was stopped by the addition

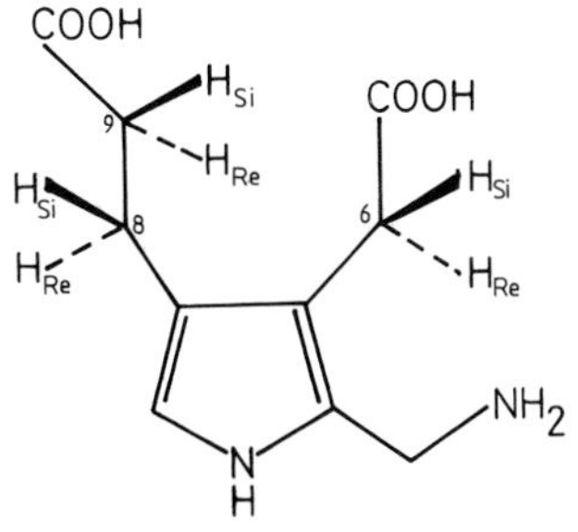

Fig. 3. PBG containing isotopic hydrogen atoms at C-6, C-8, and C-9. As described: [6R,8R,9R-³H₃]PBG(**2a**); 6H$_{Re}$,8H$_{Re}$, and 9H$_{Re}$ = ³H. Variation A: [6S,8S,9S-³H₃]PBG(**2b**); 6H$_{Si}$, 8H$_{Si}$, and 9H$_{Si}$ = ³H. Variation B: (6R,8R,9R)-[6,8,9-³H₂, 6,8,9, ²H₃]PBG(**2c**); all H$_{Re}$ = ³H and H$_{Si}$ = ²H.

of 5% hydrogen peroxide (6.6 ml). The $[2R\text{-}^3H_1]$succinate was purified by ion-exchange chromatography on a column of Dowex chloride (1-X8, 20–50 mesh; 2.5 × 10 cm). After washing the column with water (50 ml) the succinic acid was eluted with 0.1 M hydrochloric acid (approximately 20 ml) and recovered by evaporation under reduced pressure. Yield was 50 μmol (85%), specific radioactivity 11.84 μCi/μmol. This material was diluted with carrier succinic acid to give specific radioactivity as desired.

$[2R,3R\text{-}^3H_2]$Succinyl-CoA

Reagents

Dicyclohexylcarbodiimide
Acetone (dry; distilled over K_2CO_3)
Diethyl ether (freshly dried; distilled over LiAlH$_4$)
Coenzyme A (trilithium salt)
NaHCO$_3$

The labeled succinic acid (200 μmol) obtained above was dissolved in anhydrous acetone (5 ml) and dicyclohexylcarbodiimide (200 μmol) added. The reaction was stirred at room temperature for 30 min and further carbodiimide (20 μmol) added. After a further 30 min the dicyclo-hexylurea was removed by filtration and the acetone removed under reduced pressure. The succinic anhydride was extracted from the residue with dry ether (3 × 10 ml) and the ether removed under reduced pressure to give the succinic anhydride (160–170 μmol).

Succinyl-CoA was prepared immediately prior to use by dissolving the succinic anhydride (90 μmol) in a solution of sodium bicarbonate (4.5 mg) and lithium coenzyme A (10 mg) in 2.5 ml of water.[7] The solution was stirred at 0° for 30 min. Complete reaction of the thiol was confirmed by treatment of 5 μl of the solution with 500 μl of a 1 mM solution of 5,5'-dithiobis(2-nitrobenzoic acid) (for details, see Ref. 8).

$[6R,8R,9R\text{-}^3H_3]$PBG (2a; Fig. 3)

Reagents

Glycine, 400 mM in distilled water
Tris–HCl buffer, pH 6.8, 50 mM
Pyridoxal phosphate 1 mM in distilled water
ALA synthetase from $R.$ $spheroides$[8]
ALA dehydratase from bovine liver[9]

[7] E. J. Simon and D. Shemin, $J.$ $Am.$ $Chem.$ $Soc.$ **75**, 2520 (1953).
[8] P. M. Jordan and A. Laghai-Newton, this volume [52].
[9] P. M. Jordan and J. S. Seehra, this volume [51].

The following stock solutions were mixed: glycine (2 ml), buffer (1 ml), pyridoxal phosphate (1 ml), ALA synthetase[8] (200 units), and ALA dehydratase[9] (350 units) and the solution made up to 7.5 ml with distilled water. To this was added the succinyl-CoA solution (2.5 ml, see above) and the mixture incubated at 37° under nitrogen until PBG formation was complete (usually 30–60 min) as judged by the reaction with *p*-dimethylaminobenzaldehyde.[10] The mixture was adjusted to pH 7.2 and poured onto a column of Dowex acetate (2-X8, 400 mesh; 2 × 10 cm) and the resin washed with distilled water (40 ml) to remove ALA. The washings were checked for the absence of PBG. The PBG was eluted with ice cold 1 *M* acetic acid (20 ml), the solvent removed by rotary evaporation under high vacuum below 30° and the PBG further purified by thin layer chromatography on microcrystalline cellulose plates (1 mm thick, 20 × 10 cm), developed in 1-butanol : acetic acid : water, 63 : 27 : 10. The PBG was located by radioactive scanning or by spraying a portion of the plate with the modified Ehrlich reagent,[10] and the band eluted with dilute aqueous ammonia and lyophilized. The sample of [6*R*,8*R*,9*R*-³H₃]PBG was stored desiccated at −18°.

Crystallization of PBG

PBG was dissolved in 0.5 *M* aqueous ammonia (0.2 ml) and crystallized by addition of 0.5 *M* aqueous acetic acid until the pH was 5.0. The mixture was stood on ice for 3 hr. The crystals were removed by centrifugation, washed with methanol and dried under high vacuum.

Variations

[6S,8S,9S-³H₃]PBG (**2b**, *Fig. 3*). The use of unlabeled 2-oxoglutarate and an isocitrate dehydrogenase enzyme system in tritiated water (typically 2.5 Ci/ml) in a sealed tube leads to preparation of the [2*S*-³H₁]succinate, and hence [6*S*,8*S*,9*S*-³H₃]PBG.[4]

2-Oxoglutarate (68.5 µmol) was dissolved in triethanolamine/HCl buffer pH 7.4 (0.5 *M*; 0.2 ml) containing NADPH (30 m*M*) and isocitrate dehydrogenase (12 units; 0.8 ml). Tritiated water (1 ml; 5 Ci) was added and the tube sealed and incubated at 30° for 3 hr. Reaction was terminated by addition of hydrogen peroxide (1.5 ml: 30%) and after 30 min the succinate was purified as before on a column of Dowex chloride (1-X8, 20–50 mesh) and used for the preparation of PBG. In this preparation the adsorption of succinate on Dowex is performed in a well-ventilated fume cupboard and the washings are collected carefully but rapidly. The recov-

---

[10] D. Mauzerall and S. Granick, *J. Biol. Chem.* **219**, 435 (1956).

ered tritiated washings are either stored for disposal or used for recovery of tritiated water by freeze drying.

[6R,8R,9R-³H₃, 6S,8S,9S-²H₃]PBG (**2a**, *Fig. 3*). The use of 2-oxo-[3RS-³H₂]glutarate and an isocitrate dehydrogenase enzyme system exchanged into deuterated water leads to the preparation of the deuterated–tritiated species, [2R-³H₁, 2S-²H₁]succinate,[11] which was used for the preparation of [6R,8R,9R-³H₃, 6S,8S,9S-²H₃]PBG.[12]

Ethanolamine/HCl buffer, pH 7.4 (0.5 M; 10 ml) was freeze dried and redissolved in deuterated water (99% D), refreeze dried, and again made up in deuterated water. No allowances were made for the differences in pH and pD. Isocitrate dehydrogenase (50 units, 1 ml) was dialyzed against 3 × 10 ml of deuterated water at 4° for a total of 6 hr, in a sealed tube. The enzyme was used immediately as its stability under these conditions was suspect. The equilibration was run in the presence of 0.6 mM MgCl₂ and 0.6 mM NADPH as described above (stoppered under nitrogen) with monitoring of the labilization of tritium. Succinate was isolated and incorporated into [6S,8S,9S-³H₃, 6R,8R,9R-²H₃]PBG essentially as described above.

The bacterial ALA dehydratase from *R. spheroides* can also be used in PBG synthesis in a two-step procedure with isolation of the intermediate ALA.[1]

The failure of the above method to allow regioselective labeling was solved by the Cambridge and Munich groups, which synthesized[2] a stereospecifically labeled, succinic acid derivative monomethyl[2S-³H₁]succinate, which was then converted to ALA by known reactions (see Fig. 4). In principle, this later species may be converted to PBG using ALA dehydratase. The Cambridge group has also described a method for labeling the propionate side chains in which the relative, rather than absolute, stereochemistry at C-8 and C-9 is controlled.[13] Full experimental details of these syntheses are not yet available.[2,13]

Aminomethyl Side Chain Labeling[3]

During the course of our studies on the mechanism of action of ALA synthetase, it was shown[14] that the label from [2-³H₂]glycine is stereospecifically incorporated into C-5 of ALA (Fig. 2, in structure 1 H$_{Si}$ = ³H).

[11] G. F. Barnard and M. Akhtar, *J. Chem. Soc., Perkin Trans. 1* p. 2354 (1979).
[12] C. Jones and M. Akhtar, unpublished experiments (1977).
[13] A. R. Battersby, E. McDonald, H. K. W. Wurziger, and K. J. James, *J. Chem. Soc., Chem. Commun.* p. 493 (1979).
[14] M. M. Abboud, P. M. Jordan, and M. Akhtar, *J. Chem. Soc., Chem. Commun.* p. 643 (1974).

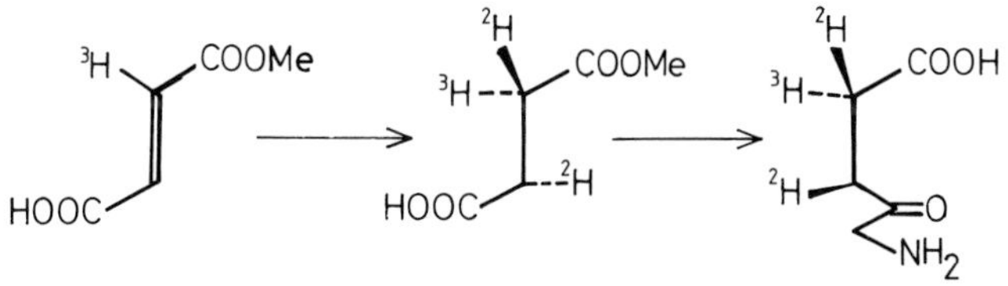

FIG. 4. Stereospecific introduction of hydrogen atoms at C-2 of ALA. In the first step, stereospecific reduction of the tritiated substrate in ^2H_2O using 2-enoate reductase gives methyl succinate which is then converted to $(2S)$-[2-^{3}H,2-^{2}H]ALA chemically.

The stereospecifically labeled ALA thus formed may be trapped to give PBG stereospecifically labeled in the aminomethyl sidechain.[3,15] The preparative procedure is essentially similar to the coupled enzyme system described above, except that incubation conditions were chosen[3] to maximize the radiochemical incorporation of glycine rather than succinyl CoA, by reducing the glycine concentration from about 130 to 3 mM (K_m about 12 mM), and the reaction was performed in phosphate buffer rather than the Tris buffer. The complications arising from the nonenzymatic racemisation of the chiral center at C-5 of ALA catalyzed by pyridoxal phosphate (required as cofactor by the ALA synthetase) were avoided by using as low a concentration of pyridoxal phosphate as possible. Furthermore the system was loaded with ALA dehydratase to reduce the life of ALA in solution. Careful control of the thiol concentration was also felt to be important (for further details, see Ref. 16). It should be noted that this method has only ever been carried out on a small scale and is not suitable for deuterium labeling (Fig. 5).

Reagents

[2RS-^{3}H$_2$]Glycine (3 μmol, freeze-dried powder)
Pyridoxal phosphate, 60 μM in distilled water
Tris–HCl buffer, 10 mM
Potassium phosphate buffer, pH 6.8, 10 mM prepared in glycerol : water (2 : 8) which was also 2 mM in 2-mercaptoethanol
Potassium phosphate buffer, pH 6.8, 10 mM which was also 20 mM in 2-mercaptoethanol
Succinyl-CoA (3.75 μmol in 150 μl prepared as in Ref. 8)
ALA synthetase and ALA dehydratase
Dowex acetate, 2-X8 (400 mesh)

[15] C. Jones, P. M. Jordan, A. G. Chaudhry, and M. Akhtar, *J. Chem. Soc., Chem. Commun.* p. 96 (1979).
[16] For further details, see C. Jones, Ph.D. Thesis, University of Southampton (1979).

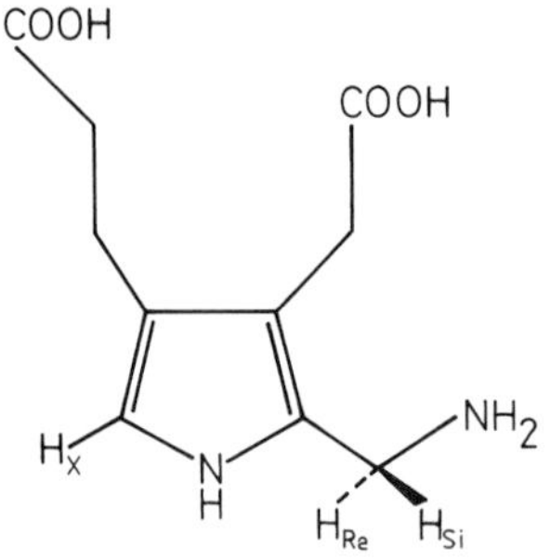

FIG. 5. PBG stereospecifically labeled at the aminomethyl group.

ALA synthetase and ALA dehydratase were prepared as described in Refs. 8 and 9. Succinyl-CoA was prepared from unlabeled succinic anhydride as described in Ref. 8. ALA synthetase[9] was dialyzed for 3 hr against 100 ml of the potassium phosphate buffer, pH 6.8 (10 mM) containing glycerol (20%) and 2-mercaptoethanol (2 mM) before use. ALA dehydratase[8] was dissolved in a small volume of 10 mM potassium phosphate buffer containing 20 mM 2-mercaptoethanol and dialyzed against this buffer for 3 hr prior to use.

The incubation mixture was prepared from [2-^{3}H$_2$]glycine (3 μmol), pyridoxal phosphate (0.1 ml), Tris–HCl buffer (0.1 ml), ALA synthetase (20 units), and ALA dehydratase (20 units) made up to 1 ml with distilled water. The final phosphate and thiol concentrations were estimated at 8 and 7 mM, respectively. The solution was warmed to 37° and reaction started by the addition of succinyl-CoA (1.25 μmol in 50 μl; for preparation see above): further aliquots were added after 10 and 20 min. After 30 min reaction was stopped by applying it to a column of Dowex acetate (2 × 8, 400 mesh; 0.5 × 6 cm) and the PBG isolated and further purified by TLC as above. The PBG from the ion exchange column was typically 98% radiochemically pure (impurities were ALA and glycine) and could often be used without further purification. Isolated radiochemical yield of the [2,11S-^{3}H$_2$]PBG was typically 30%. The enantiomeric excess of the product was about 84% as assayed using the protocol described in Ref. 3 (also see Ref. 17).

Variations

[11S-^{3}H$_1$]PBG. Dilute acid treatment of the PBG leads to exchange of the H-2 tritium in the doubly labeled PBG above. The labeled PBG (1 mg)

[17] L. Schirch, this series, Vol. 17, Part B, p. 335; P. M. Jordan and M. Akhtar, *Biochem. J.* **116**, 277 (1970).

was dissolved in 0.15 M hydrochloric acid (100 μl) and allowed to stand at room temperature for 1 hr, freeze dried, and the [11S-^{3}H$_1$]PBG purified by TLC on cellulose.

Acknowledgments

We thank SERC for their support of our work in the area of porphyrin biosynthesis.

[44] High-Performance Liquid Chromatography of Uroporphyrin and Coproporphyrin Isomers

By C. K. LIM and T. J. PETERS

Uroporphyrinogen III, the universal precursor of heme, chlorophylls, and vitamin B$_{12}$, is derived from porphobilinogen (PBG) by the sequential action of porphobilinogen deaminase (hydroxymethylbilane synthase, EC 4.3.1.8) which catalyzes the formation of hydroxymethylbilane (HMB) from four molecules of PBG and uroporphyrinogen III synthase (EC 4.2.1.75) which converts HMB into uroporphyrinogen III.[1] A small amount uroporphyrinogen I is also formed by the spontaneous chemical ring-closure of HMB. The stepwise decarboxylation of the acetic acid groups of uroporphyrinogens I and III form the corresponding coproporphyrinogen isomers[2] and is catalyzed by uroporphyrinogen decarboxylase (EC 4.1.1.37). Porphyrins are formed by oxidation of the porphyrinogens.

High-performance liquid chromatography (HPLC) has been shown to be the most effective technique for the separation of uro- and coproporphyrin isomers[3] (Fig. 1). The methyl esters of the isomers can be resolved on silica[4-6] and aminopropyl silica[7] or the porphyrin free acids can be

[1] A. R. Battersby, C. J. R. Fookes, K. E. Gustafson-Potter, E. McDonald, and G. W. J. Matcham, *J. Chem. Soc., Perkin Trans. 1*, p. 2427 (1982).
[2] A. H. Jackson, H. A. Sancovich, A. M. Ferramola, N. Evans, D. E. Games, S. A. Matlin, G. H. Elder, and S. G. Smith, *Philos. Trans. R. Soc. London, Ser. B* **273**, 191 (1976).
[3] C. K. Lim, J. M. Rideout, and D. J. Wright, *J. Chromatogr.* **282**, 629 (1983).
[4] H. Nordlov, P. M. Jordan, G. Burton, and A. I. Scot, *J. Chromatogr.* **190**, 221 (1980).
[5] A. H. Jackson, K. R. N. Rao, and S. G. Smith, *Biochem. J.* **203**, 515 (1982).
[6] A. H. Jackson, K. R. N. Rao, and S. G. Smith, *Biochem. J.* **207**, 599 (1982).
[7] I. C. Walker, M. T. Gilbert, and K. Stubbs, *J. Chromatogr.* **202**, 491 (1980).

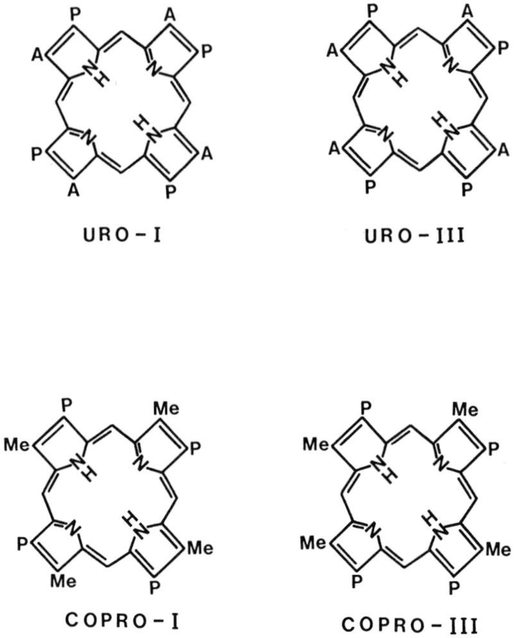

FIG. 1. Structures of uroporphyrin and coproporphyrin isomers. A, CH_2COOH; P, CH_2 CH_2COOH; Me, CH_3.

separated by reverse-phase[8–10] and reverse-phase ion-pair chromatography.[11] The separation of porphyrin free acids has the advantage of avoiding the time-consuming derivatization and extraction steps. Of the HPLC systems reported for the separation of uro- and coproporphyrin isomers, reverse-phase chromatography is superior in terms of speed, column efficiency, and resolution. A reverse-phase system with acetonitrile–ammonium acetate buffer as eluents on octadecyl silica (ODS, C_{18}) is described here.

Materials and Reagents

Uro- and coproporphyrins I and III were obtained from Sigma Chemical Co., Poole, Dorset, U.K., ammonium acetate and glacial acetic acid

[8] E. Englert, Jr., A. W. Wayne, E. E. Wales, Jr., and R. Straight, *HRC CC, J. High Resolut. Chromatogr. Chromatogr. Commun* **2**, 570 (1979).
[9] Y. Hayashi and M. Udagawa, *Talanta* **30**, 368 (1983).
[10] C. K. Lim, J. M. Rideout, and D. J. Wright, *Biochem. J.* **211**, 435 (1983).
[11] H. D. Meyer, K. Jacob, W. Vogt, and M. Knedel, *J. Chromatogr.* **199**, 339 (1980).

(AnalaR grade) from BDH Chemicals, Poole, Dorset, U.K., and acetonitrile (HPLC grade) from Rathburn Chemicals, Walkerburn, Peebleshire, U.K.

HPLC Apparatus

Isocratic separation was performed on a Pye Unicam (Cambridge, U.K.) liquid chromatograph with a variable wavelength UV-visible detector set at 404 nm. Gradient elution was carried out on a Varian Associates (Walnut Creek, CA) model 5000 liquid chromatograph with a UV-100 variable wavelength detector set at 404 nm. Sample loading (100 μl loop) was via a Rheodyne 7125 loop-valve injector.

Columns and Mobile Phases

The columns used were 25 cm × 5 mm ODS-Hypersil (Sandon Southern, Runcorn, Cheshire, U.K.), 10 cm × 5 mm ODS-Spherisorb (PhaseSep, Queensferry, Clwyd, U.K.), and 30 cm × 4.6 mm μBondapak C_{18} (Waters Associates, Milford, MA). The column packings were silica chemically bonded with octadecylsilyl groups.

The mobile phases were various proportions of acetonitrile in 1 M ammonium acetate buffer, pH 5.16. The buffer was prepared by dissolving ammonium acetate (154 g) in water (1900 ml). The pH was adjusted to 5.16 with glacial acetic acid (approximately 56 ml) and volume made up to 2 liters with water.

Uroporphyrin isomers were separated on ODS-Hypersil eluted with a mobile phase composed of 13% acetonitrile and 87% (v/v) 1 M ammonium acetate buffer, pH 5.16, or an ODS-Spherisorb or μBondapak C_{18} with 14.3% acetonitrile and 85.7% (v/v) buffer as eluent.

Coproporphyrin isomers were chromatographed on ODS-Hypersil or μBondapak C_{18} with 30% (v/v) acetonitrile in 1 M ammonium acetate buffer, pH 5.16, as the mobile phase and on ODS-Spherisorb eluted with 31% acetonitrile (v/v) in the buffer. The mobile phase flow rate was 1 ml/min.

HPLC of Uroporphyrin I and III

Figure 2a and b compares separation of uroporphyrin I and III isomers on ODS-Spherisorb and ODS-Hypersil, respectively. The difference in column selectivity is clearly demonstrated. ODS-Spherisorb was more retentive and required a higher organic modifier content for elution in a convenient time. μBondapak C_{18} behaved similarly to ODS-Spherisorb.

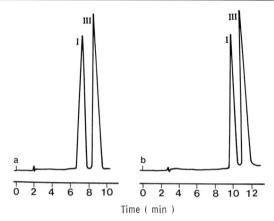

Fig. 2. Separation of uroporphyrin I and III isomers. (a) ODS-Spherisorb (10 cm × 5 mm); eluent, 14.3% (v/v) acetonitrile in 1 M ammonium acetate buffer, pH 5.16; flow rate, 1 ml/min. (b) ODS-Hypersil (25 cm × 5 mm); eluent, 13% (v/v) acetonitrile in 1 M ammonium acetate buffer, pH 5.16; flow rate, 1 ml/min.

Adequate separation of isomers can normally be obtained with 10 cm columns.

The separation is particularly sensitive to changes in pH and ammonium acetate buffer concentration.[12] Increasing the pH and/or the molar concentration of ammonium acetate buffer decreased the retention time. Optimum resolution was with 1 M ammonium acetate at pH 5.15–5.16.

The acetonitrile content in the mobile phase also significantly influenced the retention of uroporphyrins. A change of ±2% (v/v) caused excessive retention or no resolution. The optimum acetonitrile content was 12.5–13% (v/v) for separation on ODS-Hypersil and 14.3–14.5% (v/v) for chromatography on ODS-Spherisorb and μBondapak C_{18}.

Methanol alone cannot be used as an organic modifier for the separation of uroporphyrins. The extensive hydrogen bonding between the uroporphyrin carboxylic acid groups and the methanol "sorbed" onto the C_{18} stationary phase surface causing complete retention of the porphyrins.[13] Methanol containing 10% (v/v) acetonitrile[13] or 5% tetrahydrofuran,[12] however, was satisfactory as the organic modifier.

HPLC of Coproporphyrin Isomers

Figure 3a and b shows the separation of coproporphyrin I and III isomers on μBondapak C_{18} and ODS-Hypersil, respectively. The effects

[12] J. M. Rideout, D. J. Wright, and C. K. Lim, *J. Liq. Chromatogr.* **6**, 383 (1983).
[13] C. K. Lim and T. J. Peters, *Clin. Chim. Acta* **139**, 55 (1984).

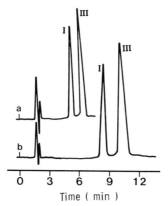

FIG. 3. Separation of coproporphyrin I and III isomers. (a) μBondapak C_{18} (30 cm × 4.6 mm) and (b) ODS-Hypersil (25 cm × 5 mm); eluent 30% (v/v) acetonitrile in 1 M ammonium acetate buffer, pH 5.16; flow rate, 1 ml/min.

of pH and buffer concentration on the retention and resolution were similar to those observed for uroporphyrins. The organic modifier concentration, however, was less critical and, depending on the separation required, 29–31% (v/v) acetonitrile in 1 M ammonium acetate, pH 5.16, was suitable for separation on all C_{18} reverse-phase columns.

The separation of coproporphyrin I and III from the II and IV isomers has been achieved on ODS-Hypersil eluted with 26% (v/v) acetonitrile in 1 M ammonium acetate, pH 5.16.[14]

Simultaneous Separation of Uroporphyrin and Coproporphyrin Isomers

The simultaneous separation of uro- and coproporphyrin isomers is best performed by gradient elution. A typical program was a 15 min linear gradient elution from 13% (v/v) acetonitrile (in 1 M ammonium acetate, pH 5.16) to 30% acetonitrile, followed by isocratic elution at 30% acetonitrile for a further 15 min (Fig. 4). The system also separated porphyrins with intermediate numbers of carboxylic acid groups between uro- and coproporphyrins.[10]

Advantages of the Ammonium Acetate Buffer System

Porphyrins are usually dissolved or extracted into acidic solutions. These solutions can be injected directly onto the column without causing

[14] D. J. Wright, J. M. Rideout, and C. K. Lim, *Biochem. J.* **209,** 553 (1983).

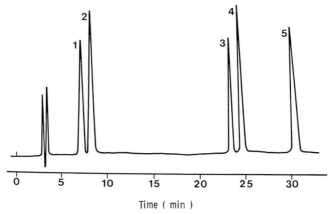

FIG. 4. Separation of uroporphyrin and coproporphyrin isomers. Column, ODS-Hypersil (25 cm × 5 mm); gradient mixtures, 1 *M* ammonium acetate buffer, pH 5.16 (a) and acetonitrile (b); elution, 15 min linear gradient from 13% (v/v) b to 30% b then 15 min isocratic elution at 30% B; flow rate, 1 ml/min. Peaks: 1, uroporphyrin I; 2, uroporphyrin III; 3, coproporphyrin I; 4, coproporphyrin III; 5, isocoproporphyrin.

damage to the stationary phase surface which is well protected by the high buffering capacity of 1 *M* ammonium acetate. The use of such a high molarity is difficult with other buffer systems because of their precipitation by acetonitrile. Ammonium acetate (1 *M*) is fully soluble in 40% (v/v) of acetonitrile which is well above that required for the separation of porphyrin isomers. The solubility can be further improved by inclusion of methanol in the mobile phase.[13]

Ammonium acetate is a good masking agent for residual accessible silanol groups on reverse-phase packing materials and is also able to accelerate the rates of proton equilibrium in the chromatographic process.[15] These properties are probably responsible for the better chromatography obtained compared to other mobile phase systems.

Preparative Separation

The systems described above are suitable for the preparative isolation of pure isomers. Ammonium acetate, however, contained metallic impurities which on prolonged contact with porphyrins forms metalloporphyrins. This problem can be overcome by adding Na_2EDTA (10 m*M*) to the mobile phase. EDTA does not alter the retention or resolution of the porphyrins.[10] Alternatively, the separated porphyrins may be collected

[15] C. K. Lim and T. J. Peters, *J. Chromatogr.* **316,** 397 (1984).

into flasks containing a little solid EDTA. Metalloporphyrins were not formed during the relatively short chromatographic separation.

Reproducibility and Recovery

The separation reproducibility was usually excellent for both uro- and coproporphyrin isomers when the same column, or columns packed with the same batch of packing material, were used. Interbatch reproducibility, however, was not as good and retention time fluctuation of up to 10% has been observed. Uroporphyrin separation was in general more sensitive to column packing variation than coproporphyrin separations. One batch of ODS-Hypersil was found to be unsuitable for separating uroporphyrin isomers while effective resolution of coproporphyrin isomers was maintained. The recoveries of uro- and coproporphyrin from the column were similar, being $98 \pm 1.5\%$ for replicated ($N = 20$) injection of 15 ng of porphyrin.

Applications

The separation of uroporphyrin isomers is especially important for assaying the enzymes hydroxymethylbilane synthase and uroporphyrino-gen III synthase.[16] The method is also useful for the rapid analysis of uro- and coproporphyrin isomers in biological and clinical materials.[10,13] Type III porphyrin standards, usually contaminated with about 5% of type 1 isomers, can be easily purified by the HPLC systems.

[16] D. J. Wright and C. K. Lim, *Biochem. J.* **213**, 85 (1983).

[45] High-Performance Liquid Chromatographic Separation of the Bilirubin Isomers

By C. K. LIM and T. J. PETERS

Bilirubin, the most important bile pigment, exists in nature as the $IX\alpha$ isomer (Fig. 1). Commercially available bilirubin preparations, however, invariably contain the $III\alpha$ and $XIII\alpha$ isomers (Fig. 1) as impurities. Thin-layer chromatography has been widely employed for the analysis and purification of bilirubin isomers, but HPLC, with its superior speed and resolving power, is being increasingly used.

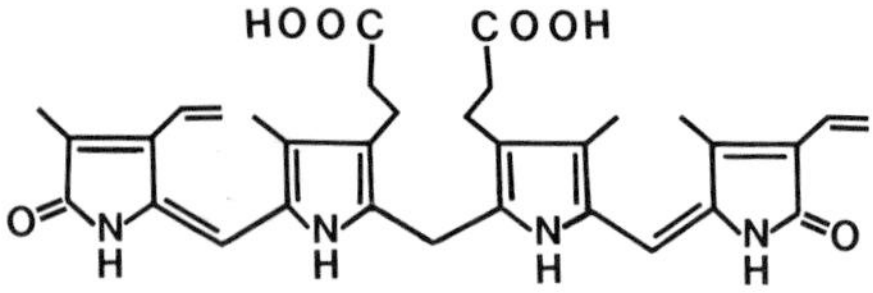

FIG. 1. Structures of bilirubin IXα, IIIα, and XIIIα isomers.

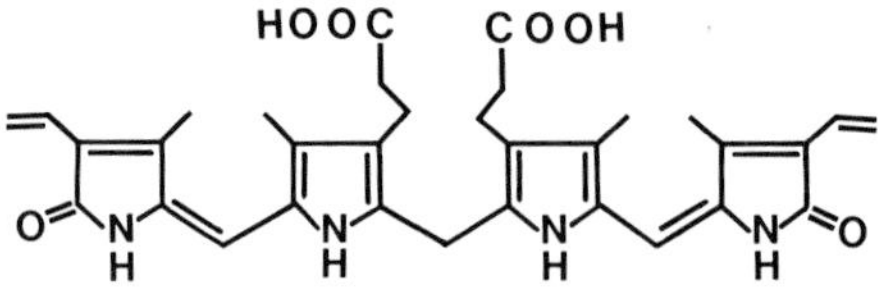

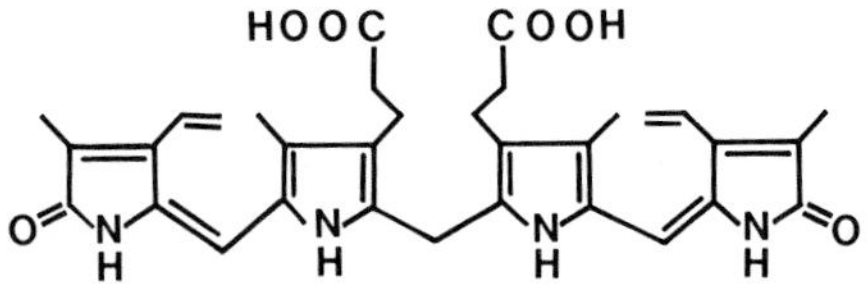

Bilirubin isomers have been separated by reverse-phase[1,2] and reverse-phase ion-pair chromatography.[3] The method described here is a modification of the reverse-phase system[2] developed by us with better control of retention times and improved resolution.

Materials and Reagents

Bilirubin was from Sigma Chemical Co., Poole, Dorset, U.K., ammonium acetate, glacial acetic acid, and dimethyl sulfoxide (DMSO) (AnalaR grade) from BDH Chemicals, Poole, Dorset, U.K., and acetonitrile (HPLC grade) from Rathburn Chemicals, Walkerburn, Peebleshire, U.K.

[1] T. A. Wooldridge and D. A. Lightner, *J. Liq. Chromatogr.* **1,** 653 (1978).
[2] C. K. Lim, R. V.A. Bull, and J. M. Rideout, *J. Chromatogr.* **204,** 219 (1981).
[3] S. Onishi, N. Kawade, S. Itoh, K. Isobe, and S. Sugiyama, *Biochem. J.* **190,** 527 (1980).

HPLC Apparatus

A Pye Unicam (Cambridge, U.K.) liquid chromatograph was used with a Pye Unicam variable-wavelength UV-visible detector set at 450 nm. Injection was via a Rheodyne 7125 injector fitted with a 100 μl loop.

Columns and Mobile Phases

The columns used were 25 cm × 5 mm ODS-Hypersil (Shandon Southern, Runcorn, Cheshire, U.K.) and 30 cm × 4.6 mm μBondapak C_{18} (Waters Associates, Milford, MA). The former is 5 μm spherical silica and the latter 10 μm irregular shape silica, chemically bonded with octadecylsilyl groups.

The mobile phases were 0.1, 0.25, and 0.5 M ammonium acetate–acetic acid buffers at pH 4.6, 5.16, and 6.8 mixed with equal volumes of acetonitrile and DMSO. The solvents were degassed in an ultrasonic bath for 10 min before used.

Sample Handling

Bilirubin must be handled under dim light. It was dissolved in DMSO containing a trace of EDTA and the solution was thoroughly purged with helium and kept in the dark. Bilirubin in biological fluids (0.5 ml) was extracted into chloroform in the presence of ascorbic acid (20 mg) and a trace of disodium EDTA (~0.1 mg). The chloroform was evaporated to dryness at 30° under nitrogen. The residue was redissolved in DMSO as described above for HPLC analysis.

HPLC of Bilirubin Isomers

The separation of bilirubin isomers on a μBondapak C_{18} and an ODS-Hypersil column is shown in Fig. 2a and b, respectively. The eluent was a mixture of 0.1 M ammonium acetate buffer, pH 4.6, acetonitrile, and DMSO in equal volumes. The mobile phase flow rate was 1 ml/min.

The longer retention and better separation achieved on ODS-Hypersil compared to that of μBondapak C_{18} (10 μm) is due to the smaller particle size packing material (5 μm). The retention and resolution, however, can be precisely controlled by manipulating the pH and concentration of ammonium acetate in the buffer and/or the organic modifier content in the mobile phase. This is an improvement on the previous method[2] where retention could only be controlled by adjusting the organic modifier concentration.

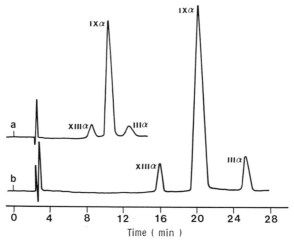

FIG. 2. Separation of bilirubin IIIα, IXα, and XIIIα isomers on (a) μBondapak C₁₈ and (b) ODS-Hypersil. Eluent, 0.1 *M* ammonium acetate, pH 4.6 : acetonitrile : DMSO (1 : 1 : 1 by vol); flow rate, 1 ml/min; detector, UV-visible 450 nm.

Solubility Consideration

The solubility of a compound in the mobile phase is an important factor in developing a HPLC separation.

Bilirubin is usually dissolved in chloroform or DMSO. DMSO was therefore chosen as one of the organic modifiers to improve the solubility of bilirubin in the mobile phase. The column back pressure generated by mixing DMSO with aqueous buffer is relatively high due to the viscosity of DMSO. Acetonitrile was thus introduced as part of the ternary mobile phase system as it considerably lowered the column pressure and also provided an extra parameter for controlling the separations.

The retention of bilirubin isomers can be effectively controlled by the relative proportions of acetonitrile, DMSO, and aqueous buffer. Retention and resolution increased with increasing DMSO and/or ammonium acetate buffer contents. Figure 3 shows the degree of separation achieved when the DMSO content in the mobile phase was increased. The mobile phase was acetonitrile : DMSO : 0.1 *M* ammonium acetate, pH 4.6 (40 : 60 : 50 by vol). The system is well suited for the preparative separation of bilirubin isomers.

Effect of pH on Retention and Resolution

The pH of ammonium acetate buffer significantly affected the retention and resolution of bilirubin isomers. A plot of the capacity ratios ($κ'$) against pH shows that $κ'$ decreased as pH increased (Fig. 4), and resolu-

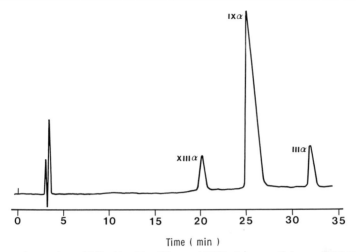

FIG. 3. Separation of bilirubin IIIα, IXα, and XIIIα isomers. Column, ODS-Hypersil; eluent, 0.1 M ammonium acetate, pH 4.6 : acetonitrile : DMSO (50 : 40 : 60 by vol); flow rate, 1 ml/min; detector, UV-visible 450 nm.

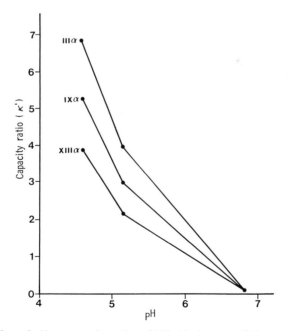

FIG. 4. Effect of pH on capacity ratios of bilirubin isomers. Column, ODS-Hypersil; eluent, 0.1 M ammonium acetate, pH 4.6, 5.16, and 6.8 : acetonitrile : DMSO (1 : 1 : 1 by vol).

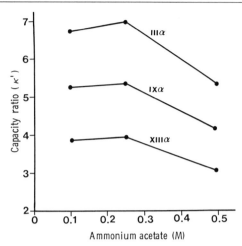

FIG. 5. Relationship between capacity ratios of bilirubin isomers and ammonium acetate buffer concentration. Column, ODS-Hypersil; eluent, 0.1, 0.25, and 0.5 M ammonium acetate, pH 4.6 : acetonitrile : DMSO (1 : 1 : 1 by vol).

tion of isomers was lost at pH 6.8. The optimum pH was between 4.5 and 5.2.

Effect of Buffer Concentration on Retention and Resolution

The effect of ammonium acetate buffer concentration on the retention and resolution of bilirubin isomers is shown in Fig. 5. There was a slight increased in κ' as the buffer concentration was increased from 0.1 to 0.25 M and a significant decrease in κ' when the concentration was increased to 0.5 M. The resolution of the isomers, however, was maintained at all buffer concentrations. The optimum buffer was between 0.1 and 0.3 M.

Reproducibility and Recovery

The reproducibility and recovery of the method were both highly acceptable. The recovery from 15 replicated injections of 10 μg of bilirubin was 95 ± 2.5%. There was no significant variation in retention time (<0.2 min).

Applications

The HPLC system is suitable for the analytical and preparative separation of bilirubin isomers. It is also applicable to the analysis of photobilirubins and bilirubin conjugates after transesterification to bilirubin mono- and dimethyl ester.[4] The separation of bilirubin mono- and diglucuronide has not been investigated but these compounds are expected to chromatograph well with the present system.

[4] N. Blanckaert, P. M. Kabra, F. A. Farina, B. E. Stafford, L. J. Morton, and R. Schmid, *J. Lab. Clin. Med.* **96**, 198 (1980).

[46] Purification of Hepatic Mitochondrial 5-Aminolevulinate Synthase

By Iain A. Borthwick, Gopesh Srivastava, Byron A. Pirola, Brian K. May, and William H. Elliott

The mitochondrial enzyme 5-aminolevulinate (ALA) synthase (EC 2.3.1.37) is the first and rate-limiting enzyme in the hepatic heme biosynthetic pathway. The enzyme is a matrix protein which appears to be loosely bound to the inner mitochondrial membrane. In the presence of the cofactor pyridoxal phosphate, ALA synthase catalyses the condensation of succinyl-CoA and glycine to give ALA.

Under normal physiological conditions, ALA synthase is present in low amounts in liver mitochondria with its synthesis being tightly regulated by the end-product heme.[1,2] This low level of hepatic ALA synthase in experimental animals, such as rats or chick embryos, is greatly increased by the administration of porphyrinogenic drugs.[1,2]

The attempted purification of hepatic ALA synthase from rats, chickens, and guinea pigs has been plagued with difficulties over the past decade due to the tendency of the enzyme to form large molecular weight aggregates and to its susceptibility to proteolytic degradation.

We describe here in detail a purification protocol for chick embryo liver mitochondrial ALA synthase. This procedure can also be used for isolation of highly purified rat liver mitochondrial ALA synthase.

Assay Method

Principle. Hepatic ALA synthase from drug-induced chick embryos and rats is detected by a colorimetric assay based on the procedure of

[1] S. Sassa and A. Kappas, *Adv. Hum. Genet.* **11**, 121 (1981).
[2] S. Granick, *J. Biol. Chem.* **241**, 1359 (1966).

Poland and Glover.[3] In this assay succinyl-CoA is generated by an endogenous system. The product ALA is converted to a pyrrole which condenses with modified Ehrlich's reagent and the resulting pink-colored ALA pyrrole complex is measured spectrophotometrically. A potential source of error in the assay is the presence of mitochondrial aminoacetone synthase which generates aminoacetone and hence aminoacetone pyrrole. For determination of rat liver ALA synthase activity, separation of aminoacetone pyrrole from ALA pyrrole is necessary. However, unlike mammalian liver, the formation of aminoacetone by chick embryo liver is insignificant and extraction of aminoacetone pyrrole is not required.

Reagents. Assay mix: 75 mM Tris–HCl, pH 7.6; 150 mM glycine, 15 mM succinate, 75 mM MgCl$_2$, 375 mM NaCl, 25.5 mM ATP, 0.375 mM pyridoxal phosphate, 15 mM EDTA, 1.5 mM dithioerythritol.

Succinyl-CoA synthase (EC 6.2.1.5): For routine purification we have found it convenient to use a crude extract of *E. coli* (Crooke's strain) prepared according to Ramaley *et al.*[4] but only taken to the stage of ammonium sulfate precipitation. This extract is stored in small aliquots at −80° and is stable for up to 2 years.

Acetylacetone in 1 M sodium acetate buffer, pH 5.5 (1 : 10 v/v).

Modified Ehrlich's reagent[4]: 5.6 g HgCl$_2$, 100 ml H$_2$O, 320 ml 70% perchloric acid, 1340 ml glacial acetic acid. This stock keeps indefinitely at 4°. To 55 ml of this stock is added *p*-dimethylaminobenzaldehyde (1 g). This solution is changed after 2 weeks.

Procedure. Immediately prior to use 15 μl of succinyl-CoA synthase (20 units) and 30 μl of 20 mM CoA is added per 1.0 ml of assay mix. Aliquots (100 μl) are then distributed in 1.5-ml Eppendorf tubes and equilibrated at 37° for 3 min prior to addition of ALA synthase activity in a total of 50 μl. The assay is stopped at an appropriate time (usually 15–30 min) by addition of 100 μl of cold 10% (w/v) trichloroacetic acid and the tubes kept on ice for 10 min. The tubes are centrifuged at 10,000 rpm for 5 min in an Eppendorf microfuge and 175 μl of each supernatant removed to a second Eppendorf tube. To this is added 87.5 μl of freshly made acetylacetone in 1 M sodium acetate buffer, pH 5.5 (1 : 10 v/v, respectively). The tubes are heated at 80° for 10 min followed by immediate addition of 262.5 μl of modified Ehrlich's reagent. (If necessary aminoacetone pyrrole can be removed by extraction with dichloromethane[3] prior to addition of modified Ehrlich's reagent.) Finally, the tubes are centrifuged at 10,000

[3] A. Poland and E. Glover, *Science* **179**, 426 (1973).
[4] R. F. Ramaley, W. A. Bridger, R. W. Moyer, and P. D. Boyer, *J. Biol. Chem.* **242**, 4287 (1976).

TABLE I
PURIFICATION OF 5-AMINOLEVULINATE SYNTHASE FROM CHICK EMBRYO
LIVER MITOCHONDRIA[a]

Step and fraction	Volume (ml)	Total protein (mg)	Total activity (units)	Specific activity (units/mg protein)	Yield (%)
Mitochondria (from 22 g of liver)	6.5	456.0	72,966	160	100
Mitoplasts	20.0	231.0	69,318	305	95
Freeze-dried supernatant	6.0	123.0	67,944	550	93
Sephacryl S-200 chromatography	16.0	31.0	59,480	1,920	88
Chromatofocusing	55.0	1.93	35,800	18,500	53
CoA-agarose affinity chromatography	10.6	0.43	14,870	35,000	22

[a] From Borthwick et al.,[5] reproduced with permission.

rpm for 3 min in an Eppendorf microfuge and the absorbance of supernatants measured at 555 nm in glass cuvettes of 1 cm optical length. The amount of ALA produced is calculated using a molar extinction coefficient of 58 mM^{-1} cm^{-1}. ALA synthase activity is expressed as units/mg protein where 1 unit is the amount of enzyme catalyzing the production of 1 nmol of ALA/hr at 37°.

Purification of Chick Embryo ALA Synthase

Chick embryo liver ALA synthase is purified 200-fold under conditions which minimize proteolytic degradation and aggregation of the enzyme.[5] A summary is given in Table I. ALA synthase is induced for 18 hr in 18-day-old chick embryos by injection of a combination of the porphyrinogenic drugs 2-allyl-2-isopropylacetamide (3 mg) and 3,5-diethoxycarbonyl-1,4-dihydrocollidine (6 mg).[6] All steps of the procedure are performed at 4°.

Isolation of Mitochondria. About 90 livers (20 g) are removed from drug-treated chick embryos and placed in approximately 4 volumes of ice cold buffer containing 0.25 M sucrose, 5 mM Tris–HCl, pH 7.6, and 0.1 mM pyridoxal phosphate. The livers are blotted dry, weighed, and to each gram of liver is added 4 ml of the buffer which now contains the protease inhibitor phenylmethylsulfonyl fluoride (1 mM) and is referred to as buffer

[5] I. A. Borthwick, G. Srivastava, J. D. Brooker, B. K. May, and W. H. Elliott, *Eur. J. Biochem.* **129,** 615 (1983).
[6] M. M. Bradford, *Anal. Biochem.* **76,** 248 (1976).

A. The livers are homogenized in a loose fitting Potter-Elvejhem glass homogenizer by 5 passes of a Teflon pestle at 300 rpm. Mitochondria are prepared by differential centrifugation as follows. The homogenate is centrifuged at 500 g for 5 min to pellet nuclei and the supernatant carefully decanted. The pellet is resuspended in half the original volume of buffer A, briefly homogenized (1 pass), and recentrifuged at 500 g for 5 min. The supernatants are pooled and centrifuged again at 500 g for 5 min to pellet remaining nuclei. Mitochondria are pelleted from the postnuclear supernatant by centrifugation in an SS34 Sorval rotor at 4300 g for 5 min, followed immediately by centrifugation at 12,000 g for 5 min. The uppermost fluffy layer which overlies the mitochondria is carefully removed by agitating and decanting. The mitochondria are gently suspended to a concentration of 35 mg protein/ml in buffer A which now contains the additional protease inhibitors leupeptin and chymostatin, each at 2.5 μg/ml. This buffer containing all three protease inhibitors is buffer B. During suspension of mitochondria care is taken not to disturb the underlying brown tightly packed pellet; protein estimations are rapidly determined using the method of Bradford.[6]

Preparation of Mitoplasts. Digitonin is used to remove the outer mitochondrial membrane together with any adsorbed protein. The suspended mitochondria are shaken at 0° for 15 min at 150 cycles/min with an equal volume of 0.6% (w/v) digitonin in buffer B. The suspension is immediately diluted 4-fold with this same buffer and centrifuged at 18,000 g for 15 min. The upper fluffy layer is carefully removed by suction and the underlying pellet of mitoplasts resuspended by gentle homogenization in the original volume of buffer B again leaving behind the small brown pellet at the bottom. The mitoplasts are pelleted at 18,000 g for 15 min and finally suspended in 20 ml of buffer containing 5 mM Tris–HCl, 0.1 mM pyridoxal phosphate, 1 mM dithioerythritol, pH 7.6, and the three protease inhibitors. The mitoplasts are freeze dried.

Sephacryl Chromatography. The freeze-dried mitoplasts are suspended in 8 ml of ice cold water containing the three protease inhibitors and centrifuged at 100,000 g for 60 min at 4°. About 90% of the ALA synthase activity is released into the supernatant. The supernatant is removed and made to 50 mM Tris–HCl, 50 mM MgSO$_4$, and 1 mM dithioerythritol, pH 7.6. Under these high ionic strength conditions we have found that ALA synthase does not aggregate into large molecular weight complexes which have been observed previously by others when the enzyme is exposed to low ionic strengh conditions. NaCl (0.8 M) can replace 50 mM MgSO$_4$ but the latter is used because it allows the ionic strength to be more readily reduced for the next step of chromatofocusing.

The mitoplast extract (about 8 ml) is loaded on to a Sephacryl S-200 column (1000 × 2 cm) equilibrated and run in 50 mM Tris–HCl, 50 mM MgSO$_4$, 0.1 mM pyridoxal phosphate, 1 mM dithioerythritol, pH 7.6. Note that once exposed to the high ionic strength conditions of Sephacryl S-200 chromatography, the enzyme remains disaggregated under subsequent low ionic strength conditions. The column is run at a flow rate of 20 ml/hr and 5 ml fractions collected. Prior to collection, fraction tubes contain appropriate amounts of the three protease inhibitors so that their final concentrations are those previously specified in buffer B.

ALA synthase activity elutes as a single peak just after the void volume. The most active fractions are pooled (20 ml).

Chromatofocusing. The pooled enzyme fraction is reduced in volume to 2.5 ml using an Amicon ultrafiltration cell fitted with a YM30 membrane and operated under nitrogen at 50 psi. The ionic strength is unaltered during this procedure. To reduce the ionic strength for chromatofocusing the extract is diluted 5-fold (to about 12.5 ml) with 5 mM Tris–HCl, pH 10 containing 1 mM dithioerythritol giving a final pH of 8.6. The diluted extract is loaded on a chromatofocusing column (60 × 1 cm) containing Pharmacia Poly Buffer Exchanger 94 equilibrated in 25 mM Tris–HCl, pH 8.6 with 1 mM dithioerythritol. The column is eluted with 350 ml of degassed Pharmacia Poly Buffer 96, pH 6.6, containing 1 mM dithioerythritol at a flow rate of 20 ml/hr and 5 ml fractions are collected. A linear pH gradient from 8.6 to 6.6 is generated and ALA synthase activity elutes as a broad peak about half-way through the gradient with the peak activity fractions eluting around pH 7.4.

We have found a significant stabilization of enzyme activity by the addition of high ionic strength buffer (together with protease inhibitors) to the fraction tubes prior to collection. To each fraction tube is added 250 μl of buffer containing 1 M Tris–HCl, 2 mM pyridoxal phosphate, and 2 M glycine, pH 7.6. In addition, 20 μl of each of the three protease inhibitors is added so that their final concentrations are those previously specified in buffer B.

Affinity Chromatography. Agarose-hexane–coenzyme A Type 5 (P-L Biochemicals, Milwaukee, WI) is an effective affinity ligand for chick embryo ALA synthase in the presence of pyridoxal phosphate and glycine. Bound enzyme can be eluted with the CoA analog 5′-AMP. Pooled fractions from chromatofocusing (about 50 ml) are slowly loaded at 5–10 ml/hr on to a CoA-agarose column (1 × 15 cm) equilibrated in buffer containing 50 mM Tris–HCl, 100 mM glycine, 0.1 mM pyridoxal phosphate, and 1 mM dithioerythritol pH 7.6 (buffer C). Throughout the elution procedure the column is run at a flow rate of 20 ml/hr and 5 ml fractions are collected. The column is protected from light by foil wrap-

TABLE II
PURIFICATION OF 5-AMINOLEVULINATE SYNTHASE FROM RAT LIVER MITOCHONDRIA[a]

Step and fraction	Volume (ml)	Total protein (mg)	Total activity (units)	Specific activity (units/mg protein)	Yield (%)
Mitochondria (from 86 g liver)	20.0	2,320.0	30,238	13	100
Mitoplasts	40.0	514.0	28,044	54	93
Extract from freeze-dried mitoplasts	12.0	294.0	25,496	87	84
Sephacryl S-200 chromatography	41.6	102.0	22,694	223	75
Chromatofocusing	37.0	3.2	21,710	6,800	72
CoA-agarose affinity chromatography	17.3	0.94	9,200	9,790	30

[a] From Srivastava et al.,[9] with permission.

ping. The column is washed successively with 50 ml of buffer C, 20 ml 0.5 M glycine in buffer C, then another 20 ml of buffer C. About 15% of ALA synthase activity is eluted with 20 ml of buffer C containing 20 mM 5'-AMP (pH 7.6) together with a number of other proteins. The remaining ALA synthase activity is eluted with 20 ml of the buffer C containing 50 mM 5'-AMP (pH 7.6). The column is washed with 1 M NaCl in buffer C prior to reuse.

Purity of Enzyme. From Table I it can be seen that the greatest purification is achieved at the chromatofocusing step. Analysis by sodium dodecyl sulfate–polyacrylamide gel electrophoresis of the finally purified enzyme showed a single protein band of minimum molecular weight 68,000. The enzyme has a specific activity of about 35,000 units/mg of protein. Electron microscopy work has confirmed the purity and shown that the native purified enzyme is a dimer of two identical monomers.[7]

We have found that the amounts of starting material can be increased at least 5-fold without affecting the final purity. In such instances we have found it necessary to load the material on to two Sephacryl S-200 columns each of dimensions 1000 × 2 cm.

Proteolytic Susceptibility. ALA synthase is particularly sensitive to proteolytic degradation during the purification. It is important to add protease inhibitors and to perform all steps as quickly as possible without storage of the extracts at any stage. A previous report[8] that chick embryo liver ALA synthase has a minimum molecular weight of 50,000 was due to the isolation of a degraded form of the enzyme.

[7] B. A. Pirola, F. Mayer, I. A. Borthwick, G. Srivastava, B. K. May, and W. H. Elliott, *Eur. J. Biochem.* **144,** 577 (1984).
[8] M. J. Whiting and S. Granick, *J. Biol. Chem.* **251,** 1340 (1976).

Enzyme Storage. Pure ALA synthase in elution buffer C containing 50 mM 5'-AMP is stable at 4° for at least 4 months without loss of activity or alteration of molecular weight as analyzed by sodium dodecyl sulfate–polyacrylamide gel electrophoresis. Protease inhibitors are not added. Alternatively the enzyme can be stored at −195°; following an initial loss of ~20% of activity due to freeze-thawing the enzyme loses no further activity for at least 6 months.

Purification of Rat ALA Synthase

This protocol has been successfully applied to the purification of rat ALA synthase from drug induced livers[9] and is outlined in Table II. The enzyme is judged to be 90% pure by sodium dodecyl sulfate–polyacryl-amide gel analysis. As with the chick enzyme it is important to have the protease inhibitors present and to carry out the purification as rapidly as possible in order to avoid proteolytic breakdown.

[9] G. Srivastava, I. A. Borthwick, J. D. Brooker, B. K. May, and W. H. Elliott, *Biochem. Biophys. Res. Commun.* **109**, 305 (1982).

[47] Purification and Characterization of Mammalian and Chicken Ferrochelatase

By HARRY A. DAILEY, JENNIE E. FLEMING, and BERTILLE M. HARBIN

In eukaryotic cells the terminal enzyme of the heme biosynthetic pathway, ferrochelatase (protoheme ferro-lyase, EC 4.99.1.1), is located in the mitochondria. This enzyme, which catalyzes the insertion of ferrous iron into protoporphyrin IX to form protoheme, is bound to the inner mitochondrial membrane and is only released by treatment by detergents. The active site of the enzyme is present on the matrix side of the inner membrane, but the protein molecule spans the entire bilayer.[1]

Ferrochelatase has recently been purified from rat,[2] beef,[3,4] and mouse liver[5] and from chicken erythrocytes.[6] The bovine enzyme is the best

[1] B. M. Harbin and H. A. Dailey, *Biochemistry* **24**, 366 (1985).
[2] S. Taketani and R. Tokunaga, *J. Biol. Chem.* **256**, 12748 (1981).
[3] H. A. Dailey and J. E. Fleming, *J. Biol. Chem.* **258**, 12748 (1983).
[4] S. Taketani and R. Tokunaga, *Eur. J. Biochem.* **127**, 443 (1983).
[5] H. A. Dailey and J. E. Fleming, unpublished data (1984).
[6] J. W. Hanson and H. A. Dailey, *Biochem. J.* **222**, 695 (1984).

characterized of these studied to date. The enzyme from all of these sources shares many common characteristics but differs substantially from the bacterial enzymes previously described.[7,8] In this chapter we describe a purification scheme that is effective for the bovine, murine, and avian enzyme as well as some characteristics of the bovine enzyme.

Ferrochelatase Assay

Ferrochelatase was assayed either using the pyridine hemochromogen procedure previously described[7] or with ^{59}Fe to quantitate product. Either assay procedure may be used in samples containing high protein concentrations, but for experiments where high levels of endogeneous heme are present, the radiolabel assay is preferable. Others have previously described ^{59}Fe assays that employ extraction of product followed by counting radioactivty.[9] The procedure developed in our laboratory and described here employs a relatively rapid extraction procedure that can be used for protoheme. Deuteroheme and mesoheme cannot be quantitated by this procedure since they are incompletely extracted under the conditions employed.

The assay mixture contains in 1 ml 50 mM Tris acetate, pH 8.1, 5 mM dithiothreitol, 0.2 mM [^{59}Fe]ferrous ammonium sulfate, 0.2% Triton X-100, 0.1 mM protoporphyrin, and enzyme preparation. All ingredients except the porphyrin are first mixed and the reaction is started by addition of the protoporphyrin. The reaction is carried out in the dark, open to air at 37°. For tissue samples low in activity and high in endogeneous oxidizing activities, the reaction should be done under argon or nitrogen atmosphere as previously described.[9,10] After 30 min the reaction is stopped by addition of 0.5 ml of 0.2 N HCl. To this is added 1 ml of methyl ethyl ketone, and the mixture is vortexed and chilled on ice until the phases separate. For samples high in protein concentration centrifugation may be helpful to separate phases. A 0.5 ml sample of the organic phase is removed and washed with a solution containing 0.5 ml of 0.2 N HCl, 1 ml H_2O, and 0.5 ml of fresh methyl ethyl ketone. After vortexing and separation of phases, 0.2 ml of the organic phase was removed and counted in a gamma counter. Samples may be counted in a scintillation counter if the solvent is evaporated and the heme decolorized by heating in the presence of 0.1 ml 30% H_2O_2. To quantitate the product formed one set of duplicate

[7] H. A. Dailey, *J. Biol. Chem* **257**, 14714 (1982).
[8] H. A. Dailey, Jr., *J. Bacteriol.* **132**, 302 (1977).
[9] J. R. Bloomer and K. V. Morton, *Enzyme* **28**, 220 (1982).
[10] H. A. Dailey and J. Lascelles, *J. Bacteriol.* **129**, 815 (1977).

assays containing no background heme is run. One of the duplicate samples is counted, and the second is quantitated by the pyridine hemochromogen method.

Purification Procedure

The procedure for purification of bovine liver and chicken erythrocyte ferrochelatase by this laboratory has been described previously.[3,6] Purification of mouse liver ferrochelatase is also accomplished with the same procedure although volumes are decreased to accommodate the smaller sample size.

The standard buffer used for all procedures is 20 mM Tris acetate, pH 8.1 containing 1 mM dithiothreitol, phenylmethylsulfonyl fluoride (PMSF) (10 μg/ml), and 20% glycerol. All operations were at 0–4°.

Beef liver obtained from a local abattoir was immediately sliced and stored on ice. Mitochondria were isolated from about 5 lb (approximately one-half of a steer liver). The liver slices were coarsely pureed in a food processor. Five hundred grams of this material was added to 4 liters of cold 0.25 M sucrose, 10 mM Tris acetate, pH 8.1, 1 mM EDTA, PMSF (10 μg/ml) and blended in a Waring blender at medium speed for 30 sec. Mitochondria were isolated by differential centrifugation. Nuclei, large membrane fragments, and unbroken cells are separated by centrifugation at 2000 g for 10 min. The supernatant fraction is then centrifuged at 10,000 g for 10 min to pellet out the mitochondrial fraction. This pellet is then suspended in sucrose buffer and centrifuged once again at 10,000 g for 10 min. This wash step is repeated once more and the mitochondria pellet is suspended in sucrose buffer. For the murine preparation 50 g of frozen liver (about 50 livers) is diced with a knife, suspended in 400 ml of buffer, and disrupted with Polytron on medium speed for 15 sec. Any large tissue fragments remaining are disrupted with a hand-held glass-Teflon tissue homogenizer. Mitochondria are then isolated by differential centrifugation. Preparation of hemoglobin free chicken erythrocyte mitochondria is detailed elsewhere.[6]

Mitochondria were disrupted by sonication for two 30 sec intervals with a Heat Systems sonicator set at 60 W and the membrane fraction isolated by centrifugation at 100,000 g for 90 min. The membrane fraction was suspended to yield a final protein concentration of 40 mg/ml in buffer containing 1.0% sodium cholate and 0.1 M KCl final concentration. The suspension was stirred for 3 hr to solubilize the enzyme and then centrifuged at 100,000 g for 90 min to pellet the remaining membrane fragments.

The solubilized enzyme was fractionated with ammonium sulfate by addition of either solid salt or a saturated solution (adjusted to pH 8.0).

Ferrochelatase remains in solution at 35% (30% for chicken) saturation and is precipitated out at 55% saturation. While both the mammalian liver forms of the enzyme yield firm pellets in this procedure, the chicken erythrocyte preparation yields a floating precipitate that must be carefully removed. Since this floating precipitate tends to adhere to glass rods and the wall of the tube, we usually remove the largest fragment by inserting a glass rod and slowly rotating while pulling the rod out. The remainder is obtained by gently rotating the tube around while removing liquid below the surface. The precipitate will then adhere to the walls of the tube and can be removed with a rubber policeman. The precipitated enzyme is dissolved in a minimal volume of buffer containing 1.0% Triton X-100 and 0.5 M KCl.

The enzyme preparation is loaded immediately onto a tandem set of columns composed of an initial Reactive Red 120 Sepharose CL-6B (2.5 × 10 cm) whose outlet is connected directly to a Reactive Blue CL-6B column (2.5 × 30 cm). For the mouse preparation the Blue column was 2.5 × 15 cm. Both columns are initially equilibrated with buffer containing 0.5 M KCl and 1.0% Triton X-100. The columns are first washed with 50 ml of equilibrating buffer followed by 500 ml of buffer containing 1% Triton X-100 and 1.0 M KCl. After this wash the Red column is disconnected, and the Blue column is further washed with 150 ml of buffer containing 0.5% sodium cholate and 1.0 M KCl. Ferrochelatase is eluted with buffer containing 1.0% sodium cholate and 1.5 M KCl. The purified enzyme is stored in 20% glycerol at $-20°$ and is stable for several weeks. Care must be taken to prevent freezing of the solution since activity of the purified enzyme drops rapidly with freezing and thawing. The purification procedure is outlined in Table I.

TABLE I
PURIFICATION OF BOVINE FERROCHELATASE[a]

Fraction	Protein concentration (mg/ml)	Specific activity (nmol/mg/min)[b]	Recovery (%)
Isolated mitochondrial membranes	42	0.49	100
Solubilized enzyme	20	0.94	75
$(NH_4)_2SO_4$ (55% pellet)	17	1.49	88
Reactive Blue column	0.04	959	49

[a] Data taken from Ref. 3.
[b] Enzyme units are nmol deuteroheme formed/mg protein/min.

Notes on the Purification Scheme

This purification procedure yields preparations of ferrochelatase that appear as a single band on an SDS gel with estimated molecular weights of 40,500, 42,000, and 40,000 for bovine, chicken, and mouse ferrochelatase, respectively. During our studies we have used livers from a variety of cattle and have found marked variations in the purity of our final enzyme preparation. Livers from Black Angus and Charolais steer yield good preparations while old bulls and dairy cows of various breeds yield mixed results.

In this laboratory we prepare our own Reactive Red 120 CL-6B following the procedure of Böhme *et al.*[11]; however, the Reactive Blue CL-6B is purchased from Pharmacia Fine Chemicals. Laboratory produced Reactive Blue did not yield a single band by SDS–gel electrophoresis but gave preparations with several minor contaminants. The reason for this is unknown, but may be due to stabilizers added to the commercial product. The Reactive Red column serves two purposes. First, since it is cheaply produced, it serves as an inexpensive guard column that protects the Reactive Blue column from becoming fouled with lipids and other extraneous material. Second, experiences in several laboratories have shown that the Red column removes a contaminating protein that otherwise coelutes from the Blue column with ferrochelatase. The Reactive Red columns are used a maximum of six times before being discarded. After each use both Red and Blue columns are regenerated by washes with 8 M urea. One additional problem is the difficulty in concentrating the purified protein. Concentration may be achieved by pressure filtration, but at considerable loss. Our experience has shown that the enzyme adsorbs to membranes of all sorts and it is difficult, if not impossible to recover active enzyme.

Properties of Bovine Ferrochelatase

The kinetic properties of the enzyme are summarized in Table II. Ferrochelatase from rat and bovine liver and chicken erythrocytes possesses similar kinetic parameters with protoporphyrin IX, the natural substrate, having the lowest measured K_m and V_{max} (Table III). Only dicarboxylate porphyrins of the IX isomer are substrates. Porphyrins with 2,4 position substitutions that are charged or larger than vinyl groups are competitive inhibitors of the enzyme with respect to the porphyrin sub-

---

[11] H.-J. Böhme, G. Kopperschlager, J. Schulz, and E. Hofmann, *J. Chromatogr.* **69**, 209 (1972).

TABLE II

KINETIC PROPERTIES OF BOVINE FERROCHELATASE[a]

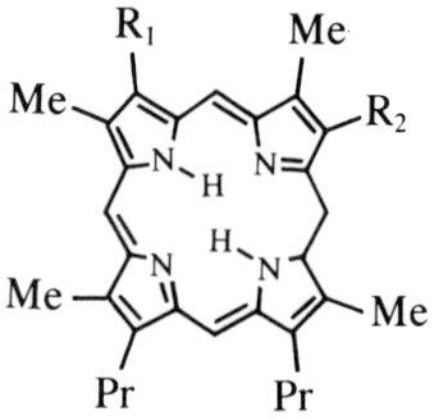

Porphyrin	R_1	R_2	K_m (μM)	K_i (μM)
Protoporphyrin[b]	CH=CH$_2$	CH=CH$_2$	11	—
Mesoporphyrin	CH$_2$CH$_3$	CH$_2$CH$_3$	34	—
Deuteroporphyrin	H	H	47	—
Hematoporphyrin	CH(OH)CH$_3$	CH(OH)CH$_3$	22	—
Hydroxyethylvinyl-deuteroporphyrin[c]	CH=CH$_2$	CH(OH)CH$_3$	23	—
O,O'-Diacetyl-hematoporphyrin	CH(OAc)CH$_3$	CH(OAc)CH$_3$	—	11
2,4-Bisacetal deuteroporphyrin	CH$_2$CH(OCH$_3$)$_2$	CH$_2$CH(OCH$_3$)$_2$	—	13
2,4-Bisglycol deuteroporphyrin	CH(OH)CH$_2$OH	CH(OH)CH$_2$OH	—	67
2,4-Disulfonic deuteroporphyrin	SO$_3^-$	SO$_3^-$	—	70
N-Methylprotoporphyrin	CH=CH$_2$	CH=CH$_2$	—	7 nM

[a] Data taken from Dailey and Smith.[13] Reprinted by permission from *Biochem. J.* **223**, 441–445. Copyright © 1984 The Biochemical Society, London. Ac, O—C—CH$_3$; Me,

$$\underset{O}{\overset{\|}{}}$$

CH$_3$; Pr, (CH$_2$)$_2$COOH.
[b] Natural substrate for this enzyme.
[c] A mixture of isomers.

TABLE III

COMPARISON OF KINETIC PARAMETERS FOR EUKARYOTIC FERROCHELATASE

Source of enzyme	Iron	K_m (μM)			Reference
		Protopor-phyrin	Mesopor-phyrin	Deuteropor-phyrin	
Bovine	80	11	34	47	3
Chicken	166	37	51	80	6
Rat	33	28	27	—	2

strate.[12,13] N-Methylprotoporphyrin is a strong competitive inhibitor with respect to porphyrin with $K_i = 7$ nM. The mammalian and chicken enzyme are also strongly inhibited by Mn^{2+} ($K_i = 15$ μM). This inhibitor is competitive with respect to ferrous iron and noncompetitive with respect to porphyrin.[3]

Reaction of a single sulfhydryl residue per monomer of purified ferrochelatase results in the total loss of activity.[14] However, the enzyme is reversibly inhibited by sodium arsenite as well as by Hg^{2+} which strongly suggests that two vicinyl sulfhydryl groups are involved in catalysis. Ferrochelatase is protected against inactivation by N-ethylmaleimide in the presence of ferrous iron, but not in the presence of porphyrin. These data have been forwarded to support a model where vicinyl sulfhydryl residues are involved in ferrous iron binding by the enzyme.[14]

Since the second substrate of ferrochelatase is a dicarboxylate porphyrin, the possible role of protein cationic residues in porphyrin binding has been examined.[15] Bovine ferrochelatase is rapidly inactivated by the arginyl reagents butanedione, cyclohexanedione, and camphorquinone 10-sulfonate. It is not inactivated by lysyl specific reagents such as trinitrobenzene sulfonate or acetimidates. The enzyme is protected against inactivation by porphyrin but not by ferrous iron. The kinetics of the modified enzyme support the model of arginyl involvement in porphyria but not iron binding. The current chemical modification and kinetic data support a model, as shown in Fig. 1, where iron binding occurs via vicinyl sulfhydryl groups prior to porphyrin binding that involves protein arginyl residue(s).

Ferrochelatase from several sources is stimulated to varying degrees by added fatty acids and phospholipids of various sorts.[2-6] There is considerable variability in the amount of stimulation, or inhibition, seen and this may reflect a difference in the content of contaminating endogenous lipids that may remain associated with the enzyme during purification.

Ferrochelatase has a reported molecular weight of approximately 40,000 as determined by SDS–polyacrylamide gel electrophoresis.[2-4,6] There is, however, some disagreement on the molecular weight of the native enzyme. Molecular weights of 200,000 and 240,000 have been reported by Taketani and Tokunaga[2,4] but 40,000 by Dailey and Fleming.[3] One explanation forwarded for this discrepancy is that dialysis used in some preparations[2,4] may lead to aggregration of the membrane protein. Resolution of this must await determination of the membrane *in situ* structure of the protein. Assuming a molecular weight of 40,000 per mol of

[12] C. L. Honeybourne, J. T. Jackson, and O. T. G. Jones, *FEBS Lett.* **98,** 207 (1979).
[13] H. A. Dailey and A. Smith, *Biochem. J.* **223,** 441 (1984).
[14] H. A. Dailey, *J. Biol. Chem.* **259,** 2711 (1984).
[15] H. A. Dailey and J. E. Fleming, in preparation.

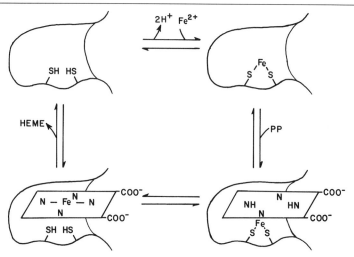

FIG. 1. Kinetic model for bovine ferrochelatase. From Dailey.[14]

enzyme, inactivation data show that 1 mol of *N*-methylprotoporphyrin binds and 1 mol of *N*-ethylmaleimide inactivates 1 mol of enzyme. So if there are multiple subunits, each is able, at least *in vitro*, to catalyze the insertion of iron.

Acknowledgments

This work was supported by Grants AM31064 and AM32303 from the National Institutes of Health. H.A.D. was the recipient of a Research Career Development Award, AM01038 from N.H. during this work.

[48] Purification and Characterization of Bacterial Ferrochelatase

By HARRY A. DAILEY

Ferrochelatase (protoheme ferrolyase, EC 4.99.1.1) is the terminal enzyme in the heme biosynthetic pathway and catalyzes the insertion of ferrous iron into protoporphyrin IX to form heme. Ferrochelatase activity has been detected in a variety of bacteria but has been highly purified from only two.[1,2] In all organisms examined the enzyme is bound to the cyto-

[1] H. A. Dailey, Jr., *J. Bacteriol.* **132,** 301 (1977).
[2] H. A. Dailey, *J. Biol. Chem.* **257,** 14714 (1982).

plasmic membrane and is only solubilized by detergents or chaotropic agents.[3,4] While the enzyme does not appear to be inducible in any prokaryotic organism studied to date, the specific activity of the enzyme varies widely among different bacterial species. This variation roughly corresponds to the ability of various organisms to produce heme with organisms such as *Paracoccus denitrificans* and *Rhodopseudomonas sphaeroides* having the highest activity and *Staphylococcus aureus* having very low activity.[5]

This chapter summarizes our work on ferrochelatase from the facultative photosynthetic organism *R. sphaeroides* and from *Aquaspirillum itersonii*. The purification scheme for the enzyme from *R. sphaeroides* is given along with details of the assay as used in our laboratory.

Ferrochelatase Assay

Our laboratory uses two different ferrochelatase assays. The most commonly used assay quantitates product formed in a predetermined period of time by measuring hemes formed as their pyridine hemochromogen as originally described by Porra.[6] The second assay procedure quantitates product radioactivity and is described elsewhere in this volume.[7]

The assay mixture is composed of 0.5 ml of 0.1 M Tris acetate, pH 8.1, 0.1 ml of 50 mM dithiothreitol (DTT), 0.05 ml of 4 mM ferrous ammonium sulfate (or ferrous citrate), 0.1 ml of 1 mM deuteroporphyrin, and enzyme extract and water to yield a final volume of 1 ml. The 1 mM deuteroporphyrin solution is prepared by weighing an appropriate amount of porphyrin (~2 mg for a 5 ml stock solution) in a glass tube. To this 0.05 ml of 4 N NH$_4$OH is added to wet the powder prior to addition of 0.5 ml of 20% (w/v) Triton X-100. This is vortexed to dissolve the porphyrin prior to the addition of 4.5 ml distilled water. The final concentration of the solution is determined spectrophotometrically in 0.1 N HCl using the extinction coefficient at (E_{mM} = 13.7) nm.[8] This stock solution is stored in a foil covered tube at 4° and may be used for routine assays for a few days. It should be emphasized that porphyrins in solution will readily aggregate, so for kinetic studies, only freshly prepared porphyrins should be used. Protoporphyrin, when used, should be made up fresh and used that day only. Mesoporphyrin and hematoporphyrin may also be used, but we

[3] H. A. Dailey, Jr. and J. Lacelles, *Arch. Biochem. Biophys.* **160**, 523 (1974).

[4] M. S. Jones and O. T. G. Jones, *Biochem. J.* **119**, 453 (1970).

[5] H. A. Dailey, Jr. and J. Lascelles, *J. Bacteriol.* **129**, 815 (1977).

[6] R. S. Porra and O. T. G. Jones, *Biochem. J.* **87**, 186 (1963).

[7] H. A. Dailey, J. E. Fleming, and B. M. Harbin, this volume [47].

[8] J.-H. Furhop and K. M. Smith, *in* "Porphyrins and Metalloporphyrins" (K. M. Smith, ed.), pp. 784 and 806. Elsevier, Amsterdam, 1976.

routinely use deuteroporphyrin since it yields the highest specific activity. Assays are run in the dark at 37°. For crude cell extracts, it may be beneficial to run the assays under an argon or nitrogen atmosphere as previously described, but our experiences with even partially purified enzyme preparations show that strict anaerobic conditions are not necessary so that assays are routinely run in open, unshaken tubes.

After 30 min incubation the reaction is stopped by addition of 0.5 ml of 50 mM iodoacetamide and vortexing. Deuteroheme formed is quantitated as its pyridine hemochrome after addition of 0.25 ml pyridine and 0.25 ml 1 N NaOH. Product is immediately quantitated by reduced minus oxidized difference spectra. For deuteroheme, a $\Delta\varepsilon_{mM}$ (545–530 nm) of 15.3 is used.[8] For proto- and mesoheme the values are 20.7 and 21.7, respectively. Since the heme is oxidized by the vortexing in air, a few grains of fresh, free flowing dithionite are added to one cuvette to reduce the heme.

One of the most common difficulties that may occur with this assay is poor reduction due to old or damp dithionite. It is best to dispense small amounts of fresh dithionite into small dark vials and store them desiccated using only one vial at a time. Another problem may occur if one tries substituting a reductant other than DTT. Ascorbate is not recommended and variable results have been found with glutathione. One difficulty that we have noted is that the use of disposable plastic cuvettes causes poor reduction or unusual spectra. Thus, we routinely use quartz or glass cuvettes. Porphyrin solutions are not prepared in dimethyl sulfoxide since this may inhibit ferrochelatase activity in some preparations.

Purification of Ferrochelatase from *R. sphaeroides*

The purification scheme described is for a 10 liter culture of *R. sphaeroides* L or 2.4.1 (ATCC).[2] For the purification of ferrochelatase (Table I), cultures were grown at 30° in 1-liter flasks containing 500 ml of yeast extract–malate–glutamate (YEMG) media at 140 rpm. A 5 ml, 16 hr culture grown in YEMG broth was used as the starter culture for each 500 ml. A total of 10 liters of culture was harvested by centrifugation at 8000 g for 10 min after 24 hr of growth. The cells were washed once in 10 mM Tris acetate, pH 8.1 and then suspended to a final volume of 250 ml in the same buffer. All harvesting operations were carried out at 4°. Cells were lysed by sonication with a Heat Systems Cell Disrupter set at 60 W. The cell suspension was kept in an ice bath and was sonicated five times for 30 sec each time with stirring for 30 sec between each sonication. This material was centrifuged at 8000 g for 10 min at 4° to remove whole cells and large cell fragments.

All operations below were carried out at 4° and, unless otherwise stated, the buffer used in all steps was 10 mM Tris acetate, pH 8.1, 0.5

TABLE I
PURIFICATION OF *R. sphaeroides* FERROCHELATASE[a]

Fraction	Protein (mg/ml)	Recovery (%)	Specific activity[b] (deutero-heme/min/mg)	Purification (fold)
Crude membranes	21.0	100	0.98	1
NaSCN-washed membranes	18.0	92	1.27	1.3
Sodium deoxycholate solubilized	6.5	86	5.13	5.2
Ammonium sulfate fraction	7.0	73	9.25	9.4
DEAE-Sephacel	0.5	64	46.20	47.1
Blue B[c]	0.5	46	1386	1414
Sephadex G-150[b]	0.4	43	1600	1640

[a] Data from Ref. 2.
[b] Specific activity is expressed as nmol of deuteroheme formed per min/mg of protein.
[c] Fractions from the Blue B and Sephadex G-150 columns were pooled and concentrated as described in the text. The recoveries shown are those obtained after these concentration steps.

mM dithiothreitol, 10 μg/ml phenylmethylsulfonyl fluoride (Buffer A). The crude cell extract (250 ml) obtained as described above was centrifuged at 100,000 g for 90 min. The soluble cytoplasmic fraction was poured off leaving a firmly packed purple-red pellet. This pellet was suspended in Buffer A to a volume of 225 ml with a glass-Teflon Potter homogenizer. Twenty-five milliliter of 5.0 M sodium thiocyanate was added to give a final concentration of 0.5 M. The membrane suspension was sonicated for 30 sec at 60 W setting of a Heat Systems Cell Disrupter before being centrifuged at 100,000 g for 90 min. The soluble salt wash was removed leaving a firmly packed pellet. These salt washed membranes were suspended to 237 ml in Buffer A as described above. The membrane suspension was brought to 0.5% (w/v) sodium deoxycholate by the addition of 12.5 ml of a 10% stock solution. This suspension was sonicated at 30 sec and centrifuged at 100,000 g for 90 min. The redish supernatant was carefully removed leaving behind a firmly packed pellet.

The volume of solubilized enzyme was usually about 220 ml, and this material was fractionated with ammonium sulfate by addition of a saturated solution of ammonium sulfate adjusted to pH 8.0. The solubilized enzyme solution was brought to 20% saturation, stirred for 10 min, and the precipitated material removed by centrifugation at 10,000 g for 10 min. The supernatant was removed and brought to 50% saturation and once again stirred for 10 min before centrifugation. The red pellet was dissolved in a minimal volume of Buffer A (about 30–40 ml) and enzyme preparation was immediately passed through a Sephadex G-25 column (3.5 × 30 cm) equilibrated with Buffer A to remove residual ammonium sulfate.

This material was applied to a DEAE Sephacel column (2.5 × 20 cm) that had been equilibrated with 1 liter of 10 mM Tris acetate, pH 8.1, 0.5 mM dithiothreitol, 10 μg/ml PMSF, and 0.2% (w/v) Brij 35 (Buffer B). After sample addition, 100 ml of this same buffer was washed through the column before addition of 500 ml of Buffer B containing 0.10 M sodium thiocyanate. Ferrochelatase was eluted with Buffer B containing 0.15 M sodium thiocyanate. The fractions containing ferrochelatase activity were pooled and brought to 10% (v/v) propylene glycol. This solution was passed slowly through a Blue B (Amicon) column (2.5 × 10 cm). The eluate was collected and passed through the column twice more. This procedure resulted in about 80% retention of ferrochelatase. The column was then washed with 300 ml of Buffer B containing 0.5 M KCl before ferrochelatase was eluted with 1.0 M sodium thiocyanate. The enzyme eluted from the Blue B column was immediately concentrated to a few milliliters volume by pressure dialysis with an Amicon YM-30 filter and chromatographed on a Sephadex G-150 column (1.5 × 100 cm) that is equilibrated with Buffer A plus 0.5% (w/v) sodium deoxycholate. The fractions containing ferrochelatase activity are pooled, brought to 10% (v/v) propylene glycol and concentrated with the Amicon YM-30 filter before being stored at 4°.

Notes Concerning the Purification of Bacterial Ferrochelatase

Above we have detailed the purification of ferrochelatase from *R. sphaeroides,* and previously we have described the 1000-fold purification of the enzyme from *A. itersonii.*[1] The experience of this laboratory with these organisms along with initial investigations with *Pseudomonas putida* and *Paracoccus denitrificans* have clearly shown that, unlike the enzyme from higher eukaryotic organisms, purification of bacterial ferrochelatase appears to require a case by case approach. The bacterial enzymes appear less stable than the mammalian ferrochelatase in storage, but are stable in the presence of chaotrophs such as NaSCN or NaClO$_4$, whereas, the activity of the mammalian enzymes are destroyed by low concentrations of these salts. The purification above works well with the wild type strain of *R. sphaeroides,* but not with some mutant strains. It also does not work with *P. denitrificans* although this organism is suggested to be similar to a nonphotosynthetic grown *R. sphaeroides.*[9]

[9] P. John and F. R. Whatley, *Nature (London)* **254,** 495 (1975).

TABLE II
KINETIC PARAMETERS FOR BACTERIAL FERROCHELATASE

Substrate	Apparent K_m (μM)	
	R. sphaeroides[a]	A. itersonii[b]
Ferrous iron	22	20
Protoporphyrin	18	47
Mesoporphyrin	20	45
Deuteroporphyrin	95	440
2,4-Disulfonic deuteroporphyrin	52	
2,4-Bisacetal deuteroporphyrin	56	

[a] Data from Refs. 2 and 13.
[b] Data from Ref. 1.

Properties of Bacterial Ferrochelatase

The molecular weight of detergent-solubilized purified R. sphaeroides ferrochelatase is 110,000 by gel filtration in the presence of 0.5% sodium deoxycholate and is 115,000 ± 5000 by SDS–polyacrylamide gel electrophoresis. This is considerably larger than the molecular weight of ferrochelatase from A. itersonii which is 50,000 as determined by gel filtration in the presence of detergents. Unlike the ferrochelatase from mammalian and avian sources, the bacterial ferrochelatases seem not to be stimulated by exogeneously added fatty acids or phospholipids although this property has not been studied extensively.

The absorption spectrum of the R. sphaeroides enzyme has a maximum at 278 nm with no absorption in the visible light range. The calculated millimolar extinction coefficient at 278 nm based upon a molecular weight of 115,000 is 76. (Previously an incorrect value of 90 had been reported.[2])

The ferrochelatase of both R. sphaeroides and A. itersonii has similar kinetic parameters and sensitivities to various metal ions. The apparent K_m are listed in Table II and metal sensitivities are shown in Table III. While these data for the two bacterial enzymes examined to date are similar, there are significant differences between them and eukaryotic ferrochelatase. The bacterial enzymes are only poorly inhibited by 50 μM Mn^{2+}, whereas, the avian and beef enzymes are strongly inhibited by 10 μM Mn^{2+} (the K_i for the bovine ferrochelatase is 15 μM).[10,11] One other

[10] J. W. Hanson and H. A. Dailey, Biochem. J. 222, 695 (1984).
[11] H. A. Dailey and J. E. Fleming, J. Biol. Chem 258, 11453 (1983).

TABLE III
EFFECT OF METALS ON BACTERIAL
FERROCHELATASE

| Metal ion[a] | Control activity (%) | |
	R. sphaeroides[b]	A. itersonii[c]
None	100	100
KCl	98	—
NaCl	97	100
MgCl$_2$	104	95
MnCl$_2$	97	72
NiCl$_2$	99	82
CuSO$_4$/CuCl$_2$	59	132
PbNo$_3$	50	33
HgCl$_2$	15	—

[a] Metal ion concentration was 50 μM in the assay mixture. Deuteroporphyrin was the substrate for R. sphaeroides and mesoporphyrin for A. itersonii.
[b] Data taken from Ref. 2.
[c] Data taken from Ref. 1.

major difference is in the ability of R. sphaeroides ferrochelatase to utilize 2,4-disulfonate and 2,4-bisacetal deuteroporphyrin as substrates. These 2,4 disubstituted compounds are competitive inhibitors of bovine ferrochelatase and are not substrates.[12]

Both enzymes are inactivated by both N-ethylmaleimide and iodacetamide suggesting that they contain sulfhydryl residues essential for activity. This is further supported by the inhibition by lead and mercury. The kinetics of inactivation of R. sphaeroides ferrochelatase by N-ethylmalemide and the stoichiometry of inactivation suggest that the enzyme is inhibited 100% by 2 mol of N-ethylmaleimide per mol of enzyme.[13] Ferrous iron, but not porphyrin, blocks this inhibition by sulfhydryl reagents.

Ferrochelatase of R. sphaeroides is also inactivated by the arginyl reagents butanedione and camphorquinone-10 sulfonic acid.[13] The kinetics of the modified enzyme suggest that arginyl residues may be involved in porphyrin binding, but not in metal binding. Similar data have been found for the bovine ferrochelatase.[14]

Both bacterial enzymes are inhibited by heme, the end product, with

[12] H. A. Dailey and A. Smith, Biochem J. **223**, 441 (1984).
[13] H. A. Dailey, J. E. Fleming, and B. M. Harbin, J. Bacteriol. 165, in press (1986).
[14] H. A. Dailey and J. E. Fleming, submitted for publication.

50% inhibition occurring at about 40–50 μM heme. Protoporhyrin concentration above 60 μM also appear inhibitory, but this may be more a reflection of the tendancy of protoporphyrin to aggregrate in solution than an actual inhibition of the enzyme since meso- and deuteroporphyrin, which do not aggregate as readily, do not show inhibition at this concentration.

Acknowledgment

Supported by Grants AM 32303 and AM 31064 from the National Institutes of Health.

[49] Uroporphyrinogen Decarboxylase Purification from Chicken Erythrocytes

By YOSHIHIKO SEKI, SHOSUKE KAWANISHI, and SEIYO SANO

Uroporphyrinogen decarboxylase (porphyrinogen carboxylyase, EC 4.1.1.37), is a cytosolic enzyme that catalyzes the decarboxylation of uroporphyrinogen I or III into coproporphyrinogen I or III through the intermediates of hepta-, hexa-, and pentacarboxylic porphyrinogens. The enzyme is of interest with regard to porphyria cutanea tarda,[1] hexachlorobenzene (HBC)-, 2,3,7,8-tetrachlorodibenzo-*p*-dioxin (TCDD)-, and polychlorinated biphenyl (PCB)-induced porphyria in humans[2] and in several animal species,[3-6] which display a massive hepatic accumulation and increased urinary excretion of uroporphyrin III and heptacarboxylic porphyrin.[7-10] In the present chapter, we describe an assay method, an improved method for the purification of uroporphyrinogen decarboxylase from chicken erythrocytes and its properties.[11]

[1] C. Cam and G. Nigogosyan, *JAMA*, **183**, 88 (1963).
[2] R. K. Ockner and R. Schmid, *Nature (London)* **189**, 499 (1961).
[3] F. De Matteis, B. E. Prior, and C. Rimington, *Nature (London)* **191**, 363 (1961).
[4] J. G. Vos and J. H. Koeman, *Toxicol. Appl. Pharmacol.* **17**, 656 (1970).
[5] J. A. Goldstein, P. Hickman, H. Bergman, and J. G. Vos, *Res. Commun. Chem. Pathol. Pharmacol.* **6**, 919 (1973).
[6] A. G. Smith, J. E. Francis, S. J. E. Kay, and J. B. Greig, *Biochem. Pharmacol.* **30**, 2825 (1981).
[7] L. C. San Martin De Viale, A. A. Viale, S. Nacht, and M. Grinstein, *Clin. Chim. Acta.* **28**, 13 (1970).
[8] J. A. Goldstein, P. Hickman, and D. L. Jue, *Toxicol. Appl. Pharmacol.* **27**, 437 (1974).
[9] S. Kawanishi, T. Mizutani, and S. Sano, *Biochim. Biophys. Acta* **540**, 83 (1978).
[10] S. Kawanishi, Y. Seki, and S. Sano, *FEBS Lett.* **129**, 93 (1981).
[11] S. Kawanishi, Y. Seki, and S. Sano, *J. Biol. Chem.* **258**, 4285 (1983).

Assay Method

Principle

Uroporphyrinogen decarboxylase activity is determined by measuring the amounts of hepta-, hexa-, and pentacarboxylic porphyrinogen III and coproporphyrinogen III formed from uroporphyrinogen III. Separation and quantification of free porphyrins with 8- to 4-carboxylic acid after oxidation of their corresponding porphyrinogens are performed using high-pressure liquid chromatograph (HPLC) equipped with a spectrofluorometeric detector.

Reagents

Assay buffer: 100 mM potassium phosphate buffer, pH 6.8, containing 0.1 mM EDTA, 1 mM GSH.

Substrate solution: uroporphyrin III was dissolved in a minimum volume of 0.01 N KOH. Uroporphyrinogen III was prepared by reduction of uroporphyrin III with freshly prepared 3% sodium amalgam under N_2 according to the method of Sano.[12]

Quinhydrone solution for oxidation of porphyrinogen: 15 mM quinhydrone in ethanol.

Solvent for HPLC: a mixture of acetonitrile and 10 mM KH_2PO_4 adjusted to pH 3.0 with H_3PO_4 (1:1, v/v).

Procedure

The standard reaction mixture contained 100 mM potassium phosphate buffer, pH 6.8, 0.1 mM EDTA, 1 mM GSH, and 10 μl of enzyme (0.4–2 μg) in a total volume of 1 ml. Each tube was fitted with a rubber serum tube cap and was sufficiently deoxygenated by bubbling with argon. Ten microliters of uroporphyrinogen III was then added through the cap using a Hamilton gas-tight syringe. The reaction was allowed to proceed at 37° for 30 min and stopped by cooling with ice. One hundred microliters of quinhydrone/ethanol was added to oxidize porphyrinogens to porphyrins.[12] After standing overnight at room temperature, the porphyrins were analyzed by HPLC.

Uroporphyrin, hepta-, hexa-, and pentacarboxylic porphyrins and coproporphyrin were separated directly using HPLC with a μBondapak C_{18} column (300 × 3.9 mm) and detected with a spectrofluorometer (excitation at 404 nm, emission at 620 nm) (Fig. 1).

---

[12] S. Sano, *J. Biol. Chem.* **241**, 5276 (1966).

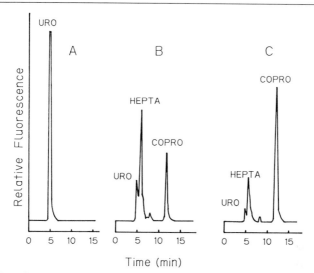

Fig. 1. HPLC separation of a mixture of porphyrins formed by uroporphyrinogen decarboxylase. The porphyrins were obtained from the oxidation of porphyrinogens formed during the decarboxylation of uroporphyrinogen III. Reaction time; 0 min (A), 15 min (B), 30 min (C). URO, Uroporphyrin; HEPTA, heptacarboxylic porphyrin; COPRO, coproporphyrin.

Enzyme Unit

One unit of uroporphyrinogen decarboxylase was defined as the amount of enzyme that catalyzes the formation of 1 nmol of coproporphyrinogen III per 1 hr in the standard assay system at 37°.

Purification Procedure

All operations were carried out at 4°. Fresh chicken blood (1 liter) containing heparin was used as an enzyme source. After centrifugation at 8000 g for 30 min, the plasma was removed and the erythrocytes were washed 3 times with cold isotonic NaCl. Packed erythrocytes were osmotically lysed by addition of 3 volumes of cold water. Stroma and unbroken cells were removed by centrifugation at 35,000 g for 60 min and the supernatant was dialyzed against 3 mM potassium phosphate buffer, pH 6.8, under N_2 overnight.

DEAE-Cellulose Chromatography

A column (6 × 20 cm) of DEAE-cellulose (DE52) was previously equilibrated with 3 mM potassium phosphate buffer, pH 6.8 containing

0.1 mM EDTA and 1 mM DTT or GSH. The dialyzed hemolysate was pumped through the column at a flow rate of 2 ml/min, and the column was washed with the same buffer until the eluates scarcely have a red color. The enzyme was eluted by a linear gradient of KCl from 0 to 0.15 M in the same buffer (2 liters) and the fractions exhibiting enzyme activity were combined.

Ammonium Sulfate Fractionation

Finely powdered ammonium sulfate was added to the eluates from DEAE-cellulose column to 45% saturation at pH 6.8, and the mixture was stirred for 30 min. The precipitate was removed by centrifugation at 12,000 g for 30 min and ammonium sulfate was added to the supernatant to 65% saturation. After 30 min stirring the enzyme was collected by centrifugation at 12,000 g for 30 min and dissolved in a minimum volume of 3 mM potassium phosphate, pH 6.8, containing 1 mM DTT.

Gel Filtration

The enzyme was applied to a Sephadex G-100 column (2.6 × 100 cm) which had been equilibrated with 50 mM potassium phosphate buffer, pH 6.8, containing 0.1 M KCl, 0.1 mM EDTA, and 1 mM DTT. The enzyme was eluted with the same solution at a flow rate of 0.4 ml/min; fractions with the highest specific activity were collected.

Chromatofocusing

The enzyme solution was then dialyzed against 25 mM imidazole–HCl buffer, pH 7.4, containing 1 mM DTT and applied to a column (1 × 18 cm) of Polybuffer Exchanger-94 previously equilibrated with the dialysis buffer. The enzyme was eluted with Polybuffer-74–HCl, pH 4.0, containing 1 mM DTT, at a flow rate of 0.4 ml/min. The purified enzyme was eluted at pH 6.2 and stored at −90°.

The table summarizes the result of a typical purification of uroporphyrinogen decarboxylase from 1 liter of chicken blood. Chromatofocusing, a new technique for separating the enzyme according to pI in self-generated pH gradients[13] was found to be a method of choice for final purification of this enzyme. The purified preparation showed a single band upon polyacrylamide slab gel electrophoresis in the presence of 0.1% SDS. The enzyme was purified approximately 5500-fold with a 12% overall yield and a specific activity of 1420 units/mg of protein.

---

[13] L. Soederberg, T. Lääs, and D. Low, *Biochem. Soc. Trans.* **9**, 302 (1981).

PURIFICATION OF UROPORPHYRINOGEN DECARBOXYLASE FROM CHICKEN ERYTHROCYTES[a]

	Volume (ml)	Total protein (mg)	Total activity (units)	Specific activity (units/mg)	Recovery (%)	Purification (fold)
Erythrocytes lysate	750	55,300	14,300	0.258	100	1
DEAE-cellulose	200	230	9,570	41.6	66.9	161
Ammonium sulfate fractionation (45–65%)	3.8	181	9,500	52.5	66.4	203
Sephadex G-100	14	18.8	4,210	224	29.4	868
Chromatofocusing	3.6	1.21	1,720	1,420	12.0	5,500

[a] One thousand milliliters.

Properties

Molecular Weight

The apparent molecular weight of uroporphyrinogen decarboxylase was estimated to be about 79,000 by Sephadex G-150 gel chromatography. On SDS–polyacrylamide gel electrophoresis, the enzyme revealed a molecular weight of 39,700. Sedimentation equilibrium analysis indicated that the enzyme had a molecular weight of about 42,000 and a tendency to self-associate at higher concentrations. Therefore, it seems likely that uroporphyrinogen decarboxylase is usually monomeric, but sometimes a dimer is formed.

Absorption Spectra

An absorption maximum was located at 282 nm with a shoulder at 293 nm. In the visible region, there was no significant spectral characteristics, suggesting the absence of pyridoxal phosphate, flavin, and heme.

Metal

Trace amounts of iron (0.15 atom/enzyme), copper (0.08 atom/enzyme), zinc (0.03 atom/enzyme), and manganese (0.01 atom/enzyme) were detected in the purified enzyme preparation. The enzyme activity was not inhibited by EDTA (5 mM) and o-phenanthroline (1 mM). These results suggest that the purified enzyme preparation contains no metal.

Most transition metals such as $MgCl_2$, $CrCl_3$, $FeSO_4$, $FeCl_3$, $Co(CH_3COO)_2$, and $ZnCl_2$ have little or no inhibitory effect on the enzyme activity at the concentration of 1 mM, but $HgCl_2$ and $CuCl_2$ produce significant inhibition at the same concentration.

Inhibitors

Uroporphyrinogen decarboxylase is completely inhibited by SH reagents (10 μM) such as p-chloromercuribenzoate, N-ethylmaleimide and 5,5'-dithiobis(2-nitrobenzoate), and diethylpyrocarbonate (1 mM). NaCl (0.3 M), KCl (0.3 M), ethanol (10%), PCB (10 μM), HCB (20 μM), SDS (0.015%), and unsaturated fatty acids (oleate, linolate, arachidonate, etc., 0.1 mM) also have strong inhibitory effect.

Activators

Detergents (Tritons, Tween, etc., 0.01%), saturated fatty acids (myristate, palmitate, stearate, etc., 0.1 mM), and phospholipids (phosphatidylcholine, lysophosphatidylcholine, etc., 0.01%) enhance the rate of reaction. For example, at a concentration of 0.01%, Triton X-100 enhances the rate of reaction 360%.

Kinetic Properties

The enzyme sequentially decarboxylates the four acetate side chains of uroporphyrinogen III to coproporphyrinogen III with the following kinetic properties.[14,15]

The first decarboxylation, i.e., the formation of heptacarboxylic porphyrinogen from uroporphyrinogen III, is faster than the second decarboxylation, the formation of coproporphyrinogen III from heptacarboxylic porphyrinogen.

The formation of coproporphyrinogen III is linear with time and the amounts of hexa- and pentacarboxylic porphyrinogens III are almost negligible.

When excess amount of substrate is used, heptacarboxylic porphyriogen III accumulates resulting in the inhibition of coproporphyrinogen III formation.

[14] R. C. Garcia, L. C. San Martin De Viale, J. M. Tomio, and M. Grinstein, *Biochim. Biophys. Acta* **309**, 203 (1973).
[15] A. G. Smith and J. E. Francis, *Biochem. J.* **195**, 241 (1981).

[50] Purification of Solubilized Chlorophyllase from *Chlorella protothecoides*

By YUZO SHIOI and TSUTOMU SASA

$$\text{Chlorophyll } a_{\text{phytol}} + \quad H_2O \rightarrow \text{chlorophyllide } a + \text{phytol} \qquad (1)$$
$$\text{Chlorophyll } a_{\text{phytol}} + \text{alcohol} \rightleftarrows \text{chlorophyll } a_{\text{alcohol}} + \text{phytol} \qquad (2)$$

Chlorophyllase (chlorophyllido-hydrolase, EC 3.1.1.14) is the chloroplast membrane enzyme and usually catalyzes both the hydrolysis [Eq. (1)] and alcohol transesterification (exchange reaction of esterifying groups) [Eq. (2)] of chlorophylls and their derivatives, but not protochlorophyll.[1] It has been previously suggested that this enzyme participates in the phytyl ester formation in the final step of chlorophyll biosyn-

[1] C. B. Jones and R. K. Ellsworth, *Plant Physiol.* **44**, 1478 (1969).

thesis. However, it has been proposed that another enzyme, chlorophyll synthase, is involved in this biosynthetic reaction.[2] The functional role of chlorophyllase in chlorophyll metabolism is still unclear.

Chlorophyllase reactions are a useful tool for preparing chlorophyllides and chlorophyll derivatives esterified with various alcohols.[3] In this chapter, a simple purification procedure for the preparation of solubilized chlorophyllase from *Chlorella protothecoides* is described.

Materials

The content of chlorophyllase in algae[4] and higher plants[5] varies greatly dependent on the plant species and their age. According to our results with several strains of *Chlorella,* only *C. protothecoides* was a rich source. The main advantage of using *Chlorella* as an enzyme source is that it does not contain phenolic compounds in contrast to higher plants. Moreover, homogeneous material is easily obtained throughout the year.

The green alga, *C. protothecoides,* was cultured in a medium containing 0.5% glycine and 0.1% potassium acetate as organic nutrients as reported previously.[6] The cells were harvested by centrifugation with a yield of about 6 g wet weight per liter of medium. After washing twice with distilled water, the cells were suspended in 10 mM potassium phosphate buffer (pH 7.2). The suspension was stored in the dark at $-18°$ until used.

Chlorophyll *a,* used as the substrate for the reaction, was extracted from spinach leaves with 80% acetone and was partially purified by precipitation with dioxane[7] and then by DEAE-Sepharose CL-6B column chromatography.[8] Chlorophyll *a* was finally dissolved in absolute acetone.

Assay method

Method 1: Hydrolytic Activity [Eq. (1)]

Principle. Chlorophyllase activity is conveniently determined by the rate of chlorophyllide *a* formation. This method was based on the parti-

---

[2] W. Rüdiger, J. Benz, and C. Guthoff, *Eur. J. Biochem.* **109,** 193 (1980).

[3] Y. Shioi, R. Fukae, and T. Sasa, *Biochim. Biophys. Acta* **722,** 72 (1983).

[4] J. Barrett and S. W. Jeffrey, *J. Exp. Mar. Biol. Ecol.* **7,** 255 (1971).

[5] M. Holden, *Photochem. Photobiol.* **2,** 175 (1963).

[6] Y. Chiba, I. Aiga, M. Idemori, Y. Satoh, K. Matsushita, and T. Sasa, *Plant Cell Physiol.* **8,** 623 (1967).

[7] K. Iriyama, N. Ogura, and A. Takamiya, *J. Biochem. (Tokyo)* **76,** 901 (1974).

[8] T. Omata and N. Murata, *Photochem. Photobiol.* **31,** 183 (1980).

tion of chlorophyll and chlorophyllide between *n*-hexane and aqueous acetone. The increase in formed product, chlorophyllide *a* concentration in the aqueous acetone layer is measured spectrophotometrically.

Reagents

40 mM phosphate buffer (pH 7.0)
0.25 M NaOH
0.6 mM chlorophyll *a*
Acetone/*n*-hexane (1:2, v/v) mixture

Procedure. A standard reaction mixture contained in 2 ml: 40 μmol of phosphate buffer (pH 7.0), 0.06 μmol of chlorophyll *a* dissolved in acetone (final acetone concentration was 25%), and the enzyme preparation. The reaction mixture was incubated at 30° for 10 min in the dark in 10-ml centrifuge tubes with stoppers and stopped by the addition of 0.2 ml of 0.25 M NaOH and 6 ml of acetone/*n*-hexane mixture (1:2, v/v). The mixture was shaken, followed by centrifugation. NaOH was added to ionize the carboxyl group for complete separation of the hydrolyzed product in the partition mixture.[9] Chlorophyllase activity was determined by measuring the absorbance at 667 nm in the aqueous acetone layer. Millimolar absorption coefficient of 74.9 at 667 nm[9] was used for the calculation of chlorophyllide *a* concentration. One unit of enzyme activity was defined as the amount of enzyme hydrolyzing 1.0 nmol of chlorophyll *a* per min.

Method 2: Alcohol Transesterifying Activity [Eq. (2)]

Principle. Alcohol transesterifying activity is determined by the use of methanol as an alcohol and by measuring the formation of methylchlorophyllide. In this reaction, chlorophyllide formation is negligible and methylchlorophyllide can be identified and quantified by thin-layer chromatography.

Reagents

40 mM phosphate buffer (pH 7.0)
0.6 mM chlorophyll *a*
Methanol
Diethyl ether
n-Hexane/benzene/acetone/methanol/water (130:70:50:20:2, v/v)
   mixture
Cellulose thin-layer plate

[9] R. F. McFeeters, C. O. Chichester, and J. R. Whitaker, *Plant Physiol.* **47**, 609 (1971).

Procedure. A standard reaction mixture comprised in 2 ml: 40 μmol phosphate buffer (pH 7.0), 0.2 μmol methanol, 0.1 μmol chlorophyll *a* dissolved in acetone (final acetone concentration was 25%), and enzyme solution. The reaction was carried out at 30° for 10 min in the dark in 10-ml centrifuge tubes with stoppers. The reaction was terminated by adding 2 ml of acetone and 5 ml of diethyl ether. The mixture was decanted into a separating funnel, and then gently shaked. The ether layer containing all the pigments was washed with distilled water several times while checking fluorescence of the washings. The ether solution was evaporated under nitrogen gas and the pigment residue was dissolved in a small volume of diethyl ether. The sample was subjected to the cellulose thin-layer chromatography using a solvent system of *n*-hexane/benzene/acetone/methanol/water (130:70:50:20:2, v/v). The concentration of each pigment band was measured spectrophotometrically and the percentage conversion was calculated. The total amount of product formed was then calculated from the known substrate concentrations used. One unit of enzyme activity was defined as the amount of enzyme that catalyzes the formation of 1.0 nmol of methylchlorophyllide *a* per min.

Purification

All procedures were carried out at 0–4°.

Step 1: Butanol Solubilization. Chlorophyllase was generally found to be bound to the membranes. Thus, soluble enzyme could be obtained only after treating the homogenate with organic solvent or detergent. Generally, Triton X-100[5,9] and cholate[10,11] are used for the solubilization of chlorophyllase. In the case of *Chlorella* enzyme, butanol treatment was convenient. Cell paste (about 50 g wet weight) was suspended in 200 ml of 10 mM phosphate buffer (pH 7.2). The suspension was treated with a 20-kHz ultrasonic oscillator for 30 min in 3-min periods or a Braun cell homogenizer for 2 min. To the homogenate, NaCl was added to 1% and *n*-butanol (−18°) added in a volume equal to that of the suspension. To obtain optimum yield during solubilization, the homogenate was further treated with a sonic oscillator for 2–3 min or a Braun cell homogenizer for 20 sec. However, prolonged butanol treatment caused a decrease in the total activity owing to denaturation of the enzyme.[12] The mixture was centrifuged at 12,000 g for 20 min and was separated into three layers; the

[10] K. Tanaka, T. Kakuno, J. Yamashita, and T. Horio, *J. Biochem. (Tokyo)* **92,** 1763 (1982).
[11] K. Shimokawa, *Phytochemistry* **21,** 543 (1982).
[12] Y. Shioi, H. Tamai, and T. Sasa, *Anal. Biochem.* **105,** 74 (1980).

upper *n*-butanol layer of green, the middle aqueous one of pale yellow, and the bottom containing cell debris. The aqueous solution from the middle layer was collected as the butanol fraction and again centrifuged at 12,000 *g* for 10 min to remove butanol. Our experiments indicate that butanol causes reversible inhibition of the enzyme activity in addition to latent enzyme inactivation. Thus, to obtain a high yield, care was taken to prevent enzyme denaturation after butanol solubilization by immediate removal of butanol by ammonium sulfate fractionation instead of by prolonged dialysis. In addition, butanol also causes the deterioration of Sephadex G-200.

Step 2: Ammonium Sulfate Fractionation. Ammonium sulfate was added with stirring to the butanol fraction to give 30% saturation. After further stirring for 20 min, the solution was centrifuged at 10,000 *g* for 20 min. The protein paste formed was collected on a nylon net and washed carefully with about 100 ml of 30% saturated ammonium sulfate solution to remove butanol. The precipitate was dissolved in a small volume of 10 m*M* phosphate buffer (pH 7.2) and clarified by centrifugation at 10,000 *g* for 10 min.

Step 3: First Sephadex G-200 Gel Filtration. The ammonium sulfate preparation was applied to a column of Sephadex G-200 (2.6 × 70 cm) equilibrated with 10 m*M* phosphate buffer (pH 7.2) containing 0.1 *M* NaCl, and eluted with the same buffer. The enzyme activity was in the void volume of the Sephadex G-200 column, indicating large molecular aggregates. Low-molecular-weight proteins were, therefore, simply removed by this molecular sieving without considerable loss of the yield. The active fractions were pooled and concentrated with Minicon B15 (Amicon Co.).

Step 4: Sepharose CL-6B Gel Filtration. Next, high-molecular-weight proteins were separated using a Sepharose CL-6B column chromatography. The concentrated enzyme solution was subjected to a column of Sepharose CL-6B (1.6 × 60 cm) equilibrated with 20 m*M* phosphate buffer (pH 7.2) containing 0.1 *M* NaCl, and eluted with the same buffer. The two protein peaks were eluted by this chromatography. Most of the enzyme activity appeared in the second peak, and sometimes a small amount was found in the first peak. These results indicate that two different molecular masses of aggregated enzyme are present. The most active fractions were collected and concentrated with Minicon B15. At the first application to the new Sepharose column, the enzyme was slightly adsorbed onto the gel. However, the amount of adsorption was lowered by repeated use of the same column.

Step 5: Second Sephadex G-200 Gel Filtration. The concentrated enzyme solution was again applied to a column of Sephadex G-200 (1.6 × 60

TABLE I
PURIFICATION OF CHLOROPHYLLASE FROM *Chlorella protothecoides*[a]

Purification step	Protein (mg)	Total activity (units)	Specific activity (units/mg protein)	Purification (fold)	Yield (%)
n-Butanol extract[b]	1105	1436	1.3	1	100
Ammonium sulfate (0–30% saturation)	12.4	1561	126	97	109
First Sephadex G-200	5.50	1359	247	190	95
Sepharose CL-6B[c]	1.31	902	689	530	63
Second Sephadex G-200	0.79	758	960	738	53

[a] Chlorophyllase was extracted from 50 g (wet wt) of cells.
[b] Enzyme assay was carried out with 40% (v/v) acetone.
[c] Sepharose CL-6B was repeatedly used to minimize loss by adsorption of the enzyme.

cm) containing 0.1 M NaCl. The most active fractions were pooled and used as the purified sample.

The enzyme was purified about 738-fold from the butanol extract with a recovery of 53%. Summary of the enzyme purification is presented in Table I.

Comments

Optimal acetone concentration for the assay of chlorophyllase depends on the degree of purification.[13] When acetone was added to the phosphate buffers, a rise of pH was always observed. Therefore, the pH of all reaction mixtures should be adjusted after addition of acetone.

Properties

Chlorophyllase prepared by the method described is homogeneous (around peak fractions) as determined by polyacrylamide gel electrophoresis. *Chlorella* enzyme is an acidic protein and does not require metal ion and thiol compound for activity. This enzyme shows high-molecular-weight aggregates similar to other enzymes.[5,14] The minimum molecular weight can be determined under dissociation conditions such as by 1% sodium lauryl sulfate at 50° for 2 hr and at 30° for 12 hr. The enzyme has a

[13] N. Ichinose and T. Sasa, *Plant Cell Physiol.* **14,** 1157 (1973).
[14] M. Kuroki, Y. Shioi, and T. Sasa, *Plant Cell Physiol.* **22,** 717 (1981).

TABLE II
CHLOROPHYLLASE DERIVED FROM VARIOUS SOURCES

Source	Molecular weight	pI	pH optimum	K_m (Chl a) (μM)	Reference
Chlorella protothecoides	38,000[a]	4.5	6.0–8.5	2	12
Chlorella vulgaris	30,000	—	7.2–7.3	70	15
Sugar beet	30,000–38,000	—	7.1	—	16
Tea leaf sprouts	38,000[a]	3.9	5.5	10	14
Greened rye seedlings	39,000	4.5	7.5	12	10
Citrus unshu (fruit peel)[b]	27,000	—	7.0	2.65	11

[a] Minimum molecular weight.
[b] Ethylene induced.

minimum molecular weight of 38,000, which falls within the range 27,000–39,000 of other chlorophyllases.

Kinetic studies with purified enzyme[13] indicated that *Chlorella* enzyme consists of at least two enzymes. One enzyme catalyzes hydrolysis of chlorophylls [Eq. (1)] and the other, alcohol transesterification of chlorophylls and its reverse reaction [Eq. (2)]. Other physicochemical and kinetical properties of the *Chlorella* enzyme and the purified enzyme from various sources are listed in Table II.[15,16]

[15] P. Böger, *Phytochemistry* **4**, 435 (1965).
[16] M. F. Bacon and M. Holden, *Phytochemistry* **9**, 115 (1970).

[51] Purification of Porphobilinogen Synthase from Bovine Liver

By PETER M. JORDAN and JASBIR S. SEEHRA

Porphobilinogen synthase (5-aminolevulinic acid dehydratase, EC 4.2.1.24), the second enzyme of the tetrapyrrole biosynthesis pathway, catalyzes the condensation between two molecules of 5-aminolevulinic acid to form the pyrrole porphobilinogen.[1] The enzyme was first described

[1] D. Shemin, *Philos. Trans. R. Soc. London, Ser. B* **273**, 109 (1976).

independently in the laboratories of Shemin[2] and Neuberger[3] and since that time 5-aminolevulinic acid dehydratases have been isolated and characterized from a wide variety of sources.[4-6] The relatively high level of 5-aminolevulinic acid dehydratase in bovine liver has made this source particularly convenient for the isolation of enzyme for the large scale enzymatic preparation of porphobilinogen. In addition, the near neutral pH optimum of the bovine enzyme makes it more attractive than the bacterial dehydratase (pH optimum 8.4[7]) for coupling with 5-aminolevulinic acid synthetase[8] in the preparation of stereospecifically labeled porphobilinogen.[9]

Assay

The product of the enzymic reaction, porphobilinogen, is estimated using modified Ehrlich's reagent.[10] Since the reaction of pyrroles with Ehrlich's reagent is affected by the presence of thiols, dithioerythritol, used to stabilize the enzyme, is removed prior to the addition of the reagent by precipitation with $HgCl_2$.

Reagents. (B.D.H., Poole, Dorset, U.K.)

0.2 *M* potassium phosphate buffer, pH 6.8, containing 20 m*M* dithioerythritol (Sigma)
50 m*M* 5-aminolevulinic acid hydrochloride (Sigma) neutralized before use to pH 7
0–0.015 units 5-aminolevulinic acid dehydratase, activated as described below
Porphobilinogen prepared as described below
10% w/v trichloroacetic acid containing 0.1 *M* $HgCl_2$
Modified Ehrlich's reagent[10] (1 g *p*-dimethylaminobenzaldehyde dissolved in 42 ml glacial acetic acid and made up to 50 ml with 8 ml perchloric acid (60–62% w/v), prepared daily

[2] R. Schmid and D. Shemin, *J. Am. Chem. Soc.* **77,** 506 (1955).
[3] K. D. Gibson, A. Neuberger, and S. Scott, *Biochem. J.* **61,** 618 (1955).
[4] D. Shemin, *in* "The Enzymes" (P. D. Boyer, ed.), 3rd ed., Vol. 7, p. 323, and references therein. Academic Press, New York, 1972.
[5] A. Cheh and J. B. Neilands, *Struct. Bonding (Berlin)* **29,** 123 and refs. therein (1976).
[6] M. Akhtar and P. M. Jordan, *Comp. Org. Chem.* **5,** 1121 (1978).
[7] D. L. Nandi, K. F. Baker-Cohen, and D. Shemin, *J. Biol. Chem.* **243,** 1224 (1968).
[8] P. M. Jordan and A. Laghai-Newton, this volume [52].
[9] M. Akhtar and C. Jones, this volume [43].
[10] D. Mauzerall and S. Granick, *J. Biol. Chem.* **219,** 435 (1956).

Method

The assay mixture for 5-aminolevulinic acid dehydratase contained in a final volume of 1 ml: 500 μl potassium phosphate buffer containing dithioerythritol, 100 μl of neutralized 5-aminolevulinic acid, distilled water and activated enzyme. After preincubation for 2 min at 37°, the reaction was started by the addition of substrate. After 10 min at 37° the reaction was terminated by the addition of 1 ml of trichloroacetic acid/HgCl$_2$. The precipitate was removed by centrifugation and 1.5 ml of supernatant was added to 1.5 ml of modified Ehrlich's reagent. The magenta complex was measured at 555 nm after allowing 10 min for color development ($\varepsilon = 60,200$ mol/liter).

One unit of enzyme catalyzes the formation of 1 μmol of porphobilinogen per hr per mg of protein under the above conditions.

Activation of 5-Aminolevulinic Acid Dehydratase

The sensitivity of mammalian 5-aminolevulinic acid dehydratase to oxygen and heavy metals and its obligatory requirement for an exogenous thiol for maximum catalytic activity are well established. In order to ensure that all experiments were conducted with fully active, reduced enzyme, 5-aminolevulinic acid dehydratase was dissolved in 0.1 M potassium phosphate buffer, pH 6.8, containing 10 mM dithioerythritol, and was incubated for 10 min at 37° or for 20 min at 20° to activate the enzyme completely. The activated enzyme was then dialyzed overnight at 0° against the same buffer to remove ammonium sulfate.

Isolation of 5-Aminolevulinic Acid Dehydratase from Bovine Liver

Initial Extract

Bovine liver (5–7 kg) was collected *immediately* after slaughter of the animal and rapidly cooled in an ice/water slurry. After removal of the fibrous portions, the remaining tissue (4 kg) was divided into eight 500 g batches and each was homogenized for 5 min in 300 ml of 0.05 M potassium phosphate buffer, pH 6.8, containing 75 mM 2-mercaptoethanol. The total volume of the homogenate was 6.5 liters.

First Heat Treatment

The homogenate was divided into three portions and each was heat treated to 60° in a 5-liter Erlenmeyer flask with continuous swirling while held in a boiling water bath over a period of 6–8 min. The temperature of

60° was maintained for 5 min after which time 500 g of "frozen" ($-15°$) 0.05 M potassium phosphate buffer, pH 6.8, containing 25 mM 2-mercaptoethanol was added to each flask. Cooling was continued until the temperature was below 7° and the denatured protein was then removed by centrifugation at 2000 g for 20 min. The enzyme was precipitated by treating the supernatant (4.7 liters) with solid ammonium sulfate to give 45% saturation. After 20 min the precipitate was collected by centrifugation at 18,000 g for 15 min.

Second Heat Treatment and Ammonium Sulfate Fractionation

The ammonium sulfate precipitate was dissolved in 1.2 liter of 0.05 M potassium phosphate buffer, pH 6.8, containing 50 mM 2-mercaptoethanol. The extract was heated with swirling to 70° over a period of 3–4 min using a boiling water bath. This temperature was maintained at 70° for a further 2 min and then the flask was rapidly cooled to 10° in an ice/water slurry. After centrifugation at 18,000 g for 15 min, the supernatant was fractionated with solid ammonium sulfate to give saturations of 0–30, 30–45, and 45–60%. The most active fraction (normally 30–45%) was dissolved in 300 ml of 0.03 M potassium phosphate buffer, pH 7.9, containing 25 mM 2-mercaptoethanol, and was dialyzed against the same buffer overnight.

DEAE BioGel Chromatography

The dialyzed enzyme was applied to a column (6.5 × 15 cm) of DEAE-BioGel, which had previously been equilibrated with 0.03 M potassium phosphate buffer, pH 7.9, containing 25 mM 2-mercaptoethanol. The column was washed with the same buffer (2 liters) to remove any loosely bound protein. A gradient of KCl (500 ml total vol; 0–0.4 M) in the same buffer was applied to elute the enzyme. The fractions containing the enzyme were pooled, treated with 55% saturated ammonium sulfate, and the precipitate was collected by centrifugation at 50,000 g for 10 min. The precipitate was dissolved in 50 ml of 0.05 M potassium phosphate buffer, pH 6.8, containing 25 mM 2-mercaptoethanol, and dialyzed against the same buffer overnight.

Hydroxyl Apatite Chromatography

The dialyzed enzyme was applied to a column of hydroxylapatite (6.5 × 8 cm) which had been equilibrated with 0.05 M potassium phosphate buffer, pH 6.8, containing 25 mM 2-mercaptoethanol. The column was washed with 1 liter of the same buffer to remove the loosely bound proteins. The strongly bound proteins were then eluted with a potassium phosphate gradient (600 ml total vol; 0.05–0.3 M, pH 6.8) containing 25 mM 2-mercaptoethanol. The fractions containing the enzyme were pooled, precipitated with solid ammonium sulfate to 55% saturation, and centrifuged at 50,000 g for 10 min.

The enzyme was reprecipitated from 0.1 M potassium phosphate buffer, pH 6.8, containing 10 mM dithioerythritol with solid ammonium sulfate to 55% saturation. The precipitate was collected by centrifugation at 50,000 g for 10 min and stored at 0° in tightly stoppered tubes. Under these conditions the enzyme retained at least 80% of its activity after 1 month. Details of the purification stages are shown in the table.

Enzymic Synthesis of Porphobilinogen

5-Aminolevulinic acid dehydratase (500 units) was incubated at 37° for 15 min in 5 ml 0.1 M potassium phosphate buffer (pH 6.8) containing 10 mM dithioerythritol and added to 2 litres 10 mM potassium phosphate buffer containing 10 mM dithioerythritol. 5-Aminolevulinic acid hydrochloride (1 g), adjusted to pH 6.8 by the dropwise addition of 0.1 M NaOH, was added and the mixture was incubated for 17 hr at 37° in the dark under an atmosphere of nitrogen. A yield of 0.71 g porphobilinogen was obtained as determined by reaction with modified Ehrlich's reagent. The pH of the incubation solution was adjusted to 7.5 using 0.1 M NaOH and slowly applied to a column (2 × 12 cm) of Dowex 1-X8 (200–400 mesh) acetate. The column was washed with 1 liter of water, the porphobilinogen was eluted with 1 M acetic acid, and was isolated by freeze-drying. The porphobilinogen was further purified by isoelectric point recrystallization (in a minimum volume) from dilute ammonia by adjusting the pH to 5.5 with 0.1 M acetic acid. The crystals were filtered and washed consecutively with ice-cold methanol and dry ether and were stored at −20° under vacuum.

Biosynthesis of Stereospecifically Labeled Porphobilinogen

Stereospecifically tritiated 5-aminolevulinic acid, prepared enzymatically using purified 5-aminolevulinic acid synthase from *Rhodopseudomonas spheroides*,[8] was converted into porphobilinogen using the dehydratase purified above. When the labeling was required in the acetic acid and propionic acid side chains of porphobilinogen, ~350 units of dehydratase was used. For the labeling of the aminomethyl side chains of porphobilinogen, 20 units of enzyme was employed. In both cases before use the enzyme was activated and dialyzed as above. These procedures are fully described in this volume.[9]

PURIFICATION OF 5-AMINOLEVULINIC ACID DEHYDRATASE FROM 4 kg BOVINE LIVER

Step	Volume (ml)	Activity (units/ml)	Protein concentration (mg/ml)	Total activity (10^3/units)	Specific activity (units/mg)	Yield (%)
First heat treatment	4500	7.78	11.1	35	0.7	—
First ammonium sulfate precipitation	1250	25.6	21.3	32	1.2	91
Second heat treatment	1200	25.8	8.6	31	3.0	89
Second ammonium sulfate (30–45% fraction)	300	93.3	17	28	5.5	80
DEAE-BioGel chromatography	70	385	18.8	27	20.5	77
Hydroxylapatite chromatography	80	310	8.1	24.7	38	71

Purity of the Enzyme

The enzyme showed a single but broad protein band (mobility 0.38) on staining with Coomassie Brilliant Blue after electrophoresis on 5% polyacrylamide gels. Assay of a duplicate gel revealed enzyme activity coincident with the protein band. Enzyme which had been S-carboxymethylated to mask all thiol groups showed a single sharp protein band on gels suggesting that during electrophoresis of the native enzyme several closely related partially oxidized species were present. The specific activity of 38 units/mg of the native enzyme is the highest reported purity of the enzyme. Altogether, the procedure yielded 650 mg of homogeneous enzyme.

Molecular Weight

The subunit molecular weight was found to be 35,000 by the method of Weber and Osborn[11] using polyacrylamide gel electrophoresis in the presence of sodium dodecyl sulfate. A molecular weight of 280,000 was determined by the native enzyme by gel filtration using BioGel A 0.5M, confirming the existence of a homooctameric structure, as previously suggested.[12]

Properties of the Enzyme

The presence of 20 mM exogenous thiol is essential to maintain the enzyme in its active reduced state. On removal of the thiol, the enzyme is rapidly and reversibly oxidized to an inactive form which can be fully reactivated by readdition of the thiol.[13] Extensive investigations on the nature of the thiol-dependent activation of the enzyme have shown that two highly reactive SH groups are present at or near the active site. Reaction with oxygen or DTNB[13] causes the formation of a disulfide bond and total but reversible loss of catalytic activity. Alkylation of the enzyme by reaction with ^{14}C-labeled iodoacetic acid or iodoacetamide[14] leads to irreversible loss of enzyme activity and the incorporation of radioactivity into the protein. Active site directed inhibitors such as 5-chlorolevulinic, 5-bromolevulinic acid, and 5-iodolevulinic acid all inactivate the enzyme at low concentrations and specifically alkylate a single thiol group.[15]

[11] K. Weber and J. Osborn, *J. Biol. Chem.* **244,** 4406 (1969).
[12] W. H. Wu, D. Shemin, K. E. Richards, and R. C. Williams, *Proc. Natl. Acad. Sci. U.S.A.* **71,** 1767 (1974).
[13] J. S. Seehra, M. G. Gore, A. G. Chaudhry, and P. M. Jordan, *Eur. J. Biochem.* **114,** 263 (1981).
[14] G. F. Barnard, R. Itoh, L. H. Hohberger, and D. Shemin, *J. Biol. Chem.* **252,** 8965 (1977).
[15] J. S. Seehra and P. M. Jordan, *Eur. J. Biochem.* **113,** 435 (1981).

Interaction with Metals

In contrast to the bacterial enzyme which requires K^+ ions for maximum activity,[4] the bovine, and indeed all mammalian dehydratases, require Zn^{2+}.[5] The metal can be removed from the enzyme by treatment with EDTA leading to almost complete loss of activity which can be restored on readdition of Zn^{2+} ions. It is not normally necessary to add Zn^{2+} ions to the enzyme since there are sufficient traces present in the buffer and thiol. The human erythrocyte dehydratase in contrast loses about 20% of the metal ion during purification and additional Zn^{2+} ions are required to maximize activity.[16] There is good evidence that at least one of the Zn^{2+} ligands is a thiol group.[17] The bovine enzyme is severely inhibited by heavy metals such as lead and mercury and an extensive literature is available on the biochemical and medical implications.[18]

Mechanism of Action

5-Aminolevulinic acid dehydratase catalyzes what is essentially a Knorr condensation reaction. Extensive studies on the bacterial enzyme established that the dehydratase could form a Schiff base with the substrate, the competitive inhibitor levulinic acid or any other γ-keto acid. These findings together with the observation that the enzyme catalyses the formation of a "mixed" pyrrole from one molecule of substrate and one molecule of levulinic acid prompted the proposal of a mechanism for the formation of porphobilinogen[19] which has been accepted for several years. More recent studies with enzyme from bovine liver,[20] human erythrocytes[21] and *R. spheroides*,[22] using a single enzyme turnover reaction technique, have led to the formulation of a new mechanism in which the first molecule of substrate which binds to the enzyme contributes to the "propionic acid side" of the product rather than the "acetic acid side" suggested by other workers.[19] Furthermore the first molecule of substrate which binds to the enzyme does so through a Schiff base with an amino group at the enzyme active site.

[16] P. N. B. Gibbs, A. G. Chaudhry, and P. M. Jordan, *Biochem. J.* **230**, 25 (1985).
[17] P. N. B. Gibbs, M. G. Gore, and P. M. Jordan, *Biochem. J.* **225**, 573 (1985).
[18] B. L. Vallee and D. D. Ulmer, *Annu. Rev. Biochem.* **41**, 91 (1972).
[19] D. L. Nandi and D. Shemin, *J. Biol. Chem.* **243**, 1236 (1968).
[20] P. M. Jordan and J. S. Seehra, *FEBS Lett.* **114**, 283 (1980).
[21] P. M. Jordan and P. N. B. Gibbs, *Biochem. J.* **227**, 1015 (1985).
[22] J. S. Seehra, Ph.D. Thesis, University of Southampton (1979).

[52] Purification of 5-Aminolevulinate Synthase

By PETER M. JORDAN and AGHDAS LAGHAI-NEWTON

$$NH_2—CH_2—COOH + CoAS—CO—CH_2—CH_2—COOH$$
$$\rightarrow NH_2—CH_2—CO—CH_2—CH_2—COOH + CoASH + CO_2$$

5-Aminolevulinate synthase (EC 2.3.1.37) catalyzes the formation of 5-aminolevulinic acid from glycine and succinyl-CoA.[1] The enzyme has been described in all living systems which biosynthesize tetrapyrroles with the exceptions of plants and algae in which glutamate appears to be the carbon precursor.[2] 5-Aminolevulinate synthase was first described in chicken erythrocytes[3] and *Rhodopseudomonas spheroides*,[4] and shown to require pyridoxal phosphate for catalytic activity. The enzyme has since been purified from a wide variety of sources,[5,6] but the higher levels in *R. spheroides* make this organism the choice for isolation of the larger amounts required for biosynthetic studies on tetrapyrroles.

Assay

Principle. 5-Aminolevulinic acid, produced enzymatically, is first condensed with acetyl acetone and the resulting pyrrole is estimated by reaction with modified Ehrlich's reagent.[7]

Reagents. (B.D.H. Poole, Dorset, U.K.)

0.25 M potassium phosphate buffer, pH 6.8
0.25 M glycine
15 mM Na$_2$EDTA
0.5 mM pyridoxal-5'-phosphate
0.05–0.2 units 5-aminolevulinate synthase
10 mM succinyl-CoA (see below)
CoA lithium salt

[1] P. M. Jordan and D. Shemin, *in* "The Enzymes" (P. D. Boyer, ed.), 3rd ed., Vol. 7, p. 339, and references therein. Academic Press, New York, 1972.

[2] P. A. Castelfranco and S. I. Beale, *Annu. Rev. Plant Physiol.* p. 241 and references therein (1983).

[3] K. D. Gibson, W. G. Laver, and A. Neuberger, *Biochem. J.* **70,** 71 (1958).

[4] G. Kikuchi, A. Kumar, P. Talmage, and D. Shemin, *J. Biol. Chem.* **233,** 1214 (1958).

[5] M. Akhtar and P. M. Jordan, *Comp. Org. Chem.* **5,** 1121 and references therein (1978).

[6] S. Granick and S. I. Beale, *Adv. Enzymol.* **46,** 33 and references therein (1978).

[7] D. Mauzerall and S. Granick, *J. Biol. Chem.* **219,** 435 (1956).

Trichloroacetic acid, 10% w/v

3 M Sodium acetate buffer, pH 4.6

Acetyl acetone (freshly distilled)

Modified Ehrlich's reagent[7] (1 g p-dimethylaminobenzaldehyde dissolved in 42 ml glacial acetic acid and made up to 50 ml with 8 ml perchloric acid (60–62% w/v) prepared daily, CARE

Succinyl-CoA. Succinyl-CoA was prepared immediately before use by stirring succinic anhydride (1 mg), sodium bicarbonate (4.5 mg), and CoA (8 mg) in 1 ml of water for 30 min at 0°.[8] Alternatively, crystalline succinyl-CoA was prepared as follows. To a round bottomed flask held in ice and fitted with a mechanical stirrer and pH electrode were added 30 mg of CoA (Boehringer) and 20 ml of distilled water. The pH was carefully adjusted to 8.5 by the dropwise addition of 1% w/v ammonia (CARE). Freshly crushed succinic anhydride (4 mg) was added and the pH was maintained at 8.3 by the further addition of a few drops of 1% ammonia. Stirring was continued for 15 min. Additional succinic anhydride was added if required and stirring was carried out for a further 10 min until the CoA had been consumed. (This was assessed by taking 5-μl aliquots from the mixture at various times and reacting them with 1 mg DTNB[9] in 1 ml of 10 mM potassium phosphate buffer, pH 6.8.) When the reaction was complete, the pH was adjusted to 5 with a few drops of dilute HCl and the succinyl-CoA was lyophilized yielding white crystals. The succinyl-CoA was stored, without further purification, at $-20°$ in a vacuum desiccator until required.

Method. The assay mixture for 5-aminolevulinate synthase contained in a final volume of 170 μl: 25 μl glycine, 25 μl buffer, 25 μl EDTA, 25 μl pyridoxal phosphate, and 50 μl enzyme/H$_2$O. After preincubation for 2 min at 37°, the reaction was started by the addition of 20 μl of succinyl-CoA. Control incubations which were terminated at zero time contained boiled enzyme or were carried out in the absence of succinyl-CoA. After incubation at 37° for 10 min, the reaction was terminated by the addition of 150 μl of 10% TCA. If a precipitate was formed (at early stages of the enzyme isolation) the solution was clarified by centrifugation. To the solution in test tubes, 300 μl of 3 M sodium acetate buffer, pH 4.6, and 25 μl of acetylacetone were added. The tubes were heated at 100° for 10 min in a boiling water bath after which time they were cooled and treated with an equal volume (645 μl) of modified Ehrlich's reagent.[7] After color development for exactly 15 min at room temperature, the optical density was measured at 553 nm ($\varepsilon = 68,000$ mol/liter). A standard curve for

[8] E. J. Simon and D. Shemin, *J. Am. Chem. Soc.* **75**, 2520 (1953).

[9] G. L. Ellman, *Arch. Biochem. Biophys.* **82**, 70 (1959).

quantitative conversion of optical densities to μmol was obtained using 5-aminolevulinic acid (Sigma) standard (0–0.02 μmol). One unit of enzyme catalyses the formation of 1 μmol of 5-aminolevulinic acid in 1 hr at 37° under the above assay conditions.

Growth and Harvesting of Bacteria

Rhodopseudomonas spheroides (Torrey Research Station, Aberdeen, Scotland) (NCIB 8253) was maintained in 15–20 ml stab cultures essentially in Lascelles medium M.S.[10] containing yeast extract (0.2% w/v) and Oxoid agar no. 3 (1.5% w/v). The Lascelles medium M.S.[10] used for growing *R. spheroides* contained per 1 liter of distilled water: sodium L-glutamate monohydrate, 1.9 g; DL-malic acid, 2.7 g; potassium dihydrogen orthophosphate, 500 mg; dipotassium hydrogen orthophosphate, 500 mg; diammonium hydrogen orthophosphate, 800 mg; magnesium sulfate heptahydrate, 400 mg; calcium chloride, 40 mg; manganese sulfate tetrahydrate, 0.85 mg; thiamin hydrochloride, 1 mg; nicotinic acid, 1 mg; and d-biotin, 0.05 mg. The pH was adjusted to 6.8 with 5 M sodium hydroxide and the medium was sterilized by autoclaving for 15 min at 15 psi. Cultures were inoculated monthly and grown semianaerobically in the light at 30–32° for 3–4 days. The cultures were then stored at 0–4° or used to inoculate 150 ml of the medium M.S. supplemented with 0.5% (w/v) yeast extract in 250-ml flasks. The flasks were incubated at room temperature under light (60-W bulb, 30 cm distant) for 3–4 days. The bacteria thus grown were used to inoculate six 4-liter lots of the medium M.S. in 5-liter flasks. Semianaerobic growth was continued for 4–5 days at 30–32° using two 60-W bulbs for illumination after which the cells were harvested by centrifugation at 50,000 rpm using a Sharples continuous flow Super Centrifuge. The bacteria were washed once with 0.1 M Tris–HCl buffer, pH 7.5, containing 10% (v/v) glycerol, 5 mM 2-mercaptoethanol, and 0.5 mM Na$_2$EDTA and recentrifuged at 10,000 g. The bacteria were stored frozen (−15°) in batches from 24-liter growths.

Enzyme Purification

The enzyme 5-aminolevulinate synthase was purified by a procedure which incorporates some stages from the original method of Warnick and Burnham.[11] All operations were carried out at 0–4° unless stated otherwise.

---

[10] J. Lascelles, *Biochem. J.* **62,** 78 (1956); **72,** 508 (1959).
[11] G. R. Warnick and B. F. Burnham, *J. Biol. Chem.* **246,** 6880 (1971).

Disruption of the Bacterial Cells. Frozen bacteria, harvested from 48 liters of Lascelles medium M.S., were thawed and suspended in 100 mM Tris–HCl buffer, pH 7.5, containing 5 mM 2-mercaptoethanol, 0.5 mM Na$_2$EDTA, and 10% (v/v) glycerol, to a final volume of 250 ml. The suspension was disrupted using an Aminco French pressure cell at 20,000 psi. The disrupted bacteria were diluted with buffer to 400 ml. If a French press is not available then sonication may be used as long as the temperature does not rise above 10°.

Salmine Sulfate Fractionation. A solution of 4% w/v salmine sulfate (55 ml), previously neutralized, was added to the above extract with rapid stirring. The mixture was centrifuged immediately at 30,000 g in 100 ml tubes for 15 min at 0°. The supernatant was retained and adjusted to 400 ml by the addition of buffer (as above).

Ammonium Sulfate Fractionation. Over a period of 20 min, 110 g of ammonium sulfate was added to 400 ml of supernatant and after gentle stirring for 20 min the solution was centrifuged at 30,000 g as before. The supernatant was discarded and the precipitate was dissolved to give a final volume of 100 ml in 10 mM Tris–HCl buffer, pH 7.5, containing 5 mM 2-mercaptoethanol, 0.5 mM Na$_2$EDTA, and 10% (v/v) glycerol.

Heat Treatment. Glycine and pyridoxal phosphate were added to the resuspended precipitate to give final concentrations of 100 and 1 mM, respectively. The solution was heated to 55° in a water bath (80°) with stirring and maintained at this temperature for 4 min. The solution was rapidly cooled in an ice/water slurry and centrifuged at 30,000 g for 15 min. The supernatant was decanted and dialyzed against 3 × 5 liters of 10 mM Tris–HCl buffer, pH 7.5, containing 5 mM 2-mercaptoethanol, 0.5 mM Na$_2$EDTA, and 10% (v/v) glycerol over a period of 60 hr after which time the enzyme was ultracentrifuged at 60,000 g for 3 hr, and the supernatant (86 ml) was retained.

DEAE Cellulose Column Chromatography. The superantant was applied to a DE-52 (Whatman) DEAE cellulose column (6.5 × 20 cm) previously equilibrated in 10 mM Tris–HCl buffer, pH 7.5, containing 5 mM 2-mercaptoethanol and 0.5 mM Na$_2$EDTA. The column was washed with 200 ml of the above buffer and the enzyme was eluted from the column with a linear gradient of KCl [total volume 600 ml; 0–0.5 M KCl in 10 mM Tris–HCl buffer, pH 7.5, containing 5 mM 2-mercaptoethanol, 0.5 mM Na$_2$EDTA, and 10% (v/v) glycerol]. The flow rate of the column was 45 ml/hr. The most active fractions were pooled, treated with 39 g/100 ml of ammonium sulfate with stirring for a period of about 2 hr, and the solution was centrifuged at 30,000 g for 15 min. The supernatant was discarded and the precipitate was dissolved to give a final volume of 25 ml in 10 mM potassium phosphate buffer, pH 6.8, containing 5 mM 2-mercaptoethanol,

0.5 mM Na$_2$EDTA, and 10% (v/v) glycerol. The enzyme was dialyzed against 2.5 liters of the same buffer for 24 hr.

Calcium Phosphate Column Chromatography. The enzyme was applied to a calcium phosphate[12] column (5 × 20 cm) equilibrated with 10 mM potassium phosphate buffer, pH 6.8, containing 5 mM 2-mercaptoethanol and 10% (v/v) glycerol and the column was washed with 100 ml of this buffer. The enzyme was eluted with a linear gradient of phosphate buffer [total volume 500 ml; 10–50 mM potassium phosphate buffer, pH 6.8, all containing 5 mM 2-mercaptoethanol and 10% (w/v) glycerol]. The flow rate of the column was 35 ml/hr. The most active fractions of enzyme were pooled and treated with 39 g/100 ml of ammonium sulfate. After stirring for 3 hr, the suspension was centrifuged at 30,000 g for 20 min and the precipitate was dissolved to give a final volume of 3.2 ml of 25 mM potassium phosphate buffer, pH 7.5, containing 5 mM 2-mercaptoethanol, 0.5 mM Na$_2$EDTA, and 20% (v/v) glycerol. The enzyme was dialyzed against 2.5 liters of the same buffer for 24 hr.

DEAE-Sephadex Column Chromatography. The dialyzed enzyme was applied to a DEAE-Sephadex A-25 column (2.5 × 25 cm) previously equilibrated with the above buffer. The column was washed with 80 ml of buffer and the enzyme was eluted with a linear gradient of KCl (400 ml total volume; 0–200 mM KCl in the same buffer). The flow rate of the column was 25 ml/hr. Active fractions were pooled and the enzyme was concentrated to a small volume (about 5 ml containing 0.5 mg protein/ml) using a 70 ml concentration cell (Amicon) fitted with a UM20E membrane. If further reduction in volume was made then a loss of enzyme activity was observed due to the concentration of glycerol. Speed was most important during the concentration stage which should not take more than 2 hr. About 320 units of enzyme activity with a specific activity of 133 was obtained from the purification representing a 4430-fold purification. The enzyme was divided into aliquots (100 μl) and stored at $-15°$, conditions under which 80% of the activity was retained after 2 months and 50% after a year. A summary of the enzyme purification is given in the table.

Biosynthesis of Stereospecifically Labeled 5-Aminolevulinic
    Acid and Porphobilinogen

The enzyme thus purified may be utilized for the enzymatic synthesis of tritiated or deuterated 5-aminolevulinic acid (at positions 2 and 3) from appropriately labeled succinic acid precursors.[13] It was not necessary to

---

[12] C. K. Matthews, F. Brown, and S. S. Cohen, *J. Biol. Chem.* **239,** 2957 (1964).
[13] M. Akhtar and C. Jones, this volume [43].

PURIFICATION OF 5-AMINOLEVULINATE SYNTHASE

Fraction	Vol (ml)	Total protein (mg)	Total activity (units)	Specific activity (units/mg protein)	Purification (fold)
Initial extract	250	18,000	540	0.03	1
Salmine sulfate fractionation	40	4,800	720	0.15	5
Ammonium sulfate fractionation	90	975	429	0.44	14.7
Heat treatment	90	606	400	0.66	22
Dialysis	86	323	1,002	3.1	103.3
DEAE-cellulose column chromatography	25	96	681	7.1	236.7
Calcium phosphate column chromatography	3.2	17	456	27	886.6
DEAE-Sephadex column chromatography	4.8	2.4	320	133	4,430

use the highest specific activity enzyme for these transformations, and hence combined enzyme fractions from the calcium phosphate column were used, after concentration to 60–70 units/ml as described above, and dialysis against 100 ml of 10 mM potassium phosphate buffer, pH 6.8, containing glycerol (20%) and 2-mercaptoethanol (2 mM) for 2 hr. Typically, 3 ml containing 200 units of enzyme was employed per incubation.[13] Although 5-aminolevulinic acid may be isolated by ion-exchange chromatography,[7] in practice it is more convenient to couple the 5-aminolevulinate synthase to porphobilinogen synthase and to isolate the labeled porphobilinogen. The purification of porphobilinogen synthase and the overall transformations to give porphobilinogen, stereospecifically tritiated or deuterated in the acid side chains, are fully described elsewhere in this volume.[13,14]

The enzymatic synthesis of 5-aminolevulinic acid stereospecifically labeled at the more sensitive C-5 position using tritiated glycine as the labeled precursor required 5-aminolevulinate synthase of specific activity at least 80 units/mg. For these transformations, enzyme which had been purified through all stages of the above method after concentration and dialysis as above was used. Typically 20 units of this highly purified

[14] P. M. Jordan and J. S. Seehra, this volume [51].

enzyme was employed. 5-Aminolevulinic acid, thus labeled, was never isolated and the 5-aminolevulinate synthase was always coupled to the porphobilinogen synthase.[14] The resulting porphobilinogen was labeled stereospecifically at C-11 (label is also incorporated at C-2). The details of these methods are also fully described in this volume.[13,14]

Purity of 5-Aminolevulinate Synthase

The specific activity of 133 μmol/min/mg of protein is one of the highest published values obtained for 5-aminolevulinate synthase. The enzyme was homogeneous as judged by polyacrylamide gel electrophoresis (7.5% gel) when a single protein band (mobility 0.67) could be detected on staining with Coomassie Brilliant Blue. Enzyme activity was determined in a duplicate gel, which was cut into 2-mm slices. Assay of the slices by the method described above (except that a preincubation of 30 min and an incubation period of 3.5 hr were used) revealed a single peak of enzyme activity (mobility 0.67), coinciding precisely with the single protein band. No evidence for multiple enzyme bands was seen with the 4430-fold purified enzyme, although at earlier stages in the purification two enzyme bands were present, consistent with the presence of two fractions noted by previous workers.[15]

Molecular Properties of 5-Aminolevulinate Synthase

The molecular weight of the native enzyme was determined by Sephadex G-200 column chromatography using the appropriate protein standards. By this method the molecular weight of 5-aminolevulinate synthase was found to be 81,000 ($\pm$7,000). Electrophoresis of the enzyme using polyacrylamide gels containing sodium dodecyl sulfate[16] revealed a single sharp protein band of molecular weight 49,000 ($\pm$5,000). The enzyme thus appears to exist as a dimer of two identical subunits, as previously suggested.[17]

Enzyme Activation

During the early part of the enzyme purification, 5-aminolevulinate synthase progressively activates in cell-free extracts. This phenomenon has been noted previously and studied extensively.[11,18] The activation rate

[15] S. Tuboi, J. K. Haeng, and G. Kikuchi, *Arch. Biochem. Biophys.* **138,** 147 (1970).
[16] K. Weber and M. Osborn, *J. Biol. Chem.* **244,** 4406 (1969).
[17] D. L. Nandi and D. Shemin, *J. Biol. Chem.* **252,** 2278 (1977).
[18] J. Marriott, A. Neuberger, and G. H. Tait, *Biochem. J.* **117,** 609 (1970).

can be accelerated by the addition of cysteine trisulfide,[19] although in practice there is enough endogenous activator present at early stages of the purification to cause full activation without the need for the addition of exogenous activator. Because of the activation, the overall yield of enzyme is therefore flatteringly high (about 60%) and if the potential units of activity are taken into consideration the realistic yield is nearer 20%.

Substrates and Cofactors

Glycine is the only amino acid which the enzyme is known to accept as a substrate (K_m 0.2 mM). There is evidence however that in addition to succinyl-CoA (K_m 5 μM) the *R. spheroides* enzyme can utilize acetyl-CoA, propionyl-CoA, and butyryl-CoA, albeit at much lower rates.[20] Pyridoxal 5′-phosphate is rather weakly bound to the synthase compared with many other pyridoxal 5′-phosphate-dependent enzymes and the cofactor can be easily removed by dialysis. The enzyme is fully active with 50 μM pyridoxal 5′-phosphate and 50% active with 5 μM cofactor. These values are consistent with those found by other groups.[1]

Inhibitors

Rhodopseudomonas spheroides 5-aminolevulinate synthase is strongly inhibited by aminomalonate[21] (K_i 22 μM). The enzyme is also sensitive to reagents such as *p*-chloromercuribenzoate and to Fe^{2+} and Cu^{2+} ions. Hematin inhibits the enzyme at μM concentrations. The inhibition is reversible and noncompetitive.[22]

Enzyme Mechanism

Steady-state kinetics studies with mutant strain Y[23] suggest an ordered binding of glycine to the pyridoxal 5′-phosphate enzyme complex followed by addition of succinyl-CoA. CoA is released before the final product ALA. Results from exchange reactions with [2RS-³H₂]glycine and [5RS-³H₂]5-aminolevulinic acid are consistent with this sequence, the enzyme catalyzing the stereospecific exchange of the *proR* hydrogen atoms from both glycine and the 5-aminolevulinic acid in the absence of suc-

---

[19] J. D. Sandy, R. C. Davis, and A. Neuberger, *Biochem. J.* **150**, 245 (1975).

[20] G. Kikuchi and D. Shemin, unpublished results.

[21] M. Matthew and A. Neuberger, *Biochem. J.* **87**, 601 (1963).

[22] B. F. Burnham and J. Lascelles, *Biochem. J.* **87**, 462 (1963).

[23] J. Clement-Metral and M. Fanica-Gaignier, *Eur. J. Biochem.* **59**, 73 (1975).

cinyl-CoA or CoA.[24] Extensive studies with glycine stereospecifically tritiated at C-2 have established the mechanistic course of the enzymatic reaction and have shown that the loss of the C-1 carboxyl group of glycine occurs after the formation of the new C—C bond between C-2 of glycine and the succinyl moiety.[25] Furthermore, additional experimental evidence has shown that the decarboxylation is also catalyzed by the enzyme[26] and that the mechanism proceeds with overall inversion of configuration.

[24] A. Laghai-Newton and P. M. Jordan, unpublished results.
[25] Z. Zaman, P. M. Jordan, and M. Akhtar, *Biochem. J.* **135,** 257 (1973).
[26] M. M. Abboud, P. M. Jordan, and M. Akhtar, *J. Chem. Soc., Chem. Commun.* p. 643 (1974).

Author Index

Numbers in parentheses are footnote reference numbers and indicate that an author's work is referred to although the name is not cited in the text.

Subject Index

B

D

Dansylamidopropylcobalamin, synthesis, 16
Dansylamidopropylcobinamide, synthesis, 16
3-Dehydroretinal, 53
extraction as oxime, 53–54
HPLC, 55–57
applications, 58–61
comparison with conventional bleaching analysis, 57–58
standard oxime, preparation, 54–55
all-*trans*-13-Demethylretinoic acid, 114
5'-Deoxy-5'-adenosylcobalamin, synthesis, 16
Diazomethane, 116
Dicumarol, assay, 234
1,25-Dihydroxyvitamin D
assay
using rabbit intestinal cytosol-binding protein, 185–190
conditions, 187–188
standard curve, 187
chromatography, 186–187
competitive binding assays, 185
cytoreceptor assay, 190–198
calculation, 195–196
expected ranges, 196
interassay variation, 196
interlaboratory reproducibility, 197–198
materials, 191
procedure, 193–195
reagents, 191–192
recovery factor, 195
extraction, 186, 192–193
formation, 190
function, 154–155
HPLC, 177
microassay, 176–185
reagents, 177–178
plasma, 177
production, quantification by competitive binding protein radioassay, 161, 165–167
radioreceptor assay, 177
advantages, 185
characteristics, 181–182
intra- and interassay coefficients of variation, 184

nonequilibrium, 180–183
results, 184
standard curve, 183
validation, 182–185
1,25-Dihydroxyvitamin D_3
assay, 163
functions, 127
1,25-Dihydroxyvitamin D_3 receptor
chicken, preparation, 200
complexes with monoclonal antibodies, immunoprecipitation analysis, 202–204
hormone labeling of, 200–201
human
biology, 209
characterization, 209
detection of variants, 209–210
radioimmunoassay, 209–210
immunoblot assay, 205–207
immunoblot characterization, 210
immunochemical similarities among species, 208–209
immunologic characteristics, 206–207
molecular mass, 210
comparisons among avian and mammalian sources, 210–211
molecular weight, 210
monoclonal antibodies to, 199–211
preparation, 200
radioligand immunoassay, 204–205
sedimentation coefficient, 209
sedimentation displacement analysis, 201–202
sources, 200
ε-N-Dimethyl-L-lysine, 291
chromatographic properties, 294
methyl-^{14}C, synthesis, 293–294
4,5-Dioxovaleric acid, 374
4,7-Diphenyl-1,10-phenanthroline-disulfonic acid, 321
Disodium 2-oxo glutarate, [3R-^{3}H$_1$], 377–378
2,4-Disulfonic acid deuteroporphyrin, 330
Dolichyl phosphate mannose
HPLC system for separation from mannose, retinyl phosphate mannose, and retinyl phosphate, 65–66
molecular function, 61–62
standard, enzymatic synthesis from exogenous dolichyl phosphate, 64–65